T0145370

Advances in Intelligent Systems and Computing

Volume 977

The series "Advances in Intelligent Systems and Computing" contains publications on theory, applications, and design methods of Intelligent Systems and Intelligent Computing. Virtually all disciplines such as engineering, natural sciences, computer and information science, ICT, economics, business, e-commerce, environment, healthcare, life science are covered. The list of topics spans all the areas of modern intelligent systems and computing such as: computational intelligence, soft computing including neural networks, fuzzy systems, evolutionary computing and the fusion of these paradigms, social intelligence, ambient intelligence, computational neuroscience, artificial life, virtual worlds and society, cognitive science and systems, Perception and Vision, DNA and immune based systems, self-organizing and adaptive systems, e-Learning and teaching, human-centered and human-centric computing, recommender systems, intelligent control, robotics and mechatronics including human-machine teaming, knowledge-based paradigms, learning paradigms, machine ethics, intelligent data analysis, knowledge management, intelligent agents, intelligent decision making and support, intelligent network security, trust management, interactive entertainment, Web intelligence and multimedia.

The publications within "Advances in Intelligent Systems and Computing" are primarily proceedings of important conferences, symposia and congresses. They cover significant recent developments in the field, both of a foundational and applicable character. An important characteristic feature of the series is the short publication time and world-wide distribution. This permits a rapid and broad dissemination of research results.

**** Indexing: The books of this series are submitted to ISI Proceedings, EI-Compendex, DBLP, SCOPUS, Google Scholar and Springerlink ****

More information about this series at http://www.springer.com/series/11156

Robert Burduk · Marek Kurzynski ·
Michał Wozniak
Editors

Progress in Computer Recognition Systems

Editors
Robert Burduk
Faculty of Electronics
Wroclaw University of Science
and Technology
Wrocław, Poland

Marek Kurzynski
Faculty of Electronics
Wroclaw University of Science
and Technology
Wrocław, Poland

Michał Wozniak
Faculty of Electronics
Wroclaw University of Science
and Technology
Wrocław, Poland

ISSN 2194-5357 ISSN 2194-5365 (electronic)
Advances in Intelligent Systems and Computing
ISBN 978-3-030-19737-7 ISBN 978-3-030-19738-4 (eBook)
https://doi.org/10.1007/978-3-030-19738-4

This Springer imprint is published by the registered company Springer Nature Switzerland AG
The registered company address is: Gewerbestrasse 11, 6330 Cham, Switzerland

Preface

Discovering nature's rules and their applications to automatize decision making is still a focus of intense research. Nowadays, there are lots of works focus on developing accurate pattern recognition algorithms. Meanwhile, such methods are implemented in the form of computer software and applied in many practical areas, including character and speech recognition, biometry, machine vision, video surveillance, computer-aided medical diagnosis, etc. This observation encourages us to report current research progress of computer recognition systems with respect to both methodology and applications. This book contains a collection of 36 carefully selected articles contributed by experts of pattern recognition, which focus on topics related to pattern recognition, image analysis, computer vision, speech recognition, feature extraction, deep learning, and compound pattern classifiers. Editors would like to express their deep thanks to authors for their valuable submissions and all reviewers for their hard work. We believe that "Progress in Computer Recognition Systems" could be a great reference tool for scientists who deal with the problems of designing computer pattern recognition systems. Although the last, not least we would like to give special thanks to our colleagues Ms. Barbara Bobowska, Dr. Paweł, Trajdos, and Mr. Pawel Zyblewski, who assisted us to finish this book. We would like also to fully acknowledge the support from the Wrocław University of Technology, especially Prof. Andrzej Kasprzak—Chair of Department of Systems and Computer Networks which has also supported this volume.

May 2019

Robert Burduk
Marek Kurzynski
Michał Wozniak

Contents

Graph Grammar Models in Syntactic Pattern Recognition

Mariusz Flasiński(✉)

Information Technology Systems Department, Jagiellonian University, Cracow 30-348, ul. prof. St. Łojasiewicza 4, Kraków, Poland
mariusz.flasinski@uj.edu.pl

Abstract. The families of graph grammars used in syntactic pattern recognition are characterized in the paper. The reasons for the intractability of the problem of graph language parsing are presented. The methodological principles for the constructing of efficient syntax analysis schemes for graph-based syntactic pattern recognition are discussed.

1 Introduction

The pattern recognition area consists of two main subareas, namely the decision-theoretic subarea [1,5] and the the syntactic/structural one, which - in turn - can be divided into the structural approach [4,29], the algebraic approach [24] and the syntactic approach [13,15,19,20]. The syntactic pattern recognition models are grouped into the string-, tree- and graph-based models. Although graph language parsers are used in various application areas of pattern recognition such as scene analysis, feature recognition for CAD/CAM, analysis of visual events and activities, OCR, picture and diagram analysis, structure analysis in bioinformatics, medicine, chemistry and process monitoring and control, the constructing of efficient graph syntax analyzers is still one of the crucial open problems in the field [15]. This problem is presented and discussed in the paper.

The standard families of graph grammars used in syntactic pattern recognition are introduced in the next section. The issue of the intractability of graph language parsing is discussed in Sect. 3. Standard graph parsing schemes for syntactic pattern recognition are presented in Sect. 4. The final section includes the summary of methodological principles for the constructing of efficient graph-based syntactic pattern recognition models.

2 Graph Grammars in Syntactic Pattern Recognition

In the theory of graph grammars two large families are distinguished: grammars with connecting embedding and grammars with gluing embedding. In syntactic pattern recognition two classes of grammars belonging to these families are used,

R. Burduk et al. (Eds.): CORES 2019, AISC 977, pp. 1–10, 2020.
https://doi.org/10.1007/978-3-030-19738-4_1

namely *edNLC* graph grammars [22] (connecting embedding) and plex grammars [18] (gluing embedding). Let us introduce these classes.

Let $EDG_{\Sigma,\Gamma}$ denote a set of directed graphs with nodes (V) labelled by alphabet Σ, edges (E) labelled by alphabet Γ, and the node-labelling function $\phi : V \rightarrow \Sigma$.

Definition 2.1. An *edge-labelled directed Node Label Controlled, edNLC, graph grammar* is a quintuple $G = (\Sigma, \Sigma_T, \Gamma, P, Z)$, where Σ is a finite, non-empty set of node labels, $\Sigma_T \subseteq \Sigma$ is a set of terminal node labels, Γ is a finite, non-empty set of edge labels, P is a finite set of productions of the form (l, D, C), in which $l \in \Sigma \setminus \Sigma_T$, $D \in EDG_{\Sigma,\Gamma}$, $C : \Gamma \times \{in, out\} \rightarrow 2^{\Sigma \times \Sigma \times \Gamma \times \{in,out\}}$ is the embedding transformation, $Z \in EDG_{\Sigma,\Gamma}$ is the start graph called the axiom.

The second class of grammars generates the so-called *plex structures* [18]. A plex structure is composed out of components called *N attaching point entities, NAPEs*. While a symbol in a string constitutes a 2 attaching-point entity (it has two attaching points - one on its left side and one on its right side), NAPE has an arbitrary number of attaching points for connecting to other components. Now, we can introduce the following definition [18].

Definition 2.2. A *(CF) plex grammar* is a sextuple $G = (\Sigma_T, \Sigma_N, P, S, I, i_0)$, where Σ_T is a finite nonempty set of terminal NAPEs, Σ_N is a finite nonempty set of nonterminal NAPEs, P is a set of productions of the form: $A\Delta_A \rightarrow X\Gamma_X\Delta_X$, in which $A \in \Sigma_N$, Δ_A is a list of external connections for A, (X, Γ_X, Δ_X) is a plex structure, $S \in \Sigma_N$ is the start NAPE, I is a finite set of symbols called identifiers, $i_0 \in I$ is a special identifier, called the null identifier.

The remaining classes of graph grammars used in syntactic pattern recognition, i.e., tree-graph grammars [33], expansive graph grammars [34] and And-Or graph grammars [36], will be introduced in Sect. 5.

3 Computational Complexity of Graph Parsers

The are two main reasons for the intractability of the graph parsing problem [11,15]: the lack of ordering in a graph structure and the complex form of a graph grammar production. We will discuss the first reason in this section.

In the last papers of Fu and his collaborators on the graph parsing for syntactic pattern recognition [33,34], one can find the methodological suggestion for the imposing of linear ordering on the set of graph nodes in order to solve the first problem. This hint was applied for the construction of the *ETPL*(k)/*ETPR*(k) graph grammar-based model [7,9,12,15]. In this model the semantic aspect of a graph representation is used for the imposing of linear ordering, i.e., we refer to a *relational structure* (in the real-world) to be represented in order to index nodes in an unambiguous way. The formalization of this issue is based on Tarski's (semantic) model theory approach in the following way [15].

Definition 3.1. Let $\mathcal{U}$ be a finite set of individual objects called a universe, $\mathcal{N}_\mathcal{U}$ be a set of their names, $\mathcal{A}_\mathcal{U}$ be a set of their attributes. Let each object $o^k, k = 1, \ldots K$ of $\mathcal{U}$ be represented by its name $n_u^k \in \mathcal{N}_\mathcal{U}$ and the set of its attributes $a_u^k \in 2^{\mathcal{A}_\mathcal{U}}$. Let $\mathcal{R} \subset 2^{\mathcal{U} \times \mathcal{U}}$ be a set of binary relations such that for a pair of objects at most one relation is established, $\mathcal{N}_\mathcal{R}$ be a set of their names, $\mathcal{A}_\mathcal{R}$ be a set of their attributes, $\mathcal{R} = \{(n_r, a_r) : n_r \in \mathcal{N}_\mathcal{R}, a_r \in 2^{\mathcal{A}_\mathcal{R}}\}$. A *relational structure* is a sextuple $\mathrm{S} = (\mathcal{U}, \mathcal{R}, \mathcal{N}_\mathcal{U}, \mathcal{N}_\mathcal{R}, \mathcal{A}_\mathcal{U}, \mathcal{A}_\mathcal{R})$.

Given a relational structure, one can define the interpretation of an *EDG* graph in the following way.

Definition 3.2. Let $h = (V, E, \Sigma, \Gamma, \phi)$ be an *EDG* graph over Σ and Γ, $\mathrm{S} = (\mathcal{U}, \mathcal{R}, \mathcal{N}_\mathcal{U}, \mathcal{N}_\mathcal{R}, \mathcal{A}_\mathcal{U}, \mathcal{A}_\mathcal{R})$ be a relational structure, $\Sigma \subset \mathcal{N}_\mathcal{U}, \Gamma \subset \mathcal{N}_\mathcal{R}$. An *interpretation* $\mathcal{I}$ of the graph h over the structure S is a pair $\mathcal{I} = (\mathrm{S}, \mathcal{F})$, where: $\mathcal{F} = (\mathcal{F}_1, \mathcal{F}_2)$ is the *denotation function* defined in the following way: $\mathcal{F}_1$ assigns an object $u \in \mathcal{U}$ having a name $a \in \mathcal{N}_\mathcal{U}$ to each graph node $v \in V, \phi(v) = a, a \in \Sigma$, and $\mathcal{F}_2$ assigns a pair of objects $(u', u'') \in r, r \in \mathcal{R}$ to each graph edge $(v, \gamma, w) \in E$, $v, w \in V, \gamma \in \Gamma$ such that $\mathcal{F}_1(v) = u'$, $\mathcal{F}_1(w) = u''$ and r has the name γ.

Then, an interpreted *EDG* graph is defined as follows.

Definition 3.3. Let h be an *EDG* graph over Σ and Γ, S be a relational structure, $\mathcal{I}$ be the interpretation of h over S. An *interpreted EDG graph* is a triple $h^\mathcal{I} = (\mathrm{S}, h, \mathcal{I})$.

In an interpreted *EDG* graph, one can index its nodes, e.g., according to the ordering of features in an object to be represented or according to the ordering of objects (in 3D) in a scene to be represented, etc. in an unambiguous way [7,9,15]. Having the graph nodes indexed in an unambiguous way, we require the edges of the graph to be directed from the node having a smaller index to the node having a greater index, assuming that there exists the edge label $\gamma^{-1} \in \Gamma$ for each label $\gamma \in \Gamma$ such that the edges connecting nodes $v, w \in V$: $(v, \gamma, w) \in E$ and $(w, \gamma^{-1}, v) \in E$ describe the same relation between the objects of the relational structure represented by these nodes.

Let us introduce formally two classes of graphs with nodes indexed and edges directed in an unambiguous way on the basis of the semantic aspect of patterns to be represented [7,9,15]. (We define two ways of the ordering of nodes: the first one for the purpose of top-down parsing and the second one for the bottom-up parsing).

Definition 3.4. Let $h^\mathcal{I} = (\mathrm{S}, h, \mathcal{I})$ be an interpreted *EDG* graph over Σ and Γ. A *(reversely) indexed edge-unambiguous graph, IE (rIE), graph* over Σ and Γ defined on the basis of the graph $h^\mathcal{I}$ is an *EDG* graph $g = (V, E, \Sigma, \Gamma, \phi)$ which is isomorphic to h up to the direction of the edges, such that the following conditions are fulfilled. (1) g contains a directed spanning tree t such that the nodes of t have been indexed due to the (Reverse) Level Order Tree Traversal, LOTT (RLOTT). (2) The nodes of g are indexed in the same way as the nodes of t. (3) Every edge in g is directed from the node having a smaller index to the node having a greater index.

(LOTT means that for each node firstly the node is visited, then its child nodes are put into the FIFO queue. The RLOTT scheme is analogous to the LOTT, however it uses a LIFO queue instead of a FIFO queue.)

4 Graph Parsing Schemes in Syntactic Pattern Recognition

Standard graph parsers for syntactic pattern recognition have been defined by the analogy to the well-known parsing schemes for string languages, namely: $LL(k)$ [25,32], $LR(k)$ [23] and the Earley scheme [6]. In this section we show the analogy between top-down parsable $LL(k)$ and $ETPL(k)$ graph languages and the analogy between bottom-up parsable $LR(k)$ and $ETPR(k)$ graph languages.

Instead of presenting the formalized preliminary definitions for the graph grammars mentioned above [7,9], we introduce the semi-formal definition of *closed (reverse) two-level production ordered, closed TLPO (rTLPO), graph grammars* as in [14].

Definition 4.1. Let G be an *edNLC* graph grammar. G is called a *closed TLPO (rTLPO) graph grammar* iff it fulfills the following conditions. (1) The start graph Z and each graph D of the right-hand side of every production is an *IE (rIE)* graph with a spanning tree of height at most 2, where the root is terminal. (2) Each graph belonging to a derivation in G is an *IE (rIE)* graph. (3) For each derivational step, a production for a nonterminal node having the least (greatest) index is applied. (4) Node indices do not change during a derivation.

A derivation fulfilling conditions (3) and (4) is called a *regular left-hand (right-hand) side derivation*, denoted $\underset{rl(G)}{\Longrightarrow}$ ($\underset{rr(G)}{\Longrightarrow}$).

Now, we can show the analogous constraints imposed on top-down parsable $LL(k)$ grammars and $ETPL(k)$ graph grammars.

Let $G = (\Sigma_N, \Sigma_T, P, S)$ be a context-free grammar, $\eta \in \Sigma^*$. $FIRST_k(\eta)$ denotes a set of all the terminal prefixes of the strings of the length k (or of a length less than k, if a terminal string shorter than k is derived from α) that can be derived from η in the grammar G.

Let us introduce $LL(k)$ grammars [25,32].

Definition 4.2. Let $G = (\Sigma_N, \Sigma_T, P, S)$ be a context-free grammar, the left-most derivation in G be denoted with $\overset{*}{\underset{l(G)}{\Longrightarrow}}$. G is called an *$LL(k)$ grammar* iff for every two leftmost derivations

$$S \overset{*}{\underset{l(G)}{\Longrightarrow}} \alpha A\theta \underset{l(G)}{\Longrightarrow} \alpha\beta\theta \overset{*}{\underset{l(G)}{\Longrightarrow}} \alpha x$$
$$S \overset{*}{\underset{l(G)}{\Longrightarrow}} \alpha A\theta \underset{l(G)}{\Longrightarrow} \alpha\gamma\theta \overset{*}{\underset{l(G)}{\Longrightarrow}} \alpha y,$$

where $\alpha, x, y \in \Sigma_T^*,\ \ \beta, \gamma, \theta \in \Sigma^*,\ \ A \in \Sigma_N,$ the following condition holds.

$$If\, FIRST_k(x) = FIRST_k(y), \text{ then } \beta = \gamma.$$

Now, we define a graph-based notion which is analogous to $FIRST_k(\eta)$. Let g be an *IE* (*rIE*) graph, t be the index of some node of g having child nodes which are indexed: $i_1, i_2, \ldots i_r$. A subgraph h of the graph g consisting of the node indexed with t, its child nodes having indices $i_{a+1}, i_{a+2}, \ldots, i_{a+k}$, $a \geq 0$, $a + k \leq r$, and edges connecting the nodes indexed with: $t, i_{a+1}, i_{a+2}, \ldots, i_{a+k}$ is denoted $h = k\text{-}TL(g, t, i_{a+1})$. (A subgraph h of the graph g consisting of the node indexed with t, its child nodes having indices $i_{a+1}, i_{a+2}, \ldots, i_r$, $a \geq 0$, and edges connecting the nodes indexed with: $t, i_{a+1}, i_{a+2}, \ldots, i_r$ is denoted $h = CTL(g, t, i_{a+1})$).

Let $\cong$ denote a graph isomorphism. *PL*(k) graph grammars are defined in the following way [9].

Definition 4.3. Let $G = (\Sigma, \Sigma_T, \Gamma, P, Z)$ be a closed *TLPO* graph grammar. The grammar G is called a *PL*(k) *graph grammar* if the following condition is fulfilled: Let

$$Z \underset{rl(G)}{\overset{*}{\Longrightarrow}} X_1 A X_2 \underset{rl(G)}{\Longrightarrow} h_1 \underset{rl(G)}{\overset{*}{\Longrightarrow}} g_1$$

and

$$Z \underset{rl(G)}{\overset{*}{\Longrightarrow}} X_1 A X_2 \underset{rl(G)}{\Longrightarrow} h_2 \underset{rl(G)}{\overset{*}{\Longrightarrow}} g_2,$$

be two regular left-hand side derivations, such that A is a characteristic description of a node indexed with t, and X_1 and X_2 are substrings of characteristic descriptions. Let max be the number of nodes of the graph $X_1 A X_2$. If

$$k\text{-}TL(g_1, t, max + 1) \cong k\text{-}TL(g_2, t, max + 1)$$

then

$$CTL(h_1, t, max + 1) \cong CTL(h_2, t, max + 1).$$

One can easily notice the analogy between both definitions presented above.

Now, we can show the analogous constraints imposed on bottom-up parsable *LR*(k) grammars and *ETPR*(k) graph grammars.

Let us introduce *LR*(k) grammars [23].

Definition 4.4. Let $G = (\Sigma_N, \Sigma_T, P, S)$ be a context-free grammar, the rightmost derivation in G be denoted with $\underset{r(G)}{\overset{*}{\Longrightarrow}}$. G is called an *LR*(k) *grammar* iff for every two rightmost derivations

$$S \underset{r(G)}{\overset{*}{\Longrightarrow}} \alpha A w \underset{r(G)}{\Longrightarrow} \alpha \beta w$$

$$S \underset{r(G)}{\overset{*}{\Longrightarrow}} \gamma B y \underset{r(G)}{\Longrightarrow} \alpha \beta x,$$

where $w, x, y \in \Sigma_T^*$, $\alpha, \beta, \gamma \in \Sigma^*$, $A, B \in \Sigma_N$, the following condition holds.

$$If\, FIRST_k(w) = FIRST_k(x)\ then\ \alpha = \gamma,\ A = B,\ x = y.$$

PR(k) graph grammars are defined in the following way [15].

Definition 4.5. Let $G = (\Sigma, \Sigma_T, \Gamma, P, Z)$ be a closed *rTLPO* graph grammar. The grammar G is called a *PR(k) graph grammar* if the following condition is fulfilled. Let

$$Z \underset{rr(G)}{\overset{*}{\Longrightarrow}} X_1 A X_2 \underset{rr(G)}{\Longrightarrow} X_1 g X_2,$$
$$Z \underset{rr(G)}{\overset{*}{\Longrightarrow}} X_3 B X_4 \underset{rr(G)}{\Longrightarrow} X_1 g X_5,$$

and

$$k\text{-}TL(X_2, 1, 2) \cong k - TL(X_5, 1, 2),$$

where A, B are characteristic descriptions of certain nodes, X_1, X_2, X_3, X_4, X_5 are substrings of characteristic descriptions, g is the right-hand side of a production: $A \longrightarrow g$. Then:

$$X_1 = X_3,\ A = B,\ X_4 = X_5.$$

Again, one can easily notice the analogy between both definitions presented above.

Finally, parsable *ETPL(k)* and *ETPR(k)* graph grammars [9,15] can be defined in a semi-formal way on the basis of *PL(k)* and *PR(k)* graph grammars, respectively, as follows [14].

Definition 4.6. Let G be a *PL(k)* (*PR(k)*) graph grammar. G is called an *ETPL(k)* (*ETPR(k)*) *graph grammar* iff the following condition is fulfilled. If (v, λ, w), where v is labelled by terminal symbol, is an edge of an *IE (rIE)* graph belonging to a certain regular left-hand (right-hand) side derivation, then this edge is preserved by all the embedding transformations applied in succeeding steps of the derivation.

For *ETPL(k)* and *ETPR(k)* graph grammars efficient parsing algorithms have been defined. Let us introduce the following theorem [9,15].

Theorem 4.1. The running time of the parsing algorithms for *ETPL(k)* and *ETPR(k)* graph grammars is $\mathcal{O}(n^2)$, where n is the number of graph nodes.

Formal properties of *ETPL(k)* (*ETPR(k)*) graph grammars were studied in [11, 15]. Let us present the following theorems [11,15].

Theorem 4.2. For a given $k \geq 0$

$$\mathcal{L}(ETPL(k)) \subsetneq \mathcal{L}(ETPL(k+1)).$$

For a given $k \geq 0$

$$\mathcal{L}(ETPR(k)) \subsetneq \mathcal{L}(ETPR(k+1)).$$

Let us note that *ETPL(k)* (*ETPR(k)*) graph grammars constitute the hierarchy.

Theorem 4.3. There exists

$$L \in \mathcal{L}(ETPR(1))$$

such that for any $k \geq 0$

$$L \notin \mathcal{L}(ETPL(k)).$$

There exists

$$L \in \mathcal{L}(ETPL(1))$$

such that for any $k \geq 0$

$$L \notin \mathcal{L}(ETPR(k)).$$

Let us note that *ETPL* and *ETPR* graph grammars are incomparable according to the Theorem 4.3.

On the basis of formal properties of *ETPL* and *ETPR* graph grammars presented above, we can introduce the following analogy between parsable string languages and parsable graph languages as shown in Fig. 1.

Two other important graph parsing algorithms for syntactic pattern recognition were defined for plex languages. The syntax analyzer presented in [3] was applied for the analysis of flowcharts, whereas the parser presented in [31] was applied for the interpretation of electronic circuits. Both parsers were constructed on the basis of the well-known Earley parser [6] for string languages.

5 Concluding Remarks

As we have mentioned in Introduction, the constructing of efficient graph language parsers is one of the crucial open problems in syntactic pattern recognition. On the basis of our considerations we can formulate the following methodological recommendations [15].

The problem of the lack of ordering in a graph structure can be solved by the imposing of linear ordering on the set of graph nodes and the unambiguous (re)directing of graph edges. These operations can be performed by referring to the semantic properties of represented objects.

In order to handle the second problem, i.e., the complex form of a graph production, one can make use of parsing schemes developed for string languages. Such a methodology has been applied in case of defining *ETPL* graph grammars (based on *LL* string grammars), *ETPR* graph grammars (based on *LR* string grammars) and two parsing schemes defined for plex languages [3,31] (based on the Earley parser scheme).

If the embedding transformation of a graph grammar to be used is too strong/complex, the problem seems to be difficult. There are three basic strategies in syntactic pattern recognition to cope with this problem [15].

The imposing of constraints which weaken the embedding transformation slightly with the preserving of the generative power of the grammar at an acceptable level is the first strategy. Such a strategy was used for the parsable *ETPL* and *ETPR* graph grammars which were strong enough to be used in various application areas such as scene analysis [7,9], distorted object recognition [8], feature recognition for CAD/CAM [10], sign language recognition [17] and process monitoring and control [16].

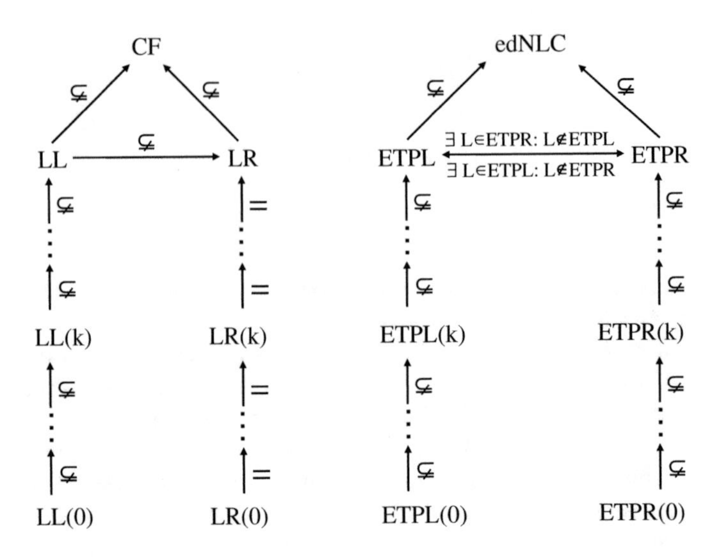

Fig. 1. The analogy between parsable *CF* string languages and parsable *edNLC* graph languages.

Table 1. Graph parsers used for syntactic pattern recognition.

Graph grammar	Parser	Running time
Tree-Graph	Sanfeliu and Fu [33]	$\mathcal{O}(n^3)$
Expansive	Shi and Fu [34]	$\mathcal{O}(n)$
edNLC	*ETPL*(*k*): Flasiński [7,9]	$\mathcal{O}(n^2)$
	ETPR(*k*): Flasiński [15]	$\mathcal{O}(n^2)$
Plex	Bunke and Haller [3]	$\mathcal{O}(n^5)$
	Peng, Yamamoto and Aoki [31]	$\mathcal{O}(n^4)$

The defining of the special classes of graph grammars for syntactic pattern recognition which are based on tree grammars and use a simplified embedding mechanism is the second strategy. This methodology was successfully used for the constructing of the Earley-based parser for tree-graph grammars in [33] and for the constructing of the parser for expansive graph languages in [34].

The constructing of the syntactic pattern recognition methods which are not based on graph grammar parsing schemes is the third strategy. This approach was used for the defining of Bunke attributed programmed graph grammars [2] which were applied for scene analysis, the interpretation of electronic circuit (EC) diagrams and flowcharts, and product family modelling in ERP systems. The And-Or graph syntactic pattern recognition model [36] was also defined within this strategy. The And-Or graph model was applied for scene analysis [21,26,27,35], the analysis of visual events and human activities [30] and the estimation of human attributes, parts and poses [28].

The summary of graph parsers used for syntactic pattern recognition is shown in Table 1.

References

1. Bishop CM (2006) Pattern recognition and machine learning. Springer, New York
2. Bunke H (1982) Attributed programmed graph grammars and their application to schematic diagram interpretation. IEEE Trans Pattern Anal Mach Intell 4:574–582
3. Bunke H, Haller B (1990) A Parser for context free plex grammars. In: Lecture Notes in Computer Science, vol 411. Springer, Heidelberg, pp 136–150
4. Bunke H, Sanfeliu A (eds) (1990) Syntactic and structural pattern recognition - theory and applications. World Scientific, Singapore
5. Duda RO, Hart PE, Stork DG (2001) Pattern classification. Wiley, New York
6. Earley J (1970) An efficient context-free parsing algorithm. Commun. ACM 13:94–102
7. Flasiński M (1988) Parsing of edNLC-graph grammars for scene analysis. Pattern Recogn 21:623–629
8. Flasiński M (1990) Distorted pattern analysis with the help of nodel label controlled graph languages. Pattern Recogn 23:765–774
9. Flasiński M (1993) On the parsing of deterministic graph languages for syntactic pattern recognition. Pattern Recogn 26:1–16
10. Flasiński M (1995) Use of graph grammars for the description of mechanical parts. Comput-Aided Des 27:403–433
11. Flasiński M (1998) Power properties of NLC graph grammars with a polynomial membership problem. Theor Comput Sci 201:189–231
12. Flasiński M (2007) Inference of parsable graph grammars for syntactic pattern recognition. Fundam Inf 80:379–413
13. Flasiński M (2016) Chapter 1: syntactic pattern recognition: paradigm issues and open problems. In: Chen, C.H. (ed.) Handbook of pattern recognition and computer vision, 5th edn. World Scientific, New Jersey/London/Singapore, pp 3–25
14. Flasiński M (2016) Introduction to artificial intelligence. Springer, Switzerland
15. Flasiński M (2019) Syntactic pattern recognition. World Scientific, New Jersey/London/Singapore
16. Flasiński M, Kotulski L (1992) On the use of graph grammars for the control of a distributed software allocation. Comput J 35:A165–A175
17. Flasiński M, Myśliński S (2010) On the use of graph parsing for recognition of isolated hand postures of Polish sign language. Pattern Recogn 43:2249–2264
18. Feder J (1971) Plex languages. Inf Sci 3:225–241
19. Fu KS (1982) Syntactic pattern recognition and applications. Prentice Hall, Upper Saddle River
20. Gonzales RC, Thomason MG (1978) Syntactic pattern recognition: an introduction. Addison-Wesley, Reading
21. Han F, Zhu SC (2009) Bottom-up/top-down image parsing with attribute grammar. IEEE Trans Pattern Anal Mach Intell 31:59–73
22. Janssens D, Rozenberg G (1980) On the structure of node-label-controlled graph languages. Inf Sci 20:191–216
23. Knuth DE (1965) On the translation of languages from left to right. Inf Control 8:607–639
24. Kulikowski JL (1971) Algebraic methods in pattern recognition. Springer, Wien

25. Lewis PM II, Stearns RE (1968) Syntax-directed transduction. J ACM 15:465–488
26. Lin L, Wu T, Porway R, Xu Z (2009) A stochastic graph grammar for compositional object representation and recognition. Pattern Recogn 42:1297–1307
27. Liu X, Zhao Y, Zhu SC (2018) Single-view 3D scene reconstruction and parsing by attribute grammar. IEEE Trans Pattern Anal Mach Intell 40:710–725
28. Park S, Nie BX, Zhu SC (2018) Attribute And-Or grammar for joint parsing of human attributes, parts and poses. IEEE Trans Pattern Anal Mach Intell 40:1555–1569
29. Pavlidis T (1977) Structural pattern recognition. Springer, New York
30. Pei M, Si Z, Yao BZ, Zhu SC (2013) Learning and parsing video events with goal and intent prediction. Comput Vis Image Underst 117:1369–1383
31. Peng KJ, Yamamoto T, Aoki Y (1990) A new parsing scheme for plex grammars. Pattern Recogn 23:393–402
32. Rosenkrantz DJ, Stearns RE (1970) Properties of deterministic top-down grammars. Inf Control 17:226–256
33. Sanfeliu A, Fu KS (1983) Tree-Graph Grammars for Pattern Recognition. In: Lecture Notes in Computer Science, vol 153. Springer, Heidelberg, pp 349–368
34. Shi QY, Fu KS (1983) Parsing and translation of attributed expansive graph languages for scene analysis. IEEE Trans Pattern Anal Mach Intell 5:472–485
35. Zarchi MS, Tan RT, Van Gemeren C, Monadjemi A, Veltkamp RC (2016) Understanding image concepts using ISTOP model. Pattern Recogn 53:174–183
36. Zhu SC, Mumford D (2006) A stochastic grammar of images. Found Trends Comput Graph Vis 2:259–362

Computer Vision Methods for Non-destructive Quality Assessment in Additive Manufacturing

Krzysztof Okarma[(✉)] and Jarosław Fastowicz

Department of Signal Processing and Multimedia Engineering,
Faculty of Electrical Engineering, West Pomeranian University of Technology,
Szczecin, 26 Kwietnia 10, 71-126, Szczecin, Poland
{okarma,jfastowicz}@zut.edu.pl

Abstract. Increasing availability and popularity of 3D printers cause growing interest in monitoring of additive manufacturing processes as well as quality assessment and classification of 3D printed objects. For this purpose various methods can be used, in some cases dependent on the type of filament, including X-ray tomography and ultrasonic imaging as well as electromagnetic methods e.g. terahertz non-destructive testing. Nevertheless, in many typical low cost solutions, utilising Fused Deposition Modelling (FDM) based technology, the practical application of such methods can be troublesome. Therefore, on-line quality assessment of the 3D printed surfaces using image analysis methods seems to be a good alternative, allowing to detect the quality decrease and stop the printing process or correct the surface in case of minor distortions to save time, energy and material. From aesthetic point of view quality assessment results may be correlated with human perception of surface quality, whereas, considering the physical issues, the presence of some surface distortions may indicate poor mechanical properties of the 3D printed object. The challenging problem of a reliable quality assessment of the 3D printed surfaces and appropriate classification of the manufactured samples can be solved using various computer vision approaches. Interesting results can be obtained assuming the appropriate location of the camera and analysis of the side view of the printed object where the linear patterns representing consecutive layers of the filament can be easily observed, especially for flat surfaces. Some exemplary experimental results of the application of texture analysis with the use of GLCM and Haralick features, Hough transform, similarity based image quality metrics, Fourier analysis and entropy are presented.

Keywords: 3D prints · Additive manufacturing · Surface quality assessment · Image analysis

1 Introduction

Rapid development of modern computer and automation technologies, which surround us everywhere, is one of the obvious symptoms of a new industrial

R. Burduk et al. (Eds.): CORES 2019, AISC 977, pp. 11–20, 2020.
https://doi.org/10.1007/978-3-030-19738-4_2

revolution towards "Industry 4.0" solutions, based on multidisciplinary integration of ICT and IT technologies with robotics, mechatronics, electrical engineering as well as transport, especially in even more popular Intelligent Transportation Systems (ITS). One of the key technologies, which may be considered as a hallmark of the new technological revolution, is the additive manufacturing technology, also referred as the 3D printing. Growing availability of low price 3D printers causes their popularity not only in industrial applications but also for home use. Nevertheless, many budget solutions and some cheaper 3D printers - usually build on user's own and based on a GPL licence - suffer from material's quality as well as assembling precision. On the other hand, a large selection and variety of different quality filaments used for manufacturing, especially popular PLA (Polyactic Acid) and ABS (Acrylonitrile Butadiene Styrene) materials, may influence the quality of the 3D printed objects as well.

Depending on specific application, related to biomedical purposes, mechanical engineering and transport, aerospace, architecture, design or even protection of cultural heritage (e.g. copies of sculptures or monuments), quality of manufactured surfaces may play a more or less important role. From aesthetic point of view some small irregularities may be accepted in some cases, however considering the mechanical properties of manufactured parts, they may indicate some distortions of their internal structure influencing some physical parameters and properties e.g. strength, durability, hardiness, etc.

As the process of additive manufacturing usually takes relatively long time, sometimes expressed in hours, particularly for larger objects, the continuation of the printing process in case of presence of some unacceptable surface distortions may be unreasonable. Hence, the necessity of using computer vision methods for the analysis of images captured by cameras mounted at the printing device, seems to be obvious. In such systems two major groups of methods may be distinguished: intended for monitoring of the printing process in terms of correctness of its progress, as well as *in situ* quality assessment of the manufactured object. The latter can be further divided into surface quality assessment and checking the compliance of the object's shape with the 3D model.

2 Vision Based Monitoring of 3D Printing Process

The idea of the application of machine vision and image analysis for fault detection and monitoring purposes for automatic assembly machines and 3D printers has relatively long history. One of the first attempts was the use of process signatures for detection of defects proposed by Fang *et al.* [6,7], whereas Cheng and Jafari [4] utilized top surface inspection for online process control during manufacturing based on correlation and dependency between two adjacent layers.

A machine vision system for fault detection in automatic assembly machines was presented by Chauhan and Surgenor [2], who proposed the machine vision performance index utilizing 5 performance factors: accuracy, processing time, noise robustness, ease of tuning and response speed. Furthermore, they analysed three major approaches based on Gaussian Mixture Models, optical flow and running average with morphological image processing [3]. A similar approach based on LabVIEW package and several webcams was proposed in the paper [31].

Another interesting example of applications of machine vision for production monitoring was the idea of defect detection in anaesthetic masks [18], whereas the use of Optical Coherence Tomography for *in situ* monitoring was examined by Gardner *et al.* [16] using the selective laser sintering process as an example. Authors of another recently published paper [25] presented the trained image analysis algorithm for anomaly detection in laser powder bed additive manufacturing devices. Some possibilities of the use of the supervised machine learning for process monitoring were analysed recently as well [5].

An interesting idea of quality assessment of electronic products was presented in the paper [32], whereas the visual inspection system preventing the possible collisions utilising machine learning and tracking based on singular points was presented recently by Makagonov, Blinova, and Bezukladnikov [20].

Nevertheless, most of these solutions are limited to fault and anomalies detection during the printing process and do not utilize any surface quality assessment methods. However, some properties of various filaments were examined in the paper [1] with the use of terahertz waves, demonstrating some possibilities of combination of THz based non-destructive evaluation with vision based surface analysis in future research.

3 Quality Assessment of 3D Printed Surfaces

3.1 Visual Feedback for Quality Monitoring

One of the first initial attempts to automatic visual quality assessment of 3D printed parts was the system developed by Straub [27] consisting of five cameras and Raspberry Pi units. Despite the encouraging results and low computational needs, the obtained results were quite unstable due to high sensitivity to even slight camera motions and changes of lighting conditions. Some improvements and further analyses were presented in some later papers, e.g. related to detection of changes of fill level [28] or microdefects [29] as well as some other attacks and security issues [30].

Recently, a multi-material 3D printing platform referred as Multifab was presented [26] where the self-calibration of printheads and visual feedback were obtained due to the use of built-in 3D scanner. Another idea related to the assessment of 3D prints was presented by Fok *et al.* [15] who used the visual information with the Ant Colony Optimization of tool paths to avoid, or at least limit, the presence of visual artifacts caused by the leaking filament during the motion of the nozzle, improving the visual quality of manufactured objects.

Holzmond and Li [17] presented the application of stereoscopic images analysis for each manufactured layer with the comparison of the printed geometry with the anticipated surface geometry with further thresholding of detected differences. Nevertheless, this approach can also be considered as defect detection rather than typical quality assessment.

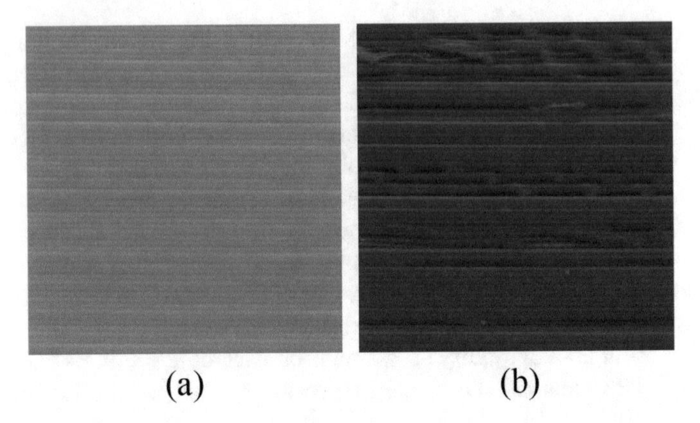

Fig. 1. Exemplary high (a) and low (b) quality samples with well visible layers and their distortions captured by a digital camera.

3.2 Description of Experimental Setup

To develop a reliable method of overall surface quality assessment, based on the analysis of images captured by cameras, some assumptions should be made. First, considering the principle of operation of the most popular Fused Deposition Modelling (FDM) devices, it can be noted that the filament is deposited layer by layer and therefore, assuming the side view, each of them should be easily visible. Hence, the most suitable is the side location of cameras. For initial experiments and verification of the usefulness of different computer vision methods some flat surfaces can be used, as was assumed in earlier research.

To verify the accuracy of classification and quality assessment a database of 3D printed flat samples was prepared, containing high and low quality samples manufactured from various colours of PLA and ABS filaments using three devices, namely Prusa i3, RepRap Pro Ormerod 2 and da Vinci Pro 3-in-1. Some of those samples contain visible distortions, forced by changing the speed of filament's delivery, parameters and configuration of stepper motors, temperature, etc. All samples were then subjectively assessed by the members of our department and divided into four major groups: high, moderately high, moderately low and low quality. Two representative examples of high and low quality samples are shown in Fig. 1. Individual samples were scanned using thee 2D flatbed scanner, photographed and additionally scanned using the GOM Atos 3D scanner.

The first experiments were made using only several PLA samples (divided into only two groups of high and low quality) scanned by a 2D flatbed scanner to avoid the influence of lighting conditions. However, most of the developed approaches were then tested using photos of larger number of samples and in some cases their modifications and extensions were necessary.

3.3 Overview of Investigated Approaches

The first group of considered methods was the application of statistical texture analysis utilizing Gray-Level Co-occurrence Matrix (GLCM) and selected Haralick features [21]. Classification of samples was based on the periodicity of chosen features determined using the vertical GLCMs (assuming the horizontal filament layers) with varying offset. Further analysis of homogeneity for scanned and photographed samples led to the proper classification of high and low quality prints with additional detection of the image type based on correlation values [11]. Nevertheless, due to the necessity of calculation of several GLCMs, those methods are rather slow and may be troublesome in practical applications.

Another investigated approach was the use of general purpose image quality assessment (IQA) methods. However, since the most universal IQA metrics are full-reference (FR) and therefore the original image without any distortions must be known, their direct application for surface quality assessment of 3D prints is not possible. Nevertheless, observing the exemplary images presented in Fig. 1, a local self-similarity can be observed mainly for the high quality sample (Fig. 1a) and therefore the calculation of the full-reference IQA metrics for the image fragments may be an interesting idea. The experiments based on Structural Similarity (SSIM), Complex Wavelet SSIM and Structural Texture Similarity (STSIM) [24] led to promising results, although the additional phase adjustments were necessary which may be based in the correlation analysis [8]. The application of some newer FR IQA approaches, such as RFSIM and Feature Similarity (FSIM), was investigated as well [22]. A significant decrease of the computation time can be obtained using the Monte Carlo method for the random choice of image regions used in comparisons applying one of the metrics, assuming the necessary phase adjustments [13]. A similar direction of experiments was the analysis of images of 3D printed surfaces in Fourier domain [9].

Encouraging results, particularly for photos, may also be obtained applying the line detection methods using Hough transform, especially for images subjected to adaptive histogram equalization using the popular CLAHE method [14]. The proposed approach with additional correction of brightness for bright filaments was based on the average length of detected lines, which should be much higher for properly manufactured samples. This approach was verified for 88 flat samples, assuming the random choice of image regions using the Monte Carlo method to prevent the influence of small geometrical distortions making it difficult to detect longer lines properly. An illustration of this approach for an exemplary sample is shown in Fig. 2. It is worth to note that the proper results obtained by such approach are determined by the presence of straight lines characteristic for flat 3D printed surfaces. In case of some other shapes, e.g. barrels or spheres, the circular Hough transform should be applied, which is more complex from computational point of view.

The presence and regularity of dominant gradient directions may also be utilised by the application of Histogram of Oriented Gradients (HOG) as shown in the paper [19]. This method, based on the local HOG features calculated for

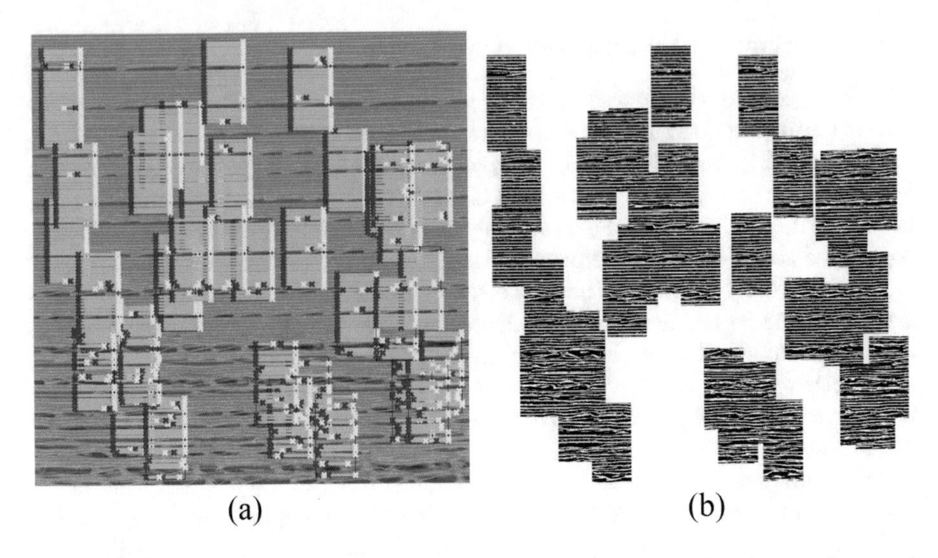

Fig. 2. Illustration of the method based on line detection using Hough transform and the Monte Carlo method [14] - exemplary low quality image with detected lines (a) and binarization results for randomly chosen regions (b).

partially overlapping blocks, utilizes the standard deviation of HOG descriptor values assuming signed orientations and 4 HOG bins.

Most of the discussed methods, including Hough transform, are strongly dependent on the colour of filament used during the manufacturing process. It means that a proper classification or quality assessment of the printed samples is possible for the predefined filament's colour and its change would require the additional colour calibration. To overcome this problem the use of image entropy was considered, which is however dependent on image brightness as well [12]. Nevertheless, the proposed combined quality indicator based on the average local entropy and its variance, calculated assuming the division of the image into blocks using the RGB and HSV colour spaces, allows for a reliable automatic quality assessment of 3D printed flat surfaces regardless of the colour of filament [23].

3.4 Discussion of the Most Essential Results

As the above method, initially tested using 18 PLA samples, was not as perfect for a bigger dataset of photographed samples, another extension of the entropy based approach, utilizing the depth maps, was proposed recently [10]. Its idea came from the assumption that slight linear changes of depth, which can be observable for each layer, looked very similar regardless of the filament's colour when the depth maps were transformed into greyscale image. Therefore all previously manufactured samples were additionally scanned using available GOM Atos 3D scanner, leading to the STL files containing the point clouds.

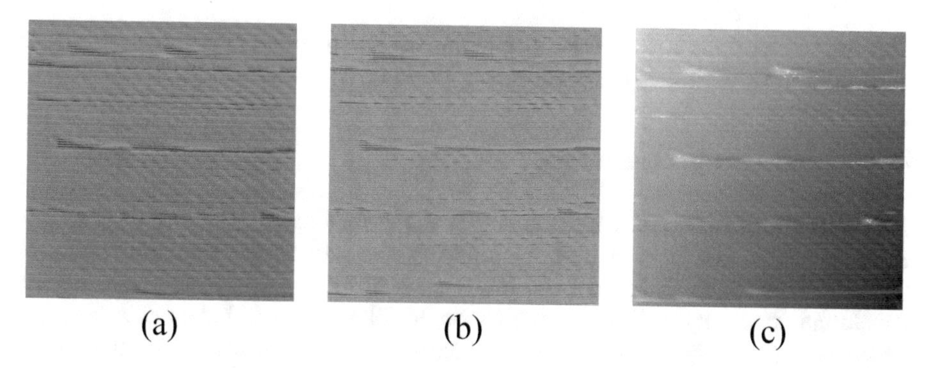

Fig. 3. Exemplary images obtained for a low quality sample - photo (a), 2D scan (b) and depth map (c).

Table 1. Experimental results obtained using selected colour independent methods.

Method	Test images	Filament	Accuracy	F-Measure
Entropy based [23]	18 (scanned)	PLA, 5 colours	1	1
HOG based [19]	74 (scanned)	PLA+ABS, 8 colours	0.838	0.867
CLAHE [14]	88 (scanned)	PLA+ABS, 9 colours	0.943	0.952
Entropy of depth map [10]	126 (3D scans)	PLA+ABS, 14 colours	0.905	0.928

After normalization and position adjustment, the depth maps were rendered into greyscale images, as shown in Fig. 3c, for which the entropy values were calculated. The finally proposed quality metric was the weighted product of the global entropy, average of 64 local entropy values and 4th root of their variance. Additionally, CLAHE algorithm was used to minimize the influence of misalignment of the scanned sample in relation to the scanner head. The proposed method was tested using 126 samples and provided the accuracy (defined as the ratio of number of properly classified samples to all samples) of 0.905, whereas the application of global entropy reached only 0.675. Detailed results are presented in the paper [10].

The experiments conducted for the improved version of the entropy based method with additional adjustment based on the deviation of the medium brightness and the division into 256 fragments proved the usefulness of the entropy based approach also for photographs, leading to proper classification of the whole dataset containing 78 samples (before 3D scanning and manufacturing some additional samples). The improvements of Hough based method led to accuracy equal to 0.795 for the dataset of 88 samples. Further extensions of the FR IQA based method reached the classification accuracy of 0.968 for the full dataset of 126 samples with the application of Feature Similarity metric for 16 blocks. However, this approach required many computations due to the necessity of numerous mutual similarity calculations followed by averaging. Nevertheless, it should be remembered that these three approaches can be applied *in situ* without the use of a high resolution 3D scanner. A short overview of some experimental results presented in original papers is shown in Table 1.

4 Concluding Remarks

Automatic surface quality assessment of 3D prints using machine vision approach is still a challenging task. Comparing the advantages and drawbacks of the presented methods, one of the main goals of further research should be the development of a hybrid metric, which would allow for even better classification accuracy, preferably regardless of the filament's colour and without the use of a 3D scanner. Nevertheless, one of the main issues may be the increase of computational complexity, particularly for combined metrics, due to the necessary calculations of their individual values.

Another direction of further research would be the development of the original database containing more subjective quality scores, similarly as in general purpose IQA databases, which would however require conducting of many perceptual experiments. This would be a necessary step to open the possibilities of converting the classification problem into optimization of the correlation between the subjective and objective scores in a similar way as for typical newly proposed image quality assessment methods. Since some of the approaches, e.g. Hough transform, directly exploit the flatness of surfaces, some additional extensive tests using the rounded shape samples should be made as well.

References

1. Busch SF, Weidenbach M, Fey M, Schäfer F, Probst T, Koch M (2014) Optical properties of 3D printable plastics in the THz regime and their application for 3D printed THz optics. J Infrared Millim Terahertz Waves 35(12):993–997
2. Chauhan V, Surgenor B (2015) A comparative study of machine vision based methods for fault detection in an automated assembly machine. Proc Manuf 1:416–428
3. Chauhan V, Surgenor B (2017) Fault detection and classification in automated assembly machines using machine vision. Int J Adv Manuf Technol 90(9):2491–2512
4. Cheng Y, Jafari MA (2008) Vision-based online process control in manufacturing applications. IEEE Trans Autom Sci Eng 5(1):140–153
5. Delli U, Chang S (2018) Automated process monitoring in 3D printing using supervised machine learning. Proc Manuf 26:865–870
6. Fang T, Jafari MA, Bakhadyrov I, Safari A, Danforth S, Langrana N (1998) Online defect detection in layered manufacturing using process signature. In: Proceedings of IEEE International Conference on Systems, Man and Cybernetics, San Diego, CA, USA, vol 5, pp 4373–4378
7. Fang T, Jafari MA, Danforth SC, Safari A (2003) Signature analysis and defect detection in layered manufacturing of ceramic sensors and actuators. Mach Vis Appl 15(2):63–75

8. Fastowicz J, Bąk D, Mazurek P, Okarma K (2018) Estimation of geometrical deformations of 3D prints using local cross-correlation and Monte Carlo sampling. In: Choraś M, Choraś RS (eds) Image Processing and Communications Challenges 9, IP&C 2017. AISC, vol 681. Springer, Cham, pp 67–74
9. Fastowicz J, Bąk D, Mazurek P, Okarma K (2018) Quality assessment of 3D printed surfaces in Fourier domain. In: Choraś M, Choraś RS (eds) Image Processing and Communications Challenges 9, IP&C 2017. AISC, vol 681. Springer, Cham, pp 75–81
10. Fastowicz J, Grudziński M, Tecław M, Okarma K (2019) Objective 3D printed surface quality assessment based on entropy of depth maps. Entropy **21**(1). Article no 97
11. Fastowicz J, Okarma K (2016) Texture based quality assessment of 3D prints for different lighting conditions. In: Chmielewski LJ, Datta A, Kozera R, Wojciechowski K (eds) Computer Vision and Graphics, ICCVG 2016. LNCS, vol 9972. Springer, Cham, pp 17–28
12. Fastowicz J, Okarma K (2017) Entropy based surface quality assessment of 3D prints. In: Silhavy R, Senkerik R, Kominkova Oplatkova Z, Prokopova Z, Silhavy P (eds) Artificial Intelligence Trends in Intelligent Systems, CSOC2017. AISC, vol 573. Springer, Cham, pp 404–413
13. Fastowicz J, Okarma K (2018) Fast quality assessment of 3D printed surfaces based on structural similarity of image regions. In: 2018 International Interdisciplinary PhD Workshop (IIPhDW), Świnoujście, Poland, pp 401–406
14. Fastowicz J, Okarma K (2019) Automatic colour independent quality evaluation of 3D printed flat surfaces based on CLAHE and Hough transform. In: Choraś M, Choraś RS (eds) Image Processing and Communications Challenges 10, IP&C 2018. AISC, vol 892. Springer, Cham, pp 123–131
15. Fok KY, Cheng C, Ganganath N, Iu H, Tse CK (2018) An ACO-based tool-path optimizer for 3D printing applications. IEEE Trans Ind Inform 15:2277–2287. https://doi.org/10.1109/TII.2018.2889740
16. Gardner MR, Lewis A, Park J, McElroy AB, Estrada AD, Fish S, Beaman JJ, Milner TE (2018) In situ process monitoring in selective laser sintering using optical coherence tomography. Opt Eng 57:57-1–57-5
17. Holzmond O, Li X (2017) In situ real time defect detection of 3D printed parts. Addit Manuf 17:135–142
18. Laucka A, Andriukaitis D (2015) Research of the defects in anesthetic masks. Radioengineering 24(4):1033–1043
19. Lech P, Fastowicz J, Okarma K (2018) Quality evaluation of 3D printed surfaces based on HOG features. In: Chmielewski LJ, Kozera R, Orłowski A, Wojciechowski K, Bruckstein AM, Petkov N (eds) Computer Vision and Graphics, vol 11114. ICCVG 2018, LNCS. Springer, Cham, pp 199–208
20. Makagonov NG, Blinova EM, Bezukladnikov II: Development of visual inspection systems for 3D printing. In: 2017 IEEE Conference of Russian Young Researchers in Electrical and Electronic Engineering (EIConRus), St. Petersburg, Russia, pp 1463–1465 (2017)
21. Okarma K, Fastowicz J: No-reference quality assessment of 3D prints based on the GLCM analysis. In: Proceedings of the 2016 21st International Conference on Methods and Models in Automation and Robotics (MMAR), Międzyzdroje, Poland, pp 788–793 (2016)
22. Okarma K, Fastowicz J: Quality assessment of 3D prints based on feature similarity metrics. In: Choraś RS (ed) Image Processing and Communications Challenges 8, IP&C 2016. AISC, vol 525, pp 104–111 (2017)

23. Okarma K, Fastowicz J (2018) Color independent quality assessment of 3D printed surfaces based on image entropy. In: Kurzynski M, Wozniak M, Burduk R (eds) Proceedings of the 10th International Conference on Computer Recognition Systems CORES 2017. AISC, vol 578. Springer, Cham, pp 308–315
24. Okarma K, Fastowicz J, Tecław M (2016) Application of structural similarity based metrics for quality assessment of 3D prints. In: Chmielewski LJ, Datta A, Kozera R, Wojciechowski K (eds) Computer Vision and Graphics, ICCVG 2016. LNCS, vol 9972, pp 244–252
25. Scime L, Beuth J (2018) Anomaly detection and classification in a laser powder bed additive manufacturing process using a trained computer vision algorithm. Addit Manuf 19:114–126
26. Sitthi-Amorn P, Ramos JE, Wangy Y, Kwan J, Lan J, Wang W, Matusik W (2015) MultiFab: a machine vision assisted platform for multi-material 3D printing. ACM Trans Graph 34(4):129-1–129-11
27. Straub J (2015) Initial work on the characterization of additive manufacturing (3D printing) using software image analysis. Machines 3(2):55–71
28. Straub J (2017) 3D printing cybersecurity: detecting and preventing attacks that seek to weaken a printed object by changing fill level. In: Proceedings of SPIE – Dimensional Optical Metrology and Inspection for Practical Applications VI, Anaheim, CA, USA, vol 10220, pp 102,200O-1–102,200O-15
29. Straub J (2017) An approach to detecting deliberately introduced defects and micro-defects in 3D printed objects. In: Proceedings of SPIE – Pattern Recognition and Tracking XXVII, Anaheim, CA, USA, vol 10203, pp 102,030L-1–102,030L-14
30. Straub J (2017) Identifying positioning-based attacks against 3D printed objects and the 3D printing process. In: Proceedings of SPIE – Pattern Recognition and Tracking XXVII, Anaheim, CA, USA, vol 10203, pp 1020,304-1–1020,304-13
31. Szkilnyk G, Hughes K, Surgenor B (2011) Vision based fault detection of automated assembly equipment. In: Proceedings of the ASME/IEEE International Conference on Mechatronic and Embedded Systems and Applications, Parts A and B, Washington, DC, USA, vol 3, pp 691–697
32. Tourloukis G, Stoyanov S, Tilford T, Bailey C (2015) Data driven approach to quality assessment of 3D printed electronic products. In: Proceedings of the 38th International Spring Seminar on Electronics Technology (ISSE), Eger, Hungary, pp 300–305

Combined kNN Classifier for Classification of Incomplete Data

Tomasz Orczyk, Rafal Doroz(✉), and Piotr Porwik

Computer Systems Department, Institute of Computer Science, University of Silesia, Bedzinska 39, 41-200 Sosnowiec, Poland
{tomasz.orczyk,rafal.doroz,piotr.porwik}@us.edu.pl

Abstract. Common problem in data classification is the incompleteness of the data, and not always it is possible to re-acquire the missing values. Another approach is to fill-in missing values using some statistical methods. This however distracts the original data and may lead to over-fit the classifier to the artificially generated values, and in consequence to over-estimate the classifier accuracy in Cross Validation tests. In this paper we propose a solution where, for a reference data consisting of complete and incomplete records, complete records serve as a reference data for a standard classifier, while the whole set serves as a reference data for single feature subspaced classifier.

Keywords: Classification · Incomplete data · kNN · Subspace

1 Introduction

Nowadays classification algorithms play significant role in many aspects of our lives. They are a vital component of decision supporting systems and risk assessment systems. In these kind of systems either one or both: the reference and classified data, are incomplete, and due to the historical character of data, missing values cannot be easily re-acquired.

If we can not re-acquire the data and we don't want to introduce any artificial data, then we must reject incomplete vectors. This leads to wasting some potentially useful data. The solution is to use feature subspacing. The most universal and ultimate subspace size is 1D. Having one subspace for each feature makes possible to create an ensemble of classifiers, where each feature is classified independently and in parallel, and there is no need to reject any existing data. This is however this occupied by loosing information about dependencies between features. To minimize this information loos we propose a combined method, where a complete records are used as a reference data for standard classifier, and all (complete and incomplete) records are used as a reference data for a single feature subspace variant. Also, complete records are classified using a standard classifier, and incomplete ones - using a single feature space variant.

R. Burduk et al. (Eds.): CORES 2019, AISC 977, pp. 21–26, 2020.
https://doi.org/10.1007/978-3-030-19738-4_3

2 Method

Proposed method is a combined classifier (later called COMB), which consists of a regular kNN classifier [8] and ensemble of Single Feature kNN classifiers. The basic concept of the ensemble of Single Feature kNN classifiers is explained on the Fig. 1.

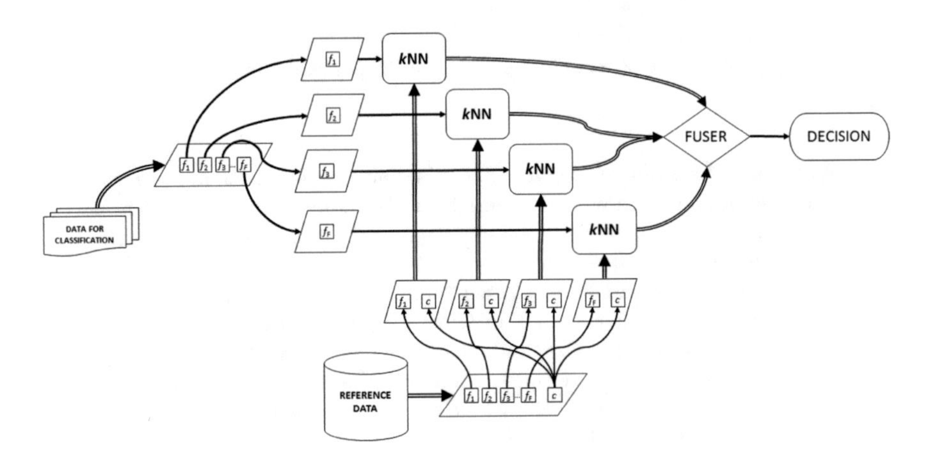

Fig. 1. Simplified architecture of ensemble of single feature kNN classifiers.

The component kNN classifiers of the Single Feature classifiers ensemble are slightly modified implementation of a regular kNN classifiers, so they return support values for all classes c, instead of a decision (single class label). When some of the f_f features from the classified data is missing (i.e. has no value), then the corresponding kNN classifier returns no decision. When same situation occurs in the reference data, such record is ignored, as distance to a non defined value (NaN) is $+\infty$, and thus it will never be considered as a nearest neighbor to the classified record. More extensive description of the SFkNN classifier can be found in [11].

The SFkNN classifier outperforms a standard kNN classifier when values are missing in either one or both of the data inputs (reference and classified), yet for a low count or no missing values at all the situation is opposite. Thus, we propose a combined classifier, which switches the classified data between SFkNN when some values are missing, and kNN when classified record is complete. Also the primary reference data set is divided between these classifiers as follows: the kNN uses only complete records from the primary set, while SFkNN uses whole primary set as is. Figure 2 depicts the simplified concept of the COMB classifier proposed in this paper.

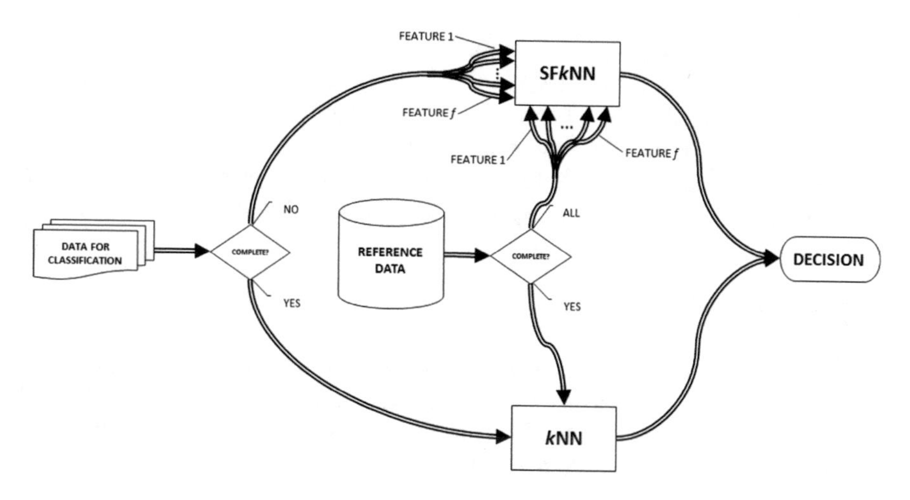

Fig. 2. Simplified architecture of the COMB classifier.

3 Experiment

For the experiment we have chosen 4 benchmark data sets from the UCI repository [6]. These are: Wine Data Set (WINE) [1], Breast Cancer Wisconsin (Diagnostic) Data Set (WDBC) [10], Cardiotocography Data Set (CTG) [4], Cryotherapy Dataset Data Set (CRYO) [9]. All these data sets consisted only of complete vectors. Table 1 shows a brief characteristics of the data used in experiments.

Table 1. Characteristics of the data used in experiments.

No.	Full name	ID	Records	Features	Classes
1.	Wine	WINE	178	13	3
2.	Wisconsin diagnostic breast cancer	WDBC	569	30	2
3.	Cardiotocography	CTG	2126	21	3
4.	Cryotherapy	CRYO	90	6	2

Missing values have been artificially introduced to these data sets according to the following scheme: each data set has been randomly divided in two subsets containing 50% of vectors each (the class balance of the original set has been reflected in these subsets); one of the subsets has been left as it was, and from other one some portion of values has been randomly removed. There were 7 thresholds of the missing value count (5%, 10%, ..., 35%), and for each threshold there were 10 variants (with different records division and different missing values placement) of data set created. Experiments were carried out in KNIME [3] environment using WEKA [7] implementation of kNN, called IBk [2]. Classification accuracy have been assessed by means of Leave-One-Out Cross Validation method, and an average classification accuracy and its standard deviation for 10 trials for each missing value thresholds have been evaluated. For a regular kNN classifier data normalization has been applied. As normalization on the whole set used for cross-validation may lead to a biased results, in this experiment for each cross-validation step, the normalisation coefficients were determined only for training subset of data, and applied to both training dataset and classified record.

4 Results

For the sake of clarity, results of the experiments have been averaged over 10 subsets for each of the missing values thresholds and presented in Tables 2, 3, 4 and 5. Additionally a standard deviation of classification accuracy has been calculated for each of the averaged value.

Table 2. Classification accuracy for the WDBC data set.

Classifier	Missing values (in incomplete 50% part of the data set).						
	5%	10%	15%	20%	25%	30%	35%
COMB	0.949 ± 0.004	0.943 ± 0.002	0.940 ± 0.004	0.938 ± 0.004	0.937 ± 0.004	0.935 ± 0.007	0.934 ± 0.004
SFkNN	0.936 ± 0.004	0.932 ± 0.003	0.930 ± 0.004	0.928 ± 0.005	0.925 ± 0.004	0.922 ± 0.007	0.921 ± 0.006
kNN	0.929 ± 0.006	0.873 ± 0.010	0.835 ± 0.011	0.787 ± 0.014	0.749 ± 0.007	0.718 ± 0.011	0.704 ± 0.008

Table 3. Classification accuracy for the CRYO data set.

Classifier	Missing values (in incomplete 50% part of the data set).						
	5%	10%	15%	20%	25%	30%	35%
COMB	0.906 ± 0.013	0.911 ± 0.022	0.901 ± 0.026	0.896 ± 0.029	0.892 ± 0.019	0.869 ± 0.019	0.874 ± 0.030
SFkNN	0.848 ± 0.011	0.841 ± 0.014	0.853 ± 0.020	0.843 ± 0.032	0.858 ± 0.027	0.848 ± 0.019	0.833 ± 0.020
kNN	0.908 ± 0.025	0.883 ± 0.014	0.884 ± 0.019	0.863 ± 0.030	0.846 ± 0.035	0.821 ± 0.031	0.839 ± 0.032

As can be seen on Fig. 3 the proposed COMB classifier gives a very stable results over a wide range of missing values number and also between each trial (low standard deviation). It also outperforms, by means of classification accuracy, both of the component classifiers in every case, what have been confirmed by evaluation of the rank based critical distance as proposed in [5]. The critical distance visualisation has been shown on Fig. 4.

Table 4. Classification accuracy for the WINE data set.

Classifier	Missing values (in incomplete 50% part of the data set).						
	5%	10%	15%	20%	25%	30%	35%
COMB	0.954 ± 0.008	0.946 ± 0.008	0.949 ± 0.005	0.929 ± 0.010	0.935 ± 0.007	0.922 ± 0.009	0.919 ± 0.014
SFkNN	0.942 ± 0.009	0.938 ± 0.007	0.942 ± 0.009	0.928 ± 0.008	0.930 ± 0.007	0.920 ± 0.010	0.917 ± 0.014
kNN	0.942 ± 0.008	0.907 ± 0.013	0.896 ± 0.014	0.878 ± 0.019	0.849 ± 0.021	0.831 ± 0.019	0.810 ± 0.035

Table 5. Classification accuracy for the CTG data set.

Classifier	Missing values (in incomplete 50% part of the data set).						
	5%	10%	15%	20%	25%	30%	35%
COMB	0.893 ± 0.003	0.889 ± 0.003	0.888 ± 0.003	0.887 ± 0.003	0.887 ± 0.002	0.885 ± 0.003	0.886 ± 0.004
SFkNN	0.870 ± 0.001	0.869 ± 0.003	0.869 ± 0.002	0.867 ± 0.003	0.868 ± 0.003	0.867 ± 0.003	0.868 ± 0.004
kNN	0.881 ± 0.004	0.867 ± 0.005	0.853 ± 0.005	0.843 ± 0.005	0.834 ± 0.004	0.828 ± 0.005	0.821 ± 0.002

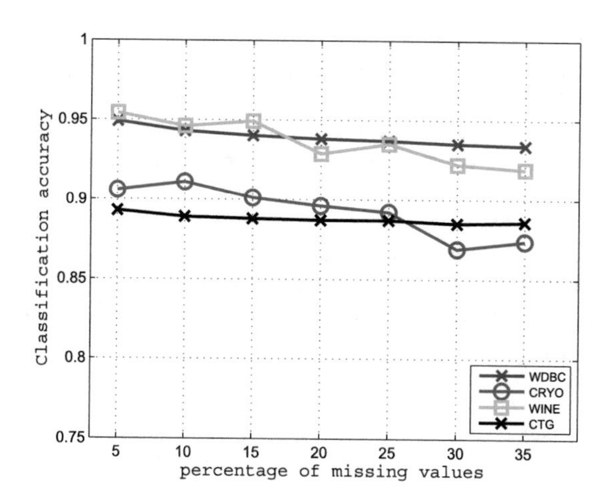

Fig. 3. Average classification accuracy for COMB classifier.

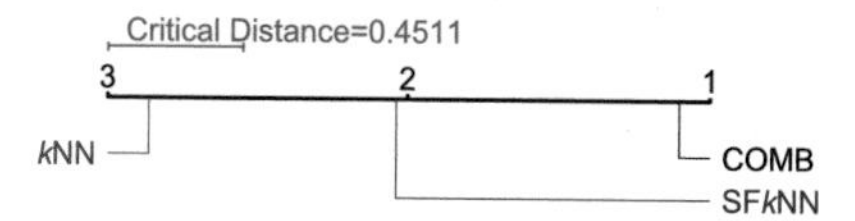

Fig. 4. Critical distance assessment for kNN, SFkNN and COMB classifier.

5 Conclusions

Proposed COMB classifier allows to improve classification accuracy, without need for reference set alternation, and without adding much computation complexity (depending on the input only one of the classifiers is executed).

References

1. Aeberhard S, Coomans D, De Vel O (1992) Comparison of classifiers in high dimensional settings. Department of Mathematics and Statistics, James Cook University, North Queensland, Australia, Technical report **92**, 02 (1992)
2. Aha D, Kibler D (1991) Instance-based learning algorithms. Mach Learn 6:37–66
3. Berthold MR, Cebron N, Dill F, Gabriel TR, Kötter T, Meinl T, Ohl P, Thiel K, Wiswedel B (2009) KNIME - the Konstanz information miner: version 2.0 and beyond. SIGKDD Explor Newsl 11(1):26–31. https://doi.org/10.1145/1656274.1656280
4. Ayres-de Campos D, Bernardes J, Garrido A, Marques-de Sa J, Pereira-Leite L (2000) SisPorto 2.0: a program for automated analysis of cardiotocograms. J Mater-Fetal Med 9(5):311–318
5. Demšar J (2006) Statistical comparisons of classifiers over multiple data sets. J Mach Learn Res 7(Jan):1–30
6. Dua D, Graff C (2017) UCI machine learning repository. http://archive.ics.uci.edu/ml
7. Eibe F, Hall M, Witten I (2016) The WEKA workbench. Online appendix for data mining: practical machine learning tools and techniques. Morgan Kaufmann
8. Jain AK, Dubes RC (1988) Algorithms for clustering data. Prentice-Hall, Inc., Upper Saddle River
9. Khozeimeh F, Alizadehsani R, Roshanzamir M, Khosravi A, Layegh P, Nahavandi S (2017) An expert system for selecting wart treatment method. Comput Biol Med 81:167–175
10. Mangasarian OL, Street WN, Wolberg WH (1995) Breast cancer diagnosis and prognosis via linear programming. Oper Res 43(4):570–577
11. Porwik P, Orczyk T, Lewandowski M, Cholewa M (2016) Feature projection k-NN classifier model for imbalanced and incomplete medical data. Biocybern Biomed Eng 36(4):644–656

Object Detection in Design Diagrams with Machine Learning

Jukka K. Nurminen[1], Kari Rainio[1](✉), Jukka-Pekka Numminen[2], Timo Syrjänen[2], Niklas Paganus[3], and Karri Honkoila[3]

[1] VTT Technical Research Centre of Finland, Espoo, Finland
{jukka.k.nurminen,kari.rainio}@vtt.fi
[2] Pöyry Oy, Vantaa, Finland
{jukka-pekka.numminen,timo.syrjanen}@poyry.com
[3] Fortum Oy, Espoo, Finland
{niklas.paganus,karri.honkoila}@fortum.com

Abstract. Over the years companies have accumulated large amounts of legacy data. With modern data mining and machine learning techniques the data is increasingly valuable. Therefore being able to convert legacy data into a computer understandable form is important. In this work, we investigate how to convert schematic diagrams, such as process and instrumentation diagrams (P&I diagrams). We use modern machine learning based approaches, in particular, the Yolo neural network system, to detect high-level objects, e.g. pumps or valves, in diagrams which are scanned from paper archives or stored in pixel or vector form. Together with connection detection and OCR this is an essential step for the reuse of old planning data. Our results show that Yolo, as an instance of modern machine learning based object detection systems, works well with schematic diagrams. In our concept, we use a simulator to automatically generate labeled training material to the system. We then retrain a previously trained network to detect the components of our interest. Detection of large components is accurate but small components with sizes below 15% of page size are missed. However, this can be worked around by dividing a big diagram into a set of smaller subdiagrams with different scales, processing them separately, and combining the results.

Keywords: Object detection · Schematic diagrams · Machine learning · Legacy data

1 Introduction

Over the years companies have accumulated large amounts of legacy data. The data can contain useful information but is difficult to take advantage of because the data is typically in paper or picture form. The goal of our research is to study ways to detect high-level objects from drawing data. As an example, a process and instrumentation diagram (P&I diagram) in drawing form is just a set of points in pixel representation, or a set of graphical objects, like lines, arcs,

R. Burduk et al. (Eds.): CORES 2019, AISC 977, pp. 27–36, 2020.
https://doi.org/10.1007/978-3-030-19738-4_4

or polylines, in vector presentation. To be useful this level is not enough. Instead we need to detect if the diagram contains higher level elements, such as pumps and valves (see Fig. 1). Also the connections between the higher level elements are important as well as the related texts. However, in this work we focus only on the detection of higher level elements. Text detection and recognition (OCR) as well as connection detection are handled with different algorithms.

Detecting high-level objects is useful for multiple reasons. First, there is a lot of excitement about AI and machine learning. A key ingredient for machine learning is data. To be useful for training machine learning systems, the data has to be at a higher conceptual level. Second, legacy data is often useful for human experts, but finding the relevant data, e.g. drawings of a specific aspect of legacy plants, can be difficult. Therefore, systems which allow users to make high-level queries, e.g. searching for drawings which have specific component configurations, are useful. Third, maintenance and upgrade of old plants (brownfield projects) can be executed more efficiently with fewer errors and lower risk if old plans in digitized form can be used as a basis. In summary, to unleash the value of legacy data, it is important to make its use easier and high-level object detection is one key enabler towards this goal. In the present work, our focus is P&I diagrams of process industry but similar issues are encountered in all kinds of two-dimensional diagrams, such as electrical circuits.

The problem of detecting high-level objects in design diagrams is not easy. The interesting symbols can appear in different sizes. They can also be rotated. Often it is not apparent which lines in the diagram are part of the object symbol and which lines are connectors or related to accompanying text. Therefore it is far easier to detect lines and other graphical primitives than the higher level objects. Old paper drawings in archives may also have corrections, stains, and other aspects which complicate the detection. In large diagrams individual components can also be very small in size.

The problem is well understood and there are various attempts to solve it. The simplest approach is to rely on human effort to convert old diagrams into an intelligent form. Some researchers suggest different kind of algorithms for the task (more details in Sect. 2). Others develop ad hoc systems combining multiple ideas for practical solutions, e.g. [3,7,20]. In general, it seems that, in spite of over 30 years of effort, there is still no good general solution to the problem.

A disadvantage of many of the approaches is that they need to be tailored to the task at hand, which makes their adoption expensive. The attractiveness of a machine learning based solution is that, ideally, the adaptation of the approach is easy. Just retrain the machine learning system to recognize new kinds of symbols. Another opportunity for machine learning is that the system can incrementally improve its detection capability by the modifications an expert makes manually to correct the machine suggested recognized components.

Machine learning has made rapid progress during the last years. One of the key new insights in supervised learning has been to avoid an extensive feature engineering phase. Earlier, experts, using domain-specific knowledge and logical thinking, tried to come up with the key features to make the detection successful.

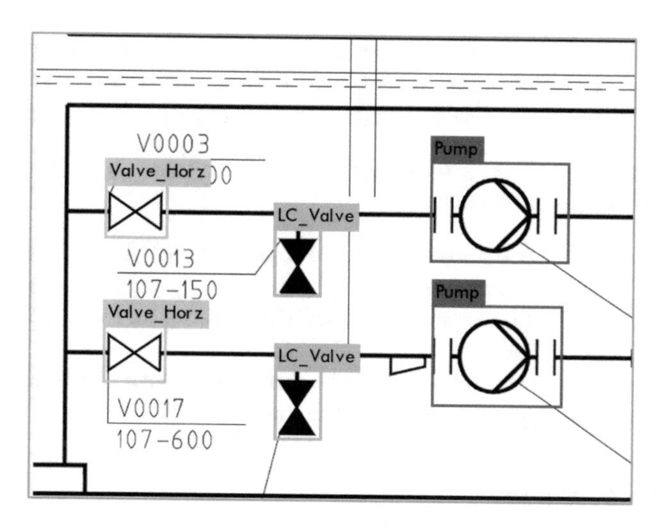

Fig. 1. Example of object detection in a P&I diagram

Instead, modern machine learning relies on huge neural networks with millions of parameters to let the system itself learn the key features. This has shifted the emphasis of machine learning to massive datasets and their use for training. Another enabler is a possibility to retrain a previously trained network. It is possible to take a network that has been trained with massive amounts of data to detect the key patterns relevant for the detection problem and use a quite small amount of data to retrain it to detect the symbols interesting for a certain use case [16].

In our work, we use Yolo, which is a state-of-the-art neural network based system for real-time object detection [17]. The design goal of Yolo has been realtime detection of objects in video streams. It can analyze tens of pictures in a second making the algorithm useful e.g. for autonomous vehicle solutions where the speed is essential. For our needs, Yolo is an overkill. Time is not hugely critical for analyzing old diagrams. Our 2D drawing data is much simpler than 3D video feeds. On the other hand, Yolo is readily available and if it works well with our object detection tasks it can be used as such. If we easily can retrain it to detect all kinds of diagram symbols it has potential to be a versatile, adaptable tool for object detection in legacy diagrams.

In comparison to object detection in a picture of regular 3D scenes the detection of schematic symbols is different

- Objects do not form regions of similar shades of color. Instead, they are just areas defined by black lines on white background. (Sometimes also other colors are used).
- Detection time does not matter unlike in video feed analysis.
- The components are always depicted from the same perspective and the lighting conditions are stable. However, there can be extra lines and text overlapping the component symbols as well as stains and dirt.

- Lines defining components, connections, and related texts are typically of the same color. Therefore, component detection has to be made by considering the whole shape of an object rather than individual lines or regions.

In this work, we wanted to experiment and see what is possible with Yolo. In this paper we present our finding:

1. We propose the use the general purpose Yolo algorithm to high-level object detection in design diagrams.
2. We report the results of our experiments of training and using Yolo to detect objects in legacy diagrams.
3. We summarize our key learnings and observations both of training and using Yolo as well as suggest ideas for further work.

2 Related Works

The problem of converting paper-based engineering drawing to intelligent form is old. The earliest references to the problem are from the 1980s. For instance, Karima et al. [12] review the technologies available for this task. Kasturi et al. [13] present a system for detecting the symbols, their relationships, and related texts in engineering drawings. A large number of papers aims for solving the problem. Jin et al. [10] apply a mixture of techniques. Adam et al. [2] apply Fourier-Mellin transformation. A lot of the work before the year 2000 is summarized in the review by Cordella and Vento [5].

Since the turn of the century new ideas include symbol detection directly from raster images without vectorization [11], using region adjacency graphs with integer optimization [4], application of dominant chain code [6]. Ruo-yu et al. [19] propose a learning-based method where the system learns to improve its recognition capability by the modification a human user makes to fix the recognized components. Ablameyko and Uchida [1] is a review of the work in early 2000 on this area. Practice-oriented solutions, e.g. [3,7], combine multiple approaches.

The classical approaches to detect objects is based on defining their characteristic features and them trying to match the picture object to those. Algorithms for this approach are summarized in [18].

The modern approach to object detection is based on supervised machine learning. Especially the object detection in video streams has attracted a lot of research. It has many use cases of significant importance, especially in autonomous driving. Therefore there has been rapid development in this area lately.

In supervised machine learning, the system is given a set of examples with known object names and their bounding boxes. The core of the system is a neural network with a vast number of parameters. The system learns the values of these parameters by trying to minimize the error in its estimations. Different machine learning systems vary in the way the neural network is structured and how the problems are represented. RCNN [22] and its variants first find regions in the picture and only then try to detect the object in each region. For schematic diagrams this is a problematic approach and finding regions in a black and white line drawing is problematic. The Yolo algorithm we use finds and classifies regions in single pass [17]. A recent detailed discussion of different machine learning based object detection algorithms is available in [14]. For our purposes the details of the algorithm do not matter much as long as the overall concept of training it with example data is not changed. Replacing Yolo with more efficient future algorithm should be simple.

3 Yolo for Object Detection in Engineering Drawings

To train Yolo to detect objects we started with a pre-trained Yolo model. The essential idea is that with pre-training it already knows how to detect essential features for object detection. With a small amount of training data we can then retrain it to detect just those components we are interested in. As Yolo has over 10 million parameters it would require a huge amount of computing and data to train it from scratch.

Having enough labeled training material, i.e. pictures with objects and their known types and bounding boxes, is frequently a problem. In our case, we came up with a way to generate training material automatically. We used Apros simulator[1] to generate the training data. Apros is intended for dynamic simulation of power plants, energy systems, and industrial processes. However, it supports user-defined components with custom symbols. We provided it with an example of each interesting component and with a small script it was then able to generate training data which contained the components in different sizes, rotations, and locations, their names and bounding boxes. This was a convenient way to rapidly generate labeled training material, a task which is commonly considered difficult in supervised machine learning.

We performed two kinds of experiments. First, we investigated Yolo training with artificial examples generated by Apros. Then, we used the component library and diagrams of Pöyry engineering company to see how the system works with real legacy data. In both experiments we used the same environment consisting of: (i) yolov3-tiny with pre-trained weights, (ii) Azure virtual machine (Standard F8 virtual machine; 8 vCPUs, 16 GB memory; Later NC6 with GPU support; 6 cores, 56 GB memory) with Ubuntu version Ubuntu 16.04.5 LTS (GNU/Linux 4.15.0-1032-azure x86_64), (iii) Apros simulator.

3.1 Artificial Apros Experiments

For these experiments we generated 2000 random Apros diagrams. To save training time we used only seven different component types. We selected these components with the aim to have some which are common and have simple symbols (e.g. ControlValve and CheckValve), some with more complicated symbols

[1] http://www.apros.fi/en/.

(e.g. Machinescreen2, LiquidEjector, GasEjector), and some with bigger symbols (Tank). The idea was to try out how the symbol size, complexity, and similarity to other symbols influence the detection.

Yolo recommends having at least 300 pictures of each object for training. To ensure this we generated 2000 random diagrams each with max 12 components. The optimum number of training examples was not in the scope of our present work but in the future it would be useful to study if a smaller amount of training examples would be adequate as the difference in symbols in 2D schematic drawings is far smaller than in video streams of 3D objects.

We performed multiple training experiments with (i) rotated and mirrored components, (ii) different component sizes, and (iii) partly visible components (components overlapping each other or located partly outside of picture area).

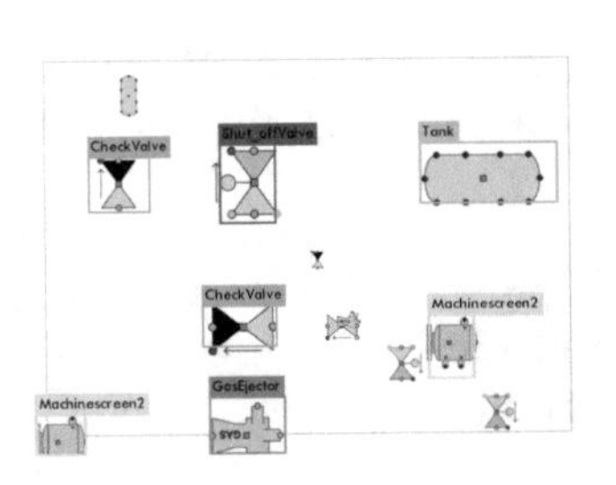

(a) Artificial test diagram

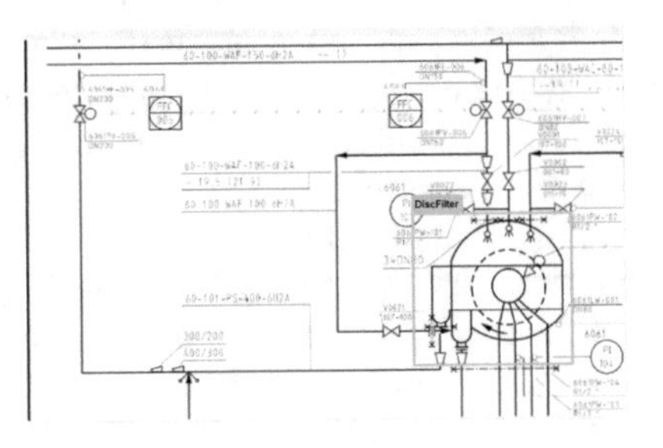

(b) Real P&I diagram

Fig. 2. Examples of detecting symbols from design diagrams

An example result is visible in Fig. 2a. The general observation is that the system is able to detect large component well. It is also able to deal with overlapping and only partly visible components as well as with rotated and mirrored components.

However, the size of the components has significant influence to the detection. It seems that components smaller than around 15% of the screen width or height are poorly detected. As can be seen in Fig. 2 the system is good at detecting the bigger components but fails to detect the smaller ones. The reason for this is that by default Yolo uses $416 \times 416 \times 3$ grid. When it scales small components to that size they lose a lot of their characteristic features and become so blurry that classifying them correctly is difficult even for a human.

To deal with this problem we have two main options. First, change the default width and height of Yolo resolution to e.g. 832×832. This increases of the neural network model a lot requiring more training and it still may not be good enough.

Alternatively, we can split the bigger diagram into smaller rectangular regions and zoom them out to make the components big enough for Yolo to detect. This assumes the resolution of the original figures is good enough to allow this kind of scaling. If the data is in vector graphics format this should be fine. For scanned data scaling is not so easy but with a high resolution scanner the diagrams should tolerate some zooming. In this approach Yolo needs to make multiple rounds to detect the components in each rectangular subdiagram. Because the detection speed of Yolo is very fast running it multiple times is not a problem. Some additional software is needed to manage the process of splitting, detecting, and combining the detection results. Naturally, this adds complexity to the component detection because the system needs to decide how to split the diagrams, which zooming ratios to apply, and how much overlap the different regions should have. Further research on making object detection algorithms better in dealing with small objects is therefore welcome as well as solutions, which automatically detect good ways to split and scale diagrams.

3.2 Experiments with Real Pöyry Data

To validate the concept with real-world data we used components from Pöyry component library. Altogether it contains around 800 different symbols. However, we limited ourselves to 200 most commonly used symbols.

We converted these symbols to Apros custom components and generated 20.000 random diagrams for training. We then tested the trained network with legacy diagrams from past Pöyry projects. Because of the large number of components and needed training instances we used Azure instances which provided support for GPUs. The use of GPUs has a dramatic effect on training speed with approximate 50× speed increase for training. We ran 250.000 training iterations which took about 24 h.

Figure 2b shows an example how the system can detect the relevant component in a complicated real design diagram. It can find the essential features and is not mislead by the incoming connections or overlapping text.

At the same time, the figure shows that in this scale the system is not able to detect the smaller components (e.g. valves above the DiscFilter) because they are too small. Therefore zooming-in to the picture and finding the components with different resolutions is essential.

4 Learnings

Small components are problematic but we can deal with them by dividing the diagram into smaller rectangles which have bigger components. This requires that the resolution of the original diagram tolerates zooming to make the components big enough in the separate diagrams. A rough rule of thumb is that each component should be around 15% of the height or width of the subdiagram.

Detecting the components works well once they are big enough. Also, the system is able to detect their bounding boxes very accurately.

Detecting small objects is a common weakness in object detection systems [15]. Some of the solutions are based on dividing the original figure into smaller subfigures and using them at different scales [15]. Another approach would be to just use Yolo to detect only large components. In the P&I diagrams smaller components, e.g. pumps and valves, have simple fixed-size symbols. Therefore they can be detected with straightforward template matching. In our case we tried this to find pumps and valves with OpenCV template matching `matchTemplate`. Because `matchTemplate` is only able to detect components of the same size we first used Tesseract OCR engine [21] to detect text and then used the text size to scale the template for `matchTemplate`. This approach works as long as the diagrams have been created with systematic rules for text and object sizes. For more flexible detection, scaling is needed. Future innovations in small object detection may provide better algorithms as well.

Depending on HW the training times vary but it will require around 20.000 iterations to show good results. Looking at the progress of learning is interesting. In early phase at smaller iteration amounts Yolo's estimates for the confidence of correct detection are small. As the training progresses the confidence grows and for many components is close to 100% in the end. Likewise, the bounding box sizes become increasingly accurate during the training. With a large number of iterations the progress stabilizes although even after 100.000 iterations some small progress was still done.

The amount of manual effort and expertise to train the system to detect new symbols is small. All that is needed are one example picture of each component together with its bounding box data. The rest is handled automatically: Apros generates the training material and Yolo uses it to train the network parameters to appropriate values. In our test some manual steps were needed to transfer data and invoke the different tools but also those steps can be automated with appropriate scripts.

In general, Yolo seems to be doing well in this task, even if is completely outside of its target domain. There is heavy competition to find improved image detection algorithms in machine learning (see e.g. [8,9] for comparison of other recent algorithms). Therefore, it is likely that we will soon see even better performing alternatives.

5 Conclusions

In this paper we have investigated the problem of converting schematic drawings into an "intelligent form". With intelligent we here mean that the system is able to understand what objects are in the drawing and how they are connected. This is an essential step for making the legacy data useful for data mining, machine learning, and brownfield design tasks.

We have used a pre-trained Yolo neural network and used data generated with our own simulator to retrain the network to detect the components we are interested in. The experiments were done with process and instrumentation diagrams of process industry but the same approach works equally well for other kinds of schematic diagrams such as drawings of electrical circuits.

The key finding is that the Yolo approach, which has been intended originally for entirely different purpose, detection of objects in video streams, works well also in schematic diagrams. It has problems in detecting small components, but this problem can be overcome by letting it detect zoomed-in parts of the picture. Because the detection speed of Yolo is very fast (multiple frames per second) we can easily afford to run tens or even hundreds of separate object detection runs to subparts of the original diagram with different zooming ratios.

Besides object detection we have also evaluated the other subtasks needed for legacy data use (detection of connections and detection and OCR of text). In our future work we plan to put the different pieces together to create a system which can process legacy diagrams and convert them automatically into a standard form such as DEXPI. Then they can be used for different kind of tasks allowing companies to gain value from their legacy data.

References

1. Ablameyko SV, Uchida S (2007) Recognition of engineering drawing entities: review of approaches. Int J Image Graph 07(04):709–733. https://doi.org/10.1142/S0219467807002878
2. Adam S, Ogier J, Cariou C, Mullot R, Labiche J, Gardes J (2000) Symbol and character recognition: application to engineering drawings. Int J Doc Anal Recogn 3(2):89–101. https://doi.org/10.1007/s100320000033
3. Arroyo E, Hoernicke M, Rodríguez P, Fay A (2016) Automatic derivation of qualitative plant simulation models from legacy piping and instrumentation diagrams. Comput Chem Eng 92:112–132
4. Bodic PL, Locteau H, Adam S, Héroux P, Lecourtier Y, Knippel A (2009) Symbol detection using region adjacency graphs and integer linear programming. In: 2009 10th international conference on document analysis and recognition. IEEE, pp 1320–1324. https://doi.org/10.1109/ICDAR.2009.202. http://ieeexplore.ieee.org/document/5277721/
5. Cordella L, Vento M (2000) Symbol recognition in documents: a collection of techniques? Int J Docu Anal Recogn 3(2):73–88. https://doi.org/10.1007/s100320000036
6. De P, Mandal S, Das A, Bhowmick P (2014) A new approach to detect and classify graphic primitives in engineering drawings. In: 2014 fourth international conference of emerging applications of information technology. IEEE, pp 243–248. https://doi.org/10.1109/EAIT.2014.33. http://ieeexplore.ieee.org/document/7052053/
7. Henderson TC (2014) Analysis of engineering drawings and raster map images. Springer
8. Huang J, Rathod V, Sun C, Zhu M, Korattikara A, Fathi A, Fischer I, Wojna Z, Song Y, Guadarrama S et al (2017) Speed/accuracy trade-offs for modern convolutional object detectors. In: IEEE CVPR, vol 4
9. Hui J (2018) Object detection: speed and accuracy comparison (faster R-CNN, R-FCN, SSD, FPN, RetinaNet and YOLOv3). https://medium.com/@jonathan_hui/object-detection-speed-and-accuracy-comparison-faster-r-cnn-r-fcn-ssd-and-yolo-5425656ae359/
10. Jin L, Zhou Z, Xiong S, Chen Y, Liu M (1998) Practical technique in conversion of engineering drawings to CAD form. In: IMTC/98 conference proceedings. IEEE instrumentation and measurement technology conference. Where instrumentation is going (Cat. No. 98CH36222) 1:8–13. https://doi.org/10.1109/IMTC.1998.679631. http://ieeexplore.ieee.org/document/679631/

11. Song J, Su F, Tai C-L, Cai S (2002) An object-oriented progressive-simplification-based vectorization system for engineering drawings: model, algorithm, and performance. IEEE Trans Pattern Anal Mach Intell 24(8):1048–1060. https://doi.org/10.1109/TPAMI.2002.1023802. http://ieeexplore.ieee.org/document/1023802/
12. Karima M, Sadhal K, McNeil T (1985) From paper drawings to computer-aided design. IEEE Comput Graph Appl 5(2):27–39. https://doi.org/10.1109/MCG.1985.276400. http://ieeexplore.ieee.org/document/4056067/
13. Kasturi R, Bow S, El-Masri W, Shah J, Gattiker J, Mokate U (1990) A system for interpretation of line drawings. IEEE Trans Pattern Anal Mach Intell 12(10):978–992. https://doi.org/10.1109/34.58870. http://ieeexplore.ieee.org/document/58870/
14. Liu L, Ouyang W, Wang X, Fieguth P, Chen J, Liu X, Pietikäinen M (2018) Deep learning for generic object detection: a survey. arXiv preprint arXiv:1809.02165
15. Meng Z, Fan X, Chen X, Chen M, Tong Y (2017) Detecting small signs from large images. In: 2017 IEEE international conference on information reuse and integration (IRI). IEEE, pp 217–224
16. Pan SJ, Yang Q et al (2010) A survey on transfer learning. IEEE Trans Knowl Data Eng 22(10):1345–1359
17. Redmon J, Farhadi A (2018) YOLOv3: an incremental improvement. arXiv
18. Roth PM, Winter M (2008) Survey of appearance-based methods for object recognition. Inst. for Computer Graphics and Vision, Graz University of Technology, Austria, Technical report ICGTR0108 (ICG-TR-01/08)
19. Ruo-yu Y, Feng S, Tong L (2010) Research of the structural-learning-based symbol recognition mechanism for engineering drawings. In: 2010 6th international conference on digital content, multimedia technology and its applications (IDC), pp 346–349
20. Smith R (2007) An overview of the Tesseract OCR engine. In: Ninth international conference on document analysis and recognition (ICDAR 2007), vol 2. IEEE, pp 629–633. https://doi.org/10.1109/ICDAR.2007.4376991. http://ieeexplore.ieee.org/document/4376991/
21. Smith R (2007) An overview of the Tesseract OCR engine. In: Ninth international conference on document analysis and recognition, ICDAR 2007, vol 2. IEEE, pp 629–633
22. Uijlings JR, Van De Sande KE, Gevers T, Smeulders AW (2013) Selective search for object recognition. Int J Comput Vis 104(2):154–171

Separation of Speech from Speech Interference Based on EGG

Lijiang Chen(✉), Lin Sun, and Xia Mao

Beihang University, Beijing 100191, China
chenlijiang@buaa.edu.cn

Abstract. In this paper, a new method based on the Electroglottography (EGG) for solving speech separation problem was proposed. By using the EGG, the mixed speech signals was segmented to silence segment S, unvoiced segment U, and voiced segment V (SUV segmentation) according to the feature of the voiced and unvoiced speech. The V-segment algorithm was based on Blind Source Separation (BSS) using FastICA. Computational Auditory Scene Analysis (CASA) and spectral subtraction methods were used to obtain the target unvoiced Ideal Binary Mask (IBM). And the waveform synthesis was then performed to obtain the target unvoiced sound for U-segment. The experimental results showed that the proposed algorithm can improve SNR, similarity coefficient and subjective evaluation with EGG.

Keywords: Speech separation · Spectral subtraction · EGG · SUV segmentation · FastICA

1 Introduction

How to distinguish the target speech from the mixed speech signals remains an important problem for speech signal processing. The research on speech separation has important theoretical and practical significance in the fields of voice communication, voice target detection, and speech signal enhancement. It is particularly important to investigate speech separation for several aspects, such as the design of hearing aids, the intelligent speaker recognition in a noisy environment, the time-frequency automatic subtitle loading system, and the artificial intelligence simultaneous interpretation system.

Electroglottography is a conventional instrument for clinical examination of modern medicine. Studies have shown that there is a direct connection between the EGG and the airflow when the voiced sound is generated [1]. It is characterized by its ability of detecting the regularity of vocal cord vibration, the opening and closing of glottis, and the different lossless vibration modes of vocal cords.

2 Related Works

Speech separation technology can be mainly divided into BSS and CASA. Most of CASA-based speech separation algorithms focus on the study of voiced sound,

R. Burduk et al. (Eds.): CORES 2019, AISC 977, pp. 37–46, 2020.
https://doi.org/10.1007/978-3-030-19738-4_5

but the separation effect of the unvoiced sound can affect intelligibility of the language because the unvoiced sound also stores part of the information in a speech signal. Hu and Wang first proposed a separate sound system [2]. Then, Wang and Hu proposed an unvoiced speech separation algorithm from nonspeech interference [3]. But it is difficult to extract useful auditory clues to separate target unvoiced sound when the noise is also a speech signal, because the unvoiced signal has no obvious features.

To solve this problem, this paper propose a speech separation algorithm based on EGG signal. Firstly, the mixture signals were separated into silence segment (S), unvoiced segment (U) and voiced segment (V) in the time domain by using the EGG of the target speech, and the FastICA algorithm [4] was then used to extract the target speech in the V-segment. The algorithm is based on spectral subtraction and IBM is used to extract the target speech in the U-segment. This research is organized as follows. In Sect. 3, we present the description of our system and the details of each stage. Section 4 is the dataset for the experiment. In Sect. 5, we report the comparison experiment result and verify the validity of the speech separation method using EGG. Finally, we conclude the paper and give further discussion in Sect. 6.

3 The Proposed System

We consider the case where the input waveform contains a recording of two people taking simultaneously in this paper. Our purpose is to extract target speech from mixture by using EGG. The framework of the system is shown as Fig. 1. The proposed system can be divided into three stages. Firstly, the EGG signal is used to separate mixture to three segments of S, U and V, which facilitates the extraction of the target speech by different methods for different speech segments. Secondly, the BSS method for the V-segment is used to extract the target voiced sound using ICA. For the U-segment, the spectral subtraction is used to obtain the unvoiced IBM. Finally, the separated unvoiced target of U-segment, voiced target of V-segment and S-segment are spliced in the time domain to obtain the final separated target speech.

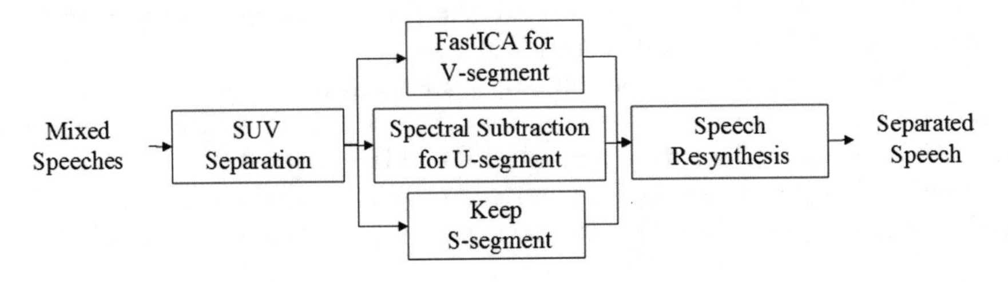

Fig. 1. Framework of the proposed system

3.1 SUV Separation from Mixture

SUV separation refers to separating and labeling of S-segment, U-segment, and V-segment from mixed speech signals. According to different response states of the EGG for unvoiced and voiced speech, the EGG of the target speech can be used to separate the V-segment, the U-segment and the S-segment from the mixed speech signal. A database of German emotional speech (EMODB) [5] is used as the original data. The data is dual-mode information where the speaker's voice information and its EGG signal are collected simultaneously during the voice recording process. This paper adopts the annotation method of Jing [6] to label the SUV manually.

3.2 Voiced Speech Separation

Each V-segment of the mixed speech signals contains the following parts: voiced target, unvoiced noise and voiced noise.

Although the harmonic characteristics and pitch characteristics are usually used as separate clues when separating the voiced sound of mixed speech signals by CASA, it is difficult to use the same frequency filtering in the case of speech interference. We use the ICA method to extract the voiced target from the V-segment. A framework for separating the voiced target is illustrated in Fig. 2.

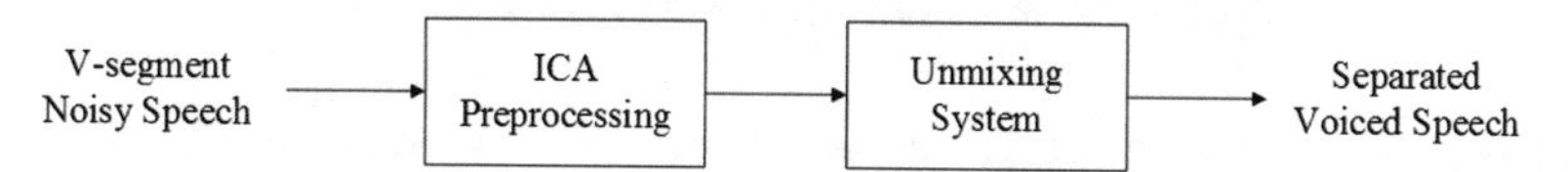

Fig. 2. Framework of the voiced target separation algorithm

The ICA model [7] can be described by the following formula:

$$\mathrm{X} = \mathrm{AS} \tag{1}$$

$$\hat{\mathrm{S}} = \mathrm{Y} = \mathrm{WX} \tag{2}$$

where $\mathrm{S} = [s_1, s_2, \ldots s_m]^T$ is an input source signals with unknown m dimension, A is an unknown $n \times m$ mixing matrix and $\mathrm{X} = [x_1, x_2, \ldots x_n]^T$ is a n-dimensional mixed speech signal. The $m \times n$ unmixing matrix is denoted as W. The ICA model estimates the value of W by using maximum likelihood estimation and maximizing non-Gaussian criteria to find the best estimated signal Y of the source signal S.

The research object in this study is a mixed speech signal composed of two isolated clean data of speakers, and it was considered that there was no additive noise in this system. m is 2 and n is 2. The first line of mixed speech is used as the research object.

The assumption of ICA is that the components of the source signal are statistically independent of each other and the average value of each component of the source signal is zero, so preprocessing is required. The mixed speech signal X is linearly transformed to a whitened vector Z.

$$Z = V \cdot (X - \bar{X}) \tag{3}$$

Z contains the whitened mixture and $E\{ZZ^T\} = I$, $V$is the whitening matrix. The whitening process is implemented by Principal Component Analysis (PCA).

FastICA refers to a rapid optimization iterative algorithm in the form of batch processing [8]. The convergence speed of FastICA is more than twice, and the step size parameter is easier to determine than the gradient search method. Based on these advantages, we use the FastICA method to extract the V-segment target voiced sound. The non-Gaussian calculation based on kurtosis is simple in theory and calculation. The probability density is required for calculation and is computationally intensive when we calculate non-Gaussian by using negative entropy. In comparison, kurtosis is used as the objective function for this purpose. The following formula is the definition of kurtosis:

$$\text{kurt}(x) = E\{x^4\} - 3\left(E\{x^2\}\right)^2 \tag{4}$$

The stronger the signal independence is, the greater the absolute value of kurtosis becomes. When the absolute value of kurtosis reaches maximum, the independent component separation can be considered successful. The objective function can be expressed as follows [9]:

$$J_{\text{kurt}}(W) = E\left\{\left(W^T Z\right)^4\right\} - 3\|W\|^4 + f\left(\|W\|^2\right) \tag{5}$$

The fixed point iterative algorithm of [10] was used to solve W. In the case of the mixed speech signal with a male voice: "She will hand it in on Wednesday." as the target speech, Fig. 3 shows the demonstration of the results of the third V-segment using the FastICA method to separate target speech.

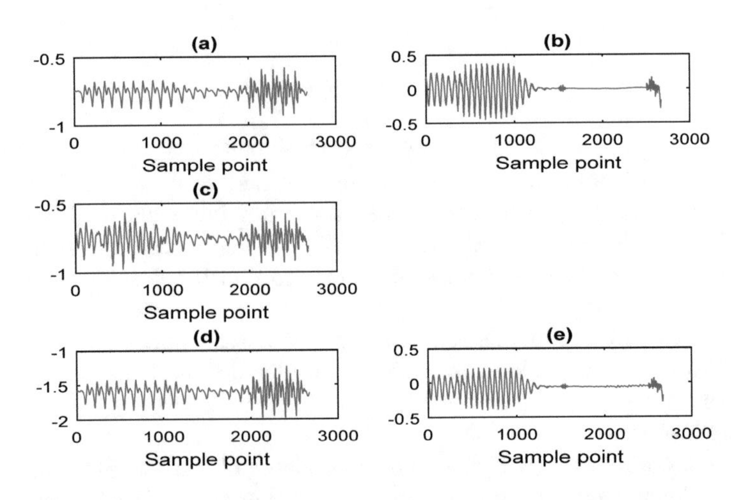

Fig. 3. Separation results of FastICA in waveform format (a) target signal (b) noise signal (c) mixed speech signal (d) output signal 1 (e) output signal 2

It can be seen from Fig. 3 that the output signal 1 is the target voiced signal. The separated speech signal in Fig. 3(d) is similar to the clean original speech in Fig. 3(a).

3.3 Unvoiced Speech Separation

Each U-segment of the mixed speech signals contains the following parts: unvoiced target, unvoiced noise and voiced noise.

A framework of the unvoiced target separation algorithm is illustrated in Fig. 4. The basic idea of the unvoiced target separation algorithm is to preprocess the whole mixed speech with CASA method to generate time-frequency (T-F) units, from which the periodic features are extracted so as to remove the voiced. Then the removed voiced segment is used to estimate the U-segment interference energy. Finally, the noise energy is subtracted from the U-segment to form the target unvoiced segment.

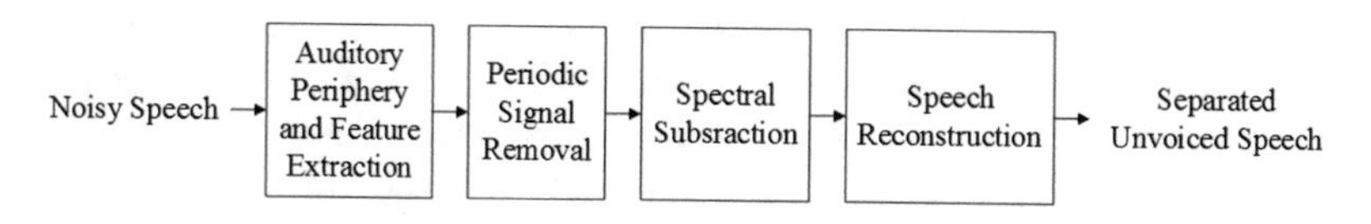

Fig. 4. Framework of the unvoiced target separation algorithm

Auditory Peripheral Analysis and Feature Extraction. The mixed speech is processed by the Gammatone filterbank and needs to be processed by the Meddis model which simulates the process of nerve fiber processing signals [11]. The speech signal in the time domain is converted into a signal in the time-frequency domain. For the output T-F signal of the auditory peripheral analysis, the signals on each channel is subjected to framing processing. The frame length is 20 ms and the frame shift is 10 ms. We use the Hamming window to frame the output signal in the time domain.

Separating clues is supposed to connect the target speech at the particular T-F unit with the target speech at the next T-F unit. It mainly extracts a series of harmonic and energy characteristics of speech including autocorrelation function (ACF) $\mathrm{A_H}$, the envelope ACF A_E, cross-channel correlation C_H, the envelope cross-channel correlation C_E and energy E.

$$\mathrm{A_H}(c,m,\tau)=\frac{1}{N}\sum_{n=0}^{N_c-1}h(c,mT-n)\,h(c,mT-n-\tau) \tag{6}$$

$$A_\mathrm{E}(c,m,\tau)=\frac{1}{N}\sum_{n=0}^{N_c-1}h_\mathrm{E}(c,mT-n)\,h_\mathrm{E}(c,mT-n-\tau) \tag{7}$$

$$C_\mathrm{H}(c,m)=\sum_{\tau=0}^{L-1}A_\mathrm{H}(c,m,\tau)\,A_\mathrm{H}(c+1,m,\tau) \tag{8}$$

$$C_{\mathrm{E}}(c,m)=\sum_{\tau=0}^{L-1}A_{\mathrm{E}}(c,m,\tau)A_{\mathrm{E}}(c+1,m,\tau) \tag{9}$$

$$\mathrm{E}(\mathrm{c},\mathrm{m})=A_{\mathrm{H}}(c,m,0) \tag{10}$$

N_{c} is the number of samples per frame. τ is the delay, and the value range is [0, 12.5 ms]. $\mathrm{h}(\mathrm{c},\mathrm{n})$ represents the value of the hair cell output at frequency channel c and time n.

Periodic Signal Removal. After the periodic signal is removed, only the unvoiced target and the unvoiced noise are left in the U-segment speech, which can reduce the error rate of the unvoiced separation. If a cell is dominated by a periodic signal, the cell will have a higher C_{H} and C_{E} [7]. So the IBM of each unit is:

$$\mathrm{L}_1(c,m)=\begin{cases}0, C_{\mathrm{H}}(c,m)>\theta_{\mathrm{R}} \text{ or } C_{\mathrm{E}}(c,m)>\theta_{\mathrm{E}}\\ 1,\ \text{else}\end{cases} \tag{11}$$

where θ_{R} and θ_{E} are the threshold of C_{H} and C_{E}, respectively. The values of θ_{R} and θ_{E} are related to the balance between the accuracy of the removal periodic signal and the integrity of the preserved unvoiced signal. It is a relatively suitable choice when θ_{R} is 0.9 and θ_{E} is 0.96 [12]. If $\mathrm{L}_1(c,m)$ is 1, this unit is dominated by unvoiced sound, and vice versa.

T-F Unit Labeling Based on Spectral Subtraction. After the periodic signal is removed, unvoiced noise is the only component of V-segment. U-segment is composed of unvoiced noise and unvoiced target. By using the two V-segments adjacent to the U-segment in order to estimate the residual noise energy of the U-segment and using the residual noise energy estimation value in order to remove the noise of the U-segment, the target unvoiced sound can be obtained. The estimated residual energy is:

$$N_{\mathrm{dB}}(c,m)=\frac{\sum_{i=m_1-l_1}^{m_1-1}E_{\mathrm{dB}}(c,i)\cdot L_1(c,i)+\sum_{i=m_2+1}^{m_2+l_2}E_{\mathrm{dB}}(c,i)\cdot L_1(c,i)}{\sum_{i=m_1-l_1}^{m_1-1}L_1(c,i)+\sum_{i=m_2+1}^{m_2+l_2}L_1(c,i)} \tag{12}$$

where $E_{\mathrm{dB}}(c,i)$ is the energy of T-F unit $U(c,i)$; $L_1(c,i)$ is IBM of $U(c,i)$, marked with 1 which means it is dominated by the unvoiced sound; m_1, m_2 are the indexes of the current U-segment initial frame and the end frame; l_1, l_2 are the lengths of the V-segment before the current U-segment. l_1, l_2 are also the length of the V-segment after the current U-segment, and are represented by the number of frames. Spectral subtraction is used to remove the unvoiced noise [3]. We first need to find the signal-to-noise ratio (dB) of the V-segment unit.

$$\delta(\mathrm{c},\mathrm{m})=10\log_{10}\left(\frac{[E(c,m)-N'(c,m)]^{+}}{N'(\mathrm{c},\mathrm{m})}\right) \tag{13}$$

$$N(c,m)=10^{N'_{\mathrm{dB}}(c,m)/10} \tag{14}$$

where $[x]^+ = \begin{cases} x, x > 0 \\ 0, x \le 0 \end{cases}$ is a positive function, and $N'(c,m)$ is the estimated noise energy of the time-frequency unit $U(c,m)$. Based on the auditory masking effect, when a unit's unvoiced target energy is greater than the unvoiced noise energy, the unit is considered to be the unvoiced target. For unit $U(c,m)$ belonging to a type-1 segment, we label it as follows.

$$L_2(c,m) = \begin{cases} 1, \delta(c,m) > 0 \text{ and } L_1(c,m) = 1 \\ \quad 0, \text{ else} \end{cases} \tag{15}$$

The flag is 1 indicating that the unit is dominated by the target unvoiced. Otherwise, the flag is 0. According to the labeling result, the separated target unvoiced is obtained by waveform synthesis.

3.4 SUV Re-synthesis

The final stage in this research is SUV re-synthesis. The voiced targets which are separated in Sect. 3.2, the unvoiced targets which are separated in Sect. 3.3 and the segment S are spliced in the time domain. The final target result of the algorithm is obtained.

4 Dataset

We use the EMODB database [5] for experimental and performance evaluation. The left channel is voice information. The right channel is EGG signal, and the sampling rate is 16000 Hz. As showed in Table 1, four male utterances and four female utterances are chosen as test data.

Table 1. The content of the speech data

Data number	Gender	Statement content
a	F	She will hand it in on Wednesday
b	F	Tonight I could tell him
c	F	In seven hours it will be
d	F	The tablecloth is lying on the fridge
e	M	What about the bags standing there under the table?
f	M	It will be in the place where we always store it
g	M	I will just discard this and then go for a drink with Karl
h	M	The tablecloth is lying on the fridge

We take 8 utterances for pairwise combination. One of the selected utterances retained its right channel which is EGG signal in each group is the target utterance, and the other utterance is interference. The left channel signal of the two utterances are combined into a mixture signal in the time domain.

5 Experiments and Results

In order to verify the effectiveness of EGG signal for target speech extraction, this paper designs a comparison experiment between group A and group B. Group B is the experimental group using the proposed system with EGG signal. Group A is the control group without EGG and zero-energy ratio is used for U-segment and V-segment discrimination [13].

5.1 Objective Evaluation Criteria

The performance of the proposed system is measured in terms of SNR and similarity coefficient e:

$$\mathrm{SNR} = 10\log\frac{s^2(n)}{\left[\hat{s}(n) - s(n)\right]^2} \tag{16}$$

$$e\left(s(n), \hat{s}(n)\right) = \frac{\left|\sum_{n=0}^{M} s(n)\,\hat{s}(n)\right|}{\sqrt{\sum_{n=0}^{M} s^2(n) \sum_{n=0}^{M} \hat{s}^2(n)}} \tag{17}$$

where $s(n)$ is the original speech and $\hat{s}(n)$ is the separated speech. The similarity coefficient $\mathrm{e}\left(s(n), \hat{s}(n)\right)$ can calculate the similarity between the separated speech and the original speech. The closer the value is to 1, the more similar the separated speech is to the original speech and the better the separation algorithm works. The SNR improvement and the similarity coefficient improvement for the proposed system is shown in Table 2. The proposed system improves the SNR with an average value of 2.70 dB and 0.93 dB as compared to input SNR and group A system respectively. From Table 2, it is proved that the proposed systems has a percentage of similarity as 88.18% which is higher than group A system. With the EGG signal, the proposed algorithm effectively uses the FastICA and spectral subtraction, to increase the quality of the speech when compared with other system.

Table 2. The SNR and similarity coefficient result table

Mixture	SNR_{M} (dB)	SNR_{A} (dB)	SNR_{B} (dB)	e_{A}	e_{B}
Af	9.886	10.1238	10.2215	0.7692	0.8896
Bd	18.4353	20.8403	23.1757	0.8741	0.9067
Eh	3.5369	5.9404	6.021	0.8446	0.9204
Ed	10.4975	12.0169	13.943	0.7213	0.8253
Fb	1.7805	2.5433	3.4912	0.8875	0.8971
Hb	−0.6991	1.6383	2.5441	0.8012	0.827
Ga	5.1785	7.8987	8.0909	0.9046	0.9071
Avg.	6.9450	8.7145	9.6410	0.8289	0.8818

5.2 Subjective Evaluation Measures

Seven volunteers are recruited for subjective audition experiments on Group A and Group B. With a total of 0, 1, 2, 3, 4, 5 points, six levels of scoring system, the subjective evaluation tables are shown in Table 3:

Table 3. Subjective evaluation result

Mixture	Group A	Group B
Af	2.6250	3.2500
Bd	3.6250	4.0000
Eh	3.5000	4.1250
Ed	2.8750	3.0000
Fb	3.2500	3.3750
Hb	3.1250	3.7500
Ga	3.6250	3.8750
Avg.	3.2321	3.6250

The average scores of speech quality in group A and group B are 3.23 and 3.63, respectively. The average score of group B is higher than that of group A. From the perspective of subjective evaluation indicators, the proposed algorithm is effective, but there is room for algorithm improvement and optimization.

6 Conclusion

This research propose an algorithm using EGG signal for separating target speech from the monaural speech of two talkers. By using EGG to separate the mixture into S-segment, U-segment and V-segment as the basis of the separation algorithm, the proposed algorithm enhances the separated speech. The performance of the proposed system is measured in terms of SNR improvement, similarity coefficient and subjective evaluation, shown in Tables 2 and 3, respectively. The experimental results show that the proposed system improves speech quality in terms of increasing SNR, similarity coefficient and subjective evaluation as compared to group A without EGG. The proposed algorithm captures most of the voiced speech components and unvoiced speech components.

Acknowledgements. This work was supported by the Fundamental Research Funds for the Central Universities (Project number: YWF-19-BJ-J-197) and the National Natural Science Foundation (Project number: 61603013). The author would like to thank the anonymous reviewers for their valuable suggestions and remarks.

References

1. Chen L, Mao X, Compare A (2013) A new method for speech synthesis combined with EGG. In: NCMMSC, pp 332–336
2. Hu G, Wang DL (2008) Segregation of unvoiced speech from nonspeech interference. J Acoust Soc Am 124(2):1306–1319
3. Hu K, Wang DL (2011) Unvoiced speech segregation from nonspeech interference via CASA and spectral subtraction. IEEE Trans Audio Speech Lang Process 19(6):1600–1609
4. Hyvärinen A (1999) Fast and robust fixed-point algorithms for independent component analysis. IEEE Trans Neural Netw 10(3):626
5. Burkhardt F, Paeschke A, Rolfes MA, Sendlmeier WF, Weiss B (2005) A database of German emotional speech. In: INTERSPEECH
6. Jing S, Mao X, Chen L, Zhang N (2015) Annotations and consistency detection for Chinese dual-mode emotional speech database. J Beijing Univ Aeronaut Astronaut 41(10):1925–1934
7. Liu Z, Miao Z, Wan L (2015) Speech blind signal separation with FastICA and Markov Chain combination. In: International conference on signal processing, pp 541–544
8. Saruwatari H, Kawamura T, Nishikawa T, Lee A, Shikano K (2006) Blind source separation based on a fast-convergence algorithm combining ICA and beamforming. IEEE Trans Audio Speech Lang Process 14(2):666–678
9. Hyvärinen A (1997) A family of fixed-point algorithms for independent component analysis. In: IEEE international conference on acoustics
10. Hyvärinen A, Oja E (2000) A fast fixed-point algorithm for independent component analysis. Int J Neural Syst 10(01):1–8
11. Strahl S, Mertins A (2009) Analysis and design of gammatone signal models. J Acoust Soc Am 126(5):2379
12. Wang DL, Brown GJ (1999) Separation of speech from interfering sounds based on oscillatory correlation. IEEE Trans Neural Netw 10(3):684–697
13. Bachu RG, Kopparthi S, Adapa B, Barkana BD (2010) Voiced/unvoiced decision for speech signals based on zero-crossing rate and energy. In: Advanced techniques in computing sciences and software engineering, pp 279–282

Improving the Quality of Clustering-Based Diagnostic Rules by Lowering Dimension of the Cluster Prototypes

Sebastian Porebski(✉) and Ewa Straszecka

Institute of Electronics,
Faculty of Automatic Control, Electronics and Computer Science,
Silesian University of Technology, Gliwice, Poland
{sebastian.porebski,ewa.straszecka}@polsl.pl

Abstract. The study concerns an evaluation of rule premise conditions generated from cluster prototypes. Different clustering methods are used as the first step of training the rule-based classifiers. If rules are generated directly from cluster prototypes, their premise complexity (the number of conditions) is equal to the dimension of considered training data. This paper describes an idea of detecting an appropriate part of the clustering-based rule premise and refinement of the rule set. It is especially valuable in medical domain applications of rule-based classifiers in the view of their generalization quality and interpretability.

Keywords: Clustering · Rule extraction · Rule set tuning · Fuzzy rule classifiers

1 Introduction

The most popular classifiers reach high efficiency due to computational capabilities of computer systems e.g. multilayer neural networks, support vector machines [16]. However, rule-based classifiers are used when classifier cannot be used as black-box system. The main advantage and reason of investigating the rule-based classifiers is their interpretability [4]. This feature is inevitable in areas in which classification or decision is the domain of people. For instance, computers cannot substitute human especially in the medical domain, though human diagnosis is based on measurements of sophisticated medical equipment.

Diagnostic rules can be created automatically (using heuristic, genetic or statistic methods) [6]. One of the popular ways is a generation of inference rules (their premises' conditions) based on the training data clustering result. Obtained clusters usually are described by cluster prototypes which are the best representatives of the cluster. These prototypes are vectors of the same dimension as clustered data. When rule premises are described by cluster prototypes, the number of their conditions equals data dimension [7,11]. It causes generation

R. Burduk et al. (Eds.): CORES 2019, AISC 977, pp. 47–56, 2020.
https://doi.org/10.1007/978-3-030-19738-4_6

of unreadable rules when the dimension of data is not small enough [4]. Different rule extraction ideas [12] show that rules does not have to be so complex hence we do not always need to use a full-scale prototype to distinguish a cluster from others. There are some, but very few works tackling the problem of redundant or similar fuzzy sets in extracted rule sets [2,18]. Our idea is different, we evaluate the fitness of every single fuzzy set in the rule premises to data cases and show that it is enough to reduce rule set complexity significantly.

This paper presents experimental results showing that reduction of the dimension of the clusters' prototypes allows improving clustering-based rule set interpretability keeping satisfactory diagnosis quality of the fuzzy rule-based classifier. A couple of popular and comprehensive clustering algorithms are chosen for a clear presentation of the study [5]. Ideas presented in this work do not depend on any clustering method because we operate with its result. Our ideas can be used in studies on clustering-based rule classifiers.

2 Rule-Based Diagnosis Support

Let us suppose that medical data set is the collection of p-dimensional vectors that are values of p medical tests performed on N patients. Exemplar patient's (k-th patient, $k \in 1, \cdots, N$) data can be described in the following way:

$$\boldsymbol{x}_k = \left[x_{k1}\ x_{k2} \cdots x_{kj} \cdots x_{kp}\right], \tag{1}$$

where x_{kj} is the value of j-th symptom. It can be a test, an anamnesis or a measurement result. Diagnostic (classification in general) "if-then" rule $\mathcal{R}$ can be the following statement: "If symptom $\mathcal{A}$ and symptom $\mathcal{B}$ and symptom **C** then diagnosis $\mathcal{D}$". Symptoms $\mathcal{A}$, $\mathcal{B}$ and $\mathcal{C}$ are linguistic values e.g. "low T3RU test result", "high thyroxine level", and $\mathcal{D}$ is the name of the diagnosis, e.g. "high thyroid activity". Premise of the i-th rule of l-th diagnosis $\mathcal{R}_i^{(l)}$ ($i = 1, \cdots, c, l = 1, \cdots, d$) can be simple (considering one symptom):

$$X_j \text{ is } \mathcal{A}_{ij}^{(l)}, \tag{2}$$

or complex (e.g. considering all symptoms):

$$X_1 \text{ is } \mathcal{A}_{i1}^{(l)} \text{ and } \cdots \text{ and } X_j \text{ is } \mathcal{A}_{ij}^{(l)} \text{ and } \cdots \text{ and } X_p \text{ is } \mathcal{A}_{ip}^{(l)}, \tag{3}$$

where X_j is the variable related to j-th symptom and $\mathcal{A}_{ij}^{(l)}$ is a fuzzy set [20] described by the fuzzy membership function $\mu_{ij}^{(l)}$ modelling symptom of diagnosis $\mathcal{D}$. Decision based on a set of rules is a matter of choice because plenty of inference mechanisms are proposed in the literature [7,11,19]. In the study, according to our former experiments [13], fuzzy belief measure from the Dempster-Shafer theory [15] (DST) is calculated according to all knowledge about l-th diagnosis saved in c rules. The rule premises are treated in the DST as fuzzy focal elements, because each premise "focuses part of the knowledge about diagnosis"

[15]. Activation degree of the rule $\mathcal{F}_i^{(l)}$ can be calculated as:

$$\mathcal{F}_i^{(l)}(\boldsymbol{x}_k) = \min_{j \in \mathcal{J}_i^{(l)}} \mu_{ij}^{(l)}(x_{kj}), \tag{4}$$

where $\mathcal{J}_i^{(l)}$ is the set of symptoms included in the i-th rule ($\mathcal{J}_i^{(l)} \subset \{1, \cdots, p\}$). If $\mathcal{F}_i^{(l)}$ are calculated for all rules of l-th diagnosis, then belief measure of this diagnosis can be calculated [13]:

$$Bel^{(l)}(\boldsymbol{x}_k) = \sum_{i=1}^{c} \mathcal{F}_i^{(l)}(\boldsymbol{x}_k) \cdot w(\mathcal{R}_i^{(l)}), \tag{5}$$

where $w(\mathcal{R}_i^{(l)})$ is the rule weight. This weight states the diagnosis belief degree when $\mathcal{R}_i^{(l)}$ is true. In the DST it is called the basic probability value [15] and it is conditioned by: $w(f) = 0$ (value equal zero for false premise f) and $\sum_i^c w(\mathcal{R}_i^{(l)}) = 1$. When $Bel^{(l)}(\boldsymbol{x}_k)$ for d diagnoses are calculated, final diagnosis for $\boldsymbol{x}_k$ is the diagnosis for which maximum belief value is obtained [17].

3 Prototype-Based Rule Premise Conditions

As the result of every clustering method we obtain some number of revealed clusters within the considered data [5]. Let us denote this number as c. Depending on clustering methods, c can be the result of the algorithm or can be a parameter set by a user. Every cluster is described by its prototype which can be one of the original data cases or a new calculated vector. Anyway, it is the vector that best represents its cluster. It is described in the following way:

$$\boldsymbol{v}_i = \left[v_{i1}\ v_{i2}\ \cdots\ v_{ij}\ \cdots\ v_{ip}\right], \tag{6}$$

where v_{ij} is the j-th value of the i-th cluster prototype ($i = 1, \cdots, c$). A determination of the cluster prototype depends on the clustering method [5]. In this paper four different clustering algorithms are used: k-means algorithm [10], fuzzy c-means algorithm [1], affinity propagation [3], Gaussian mixture models [9]. In the study, above methods are not discussed because they are rather classic and widely described [5]. They are used to obtain cluster prototypes and for the rule set generation. Usually prototype vectors are used to model fuzzy membership functions in the rule premises [7,11]. The membership function $\mu_{ij}^{(l)}$ is related to j-th value of the i-th cluster prototype in l-th diagnosis data subset. In this way c is the number of clusters and the number of rules of l-th diagnosis. In the study, v_{ij} is used as the parameter of the Gaussian membership function:

$$\mu_{ij}^{(l)} = \exp\left[-\frac{1}{2}\left(\frac{x_j - v_{ij}}{\sigma_{ij}}\right)^2\right], \tag{7}$$

where σ_{ij} is the standard deviation of the j-th symptom in the i-th cluster. If $\sigma_{ij} = 0$, then it can be calculated according to j-th symptom in all data of l-th diagnosis. When rule premises are created directly from cluster prototypes then each of c rules has p premise conditions. The example presented below reveals that some of the premise conditions are not helpful to distinguish the i-th cluster. The $\mu_{ij}^{(l)}$ membership function is appropriate if only the fuzzy set $A_{ij}^{(l)}$ matches the data set. If $\mu_{ij}^{(l)}(x_{kj}) = 0$ then we consider that j-th symptom value of k-th data case does not fit the premise condition and otherwise when $\mu_{ij}^{(l)} > 0$. However, to avoid such zero-one interpretation and keep fuzzy processing the study exact values of $\mu_{ij}^{(l)}$ are used to evaluate the j-th condition in the i-th rule premise. The Mathews Correlation Coefficient [14] can be applied for evaluation of each rule premise condition:

$$MCC_{ij}^{(l)} = \frac{tp_{ij}^{(l)} \cdot tn_{ij}^{(l)} - fp_{ij}^{(l)} \cdot fn_{ij}^{(l)}}{\sqrt{(tp_{ij}^{(l)} + fp_{ij}^{(l)}) \cdot (tp_{ij}^{(l)} + fn_{ij}^{(l)}) \cdot (fn_{ij}^{(l)} + fp_{ij}^{(l)}) \cdot (tn_{ij}^{(l)} + fn_{ij}^{(l)})}}, \tag{8}$$

where

$$tp_{ij}^{(l)} = \frac{1}{N_l} \sum_{i=1}^{N_l} \mu_{ij}^{(l)}(x_{kj}^{(l)}), \quad fp_{ij}^{(l)} = \frac{1}{N_{\neq l}} \sum_{i=1}^{N_{\neq l}} \mu_{ij}^{(l)}(x_{kj}^{(\neq l)}), \tag{9}$$

$$tn_{ij}^{(l)} = \frac{1}{N_{\neq l}} \sum_{i=1}^{N_{\neq l}} \left[1 - \mu_{ij}^{(l)}(x_{kj}^{(\neq l)})\right], \quad fn_{ij}^{(l)} = \frac{1}{N} \sum_{i=1}^{N_l} \left[1 - \mu_{ij}^{(l)}(x_{kj}^{(l)})\right], \tag{10}$$

where N_l means the number of data of l-th diagnosis and $N_{\neq l}$ means the number of data related to other diagnoses. In this way MCC evaluates the appropriateness of j-th condition in the i-th rule premise in detail because it allows checking all in which the fuzzy set $\mathcal{A}_{ij}^{(l)}$ matches all data cases from all diagnoses. Particularly, $MCC = 1$ when the $\mathcal{R}_i^{(l)}$ fits perfectly all training data ($tp_{ij}^{(l)}$ and $tn_{ij}^{(l)}$ are maximal), $MCC \approx 0$ when $\mathcal{R}_i^{(l)}$ fits the proper diagnosis data equally well as other diagnoses data ($tp_{ij}^{(l)} \approx fn_{ij}^{(l)}$, $tn_{ij}^{(l)} \approx fp_{ij}^{(l)}$) and $MCC = -1$ when $\mathcal{R}_i^{(l)}$ is useless ($fp_{ij}^{(l)}$ and $fn_{ij}^{(l)}$ are maximal).

4 Experiments

UCI benchmark used in the study is related to thyroid gland functioning. In this example, a rule set is generated with FCM method [1,5] proceeded on each diagnosis subset separately. The data set is related to three exclusive diagnoses: euthyroid, hyperthyroidism and hypothyroidism. If we choose to obtain five clusters ($c = 5$) we obtain $c \cdot d = 15$ clusters of the considered data. Since each of 215 patients in the data set is described by five hormone tests ($p = 5$), $c \cdot d \cdot p = 75$ fuzzy membership functions are generated. Figure 1 shows this rule set which is generated processing all cases of considered data set.

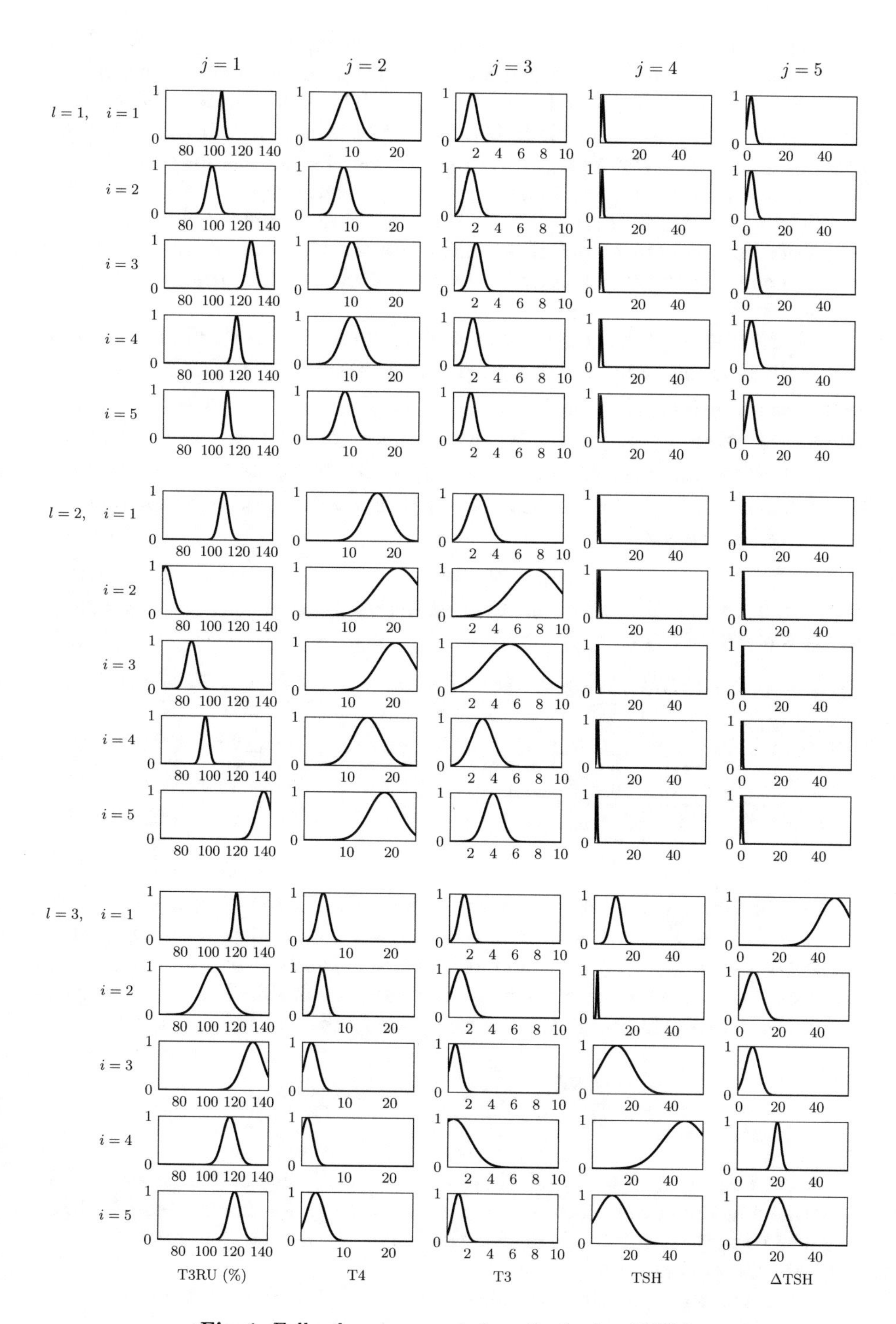

Fig. 1. Full rule set generated on the basis of FCM

Visual interpretation of the clustering-based fuzzy rules leads to the observation that there are similar premise conditions for several premises within the diagnosis, e.g. for $l = 1$, $j = 4$, $k = 1, 2, 3, 4, 5$. This observation indicates that the fourth ($j = 4$) symptom is important to perform efficient diagnosis. However, for the same symptom ($j = 4$) we can see that the premise condition is very similar also for another diagnosis ($l = 3$). Therefore, our belief about the fourth symptom importance is false. Membership functions related to the fourth symptom do not help to obtain satisfying knowledge base. When almost the same membership functions describe mutually exclusive diagnoses the symptom is ambiguous.

It is quite natural that the ambiguous part of the rule set should be detected. $MCC_{ij}^{(l)}$ (8) is calculated for this purpose. Table 1 presents $MCC_{ij}^{(l)}$ values for each premise condition described by $\mu_{ij}^{(l)}$ that Fig. 1 presents. It is necessary to choose the required quality of the rule premise condition to decide about its removal or leaving in the $\mathcal{R}_i^{(l)}$. For this purpose let us denote ϵ as required value of MCC. If $MCC_{ij}^{(l)} > \epsilon$, evaluation of $\mu_{ij}^{(l)}$ is positive and it is not removed from original rule set.

Table 1. Mathews Correlation Coefficient calculated for all premise conditions in the rule sets presented in Fig. 1

j	$\mathcal{R}_1^{(1)}$	$\mathcal{R}_2^{(1)}$	$\mathcal{R}_3^{(1)}$	$\mathcal{R}_4^{(1)}$	$\mathcal{R}_5^{(1)}$	$\mathcal{R}_1^{(2)}$	$\mathcal{R}_2^{(2)}$	$\mathcal{R}_3^{(2)}$	$\mathcal{R}_4^{(2)}$	$\mathcal{R}_5^{(2)}$	$\mathcal{R}_1^{(3)}$	$\mathcal{R}_2^{(3)}$	$\mathcal{R}_3^{(3)}$	$\mathcal{R}_4^{(3)}$	$\mathcal{R}_5^{(3)}$
1	0.19	0.03	−0.05	0.11	0.18	−0.22	0.25	0.35	0.18	0.06	0.27	−0.33	0.34	0.07	0.27
2	0.65	0.56	0.56	0.60	0.62	0.48	0.61	0.58	0.27	0.57	0.43	0.43	0.62	0.53	0.65
3	0.40	0.40	0.36	0.41	0.40	−0.23	0.39	0.38	0.04	0.39	−0.10	0.19	0.43	0.31	0.21
4	0.30	0.29	0.23	0.28	0.32	0.21	0.05	0.20	0.12	0.16	0.44	−0.23	0.27	0.20	0.19
5	0.31	0.40	0.46	0.43	0.44	0.59	0.67	0.65	0.68	0.70	0.25	−0.09	−0.03	0.26	0.39

As the first experiment, we perform diagnosis of all thyroid data cases for $\epsilon \in [-0.35, 0.7]$ changed with step 0.05. This range is determined by values presented in Table 1. Figure 2 presents diagnosis efficiency and reduction of the rule premise conditions for different ϵ. Efficiency is calculated as the percentage number of correct diagnoses and the reduction of the rule set is calculated as:

$$\Delta = \frac{100\%}{d \cdot c \cdot p} \cdot \sum_{l=1}^{d} \sum_{i=1}^{c} \sum_{j=1}^{p} I(MCC_{ij}^{(l)} \leq \epsilon). \tag{11}$$

According to Fig. 2 we can see that when $\epsilon = 0.35$ then 53.3% (40 of 75) rule conditions are removed, but efficiency is the same as for the full set (95.35%). Figure 3 presents how full rule set is refined after the reduction of the ambiguous part of rule set conditions.

According to exploratory results we are sure that rule complexity can be lowered, but we need to verify if this reduction performed in the training step will not result in poor generalization quality in the testing step of the

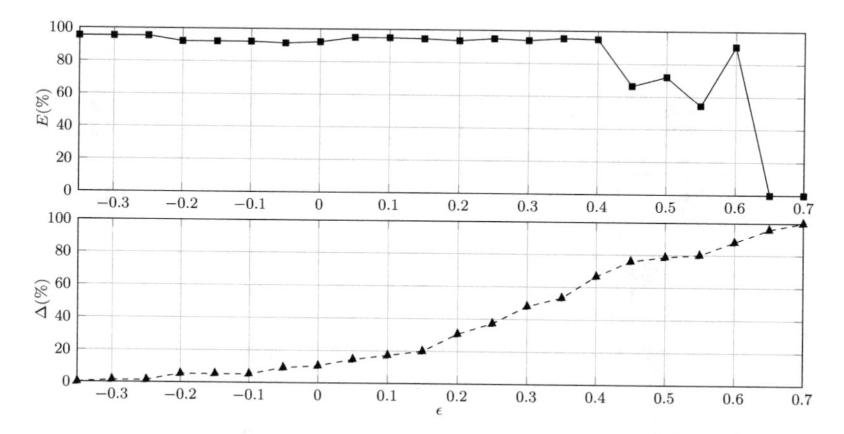

Fig. 2. Diagnosis efficiency and reduction of rule set antecedents according to different evaluation condition

rule-based diagnosis support. For this purpose 10-fold cross-validation (10FCV) can be performed. According to reference [8], it is good to perform 10FCV with optimally balanced distribution of all folds. In this point, four clustering methods [1,3,9,10] are used to generate cluster prototypes. 10FCV is also repeated for different values of the ϵ related to the quality of the rule premise. The value of ϵ is changed between -0.5 (for each experiment $MCC_{ij}^{(l)}$ is greater) and $+1$ (maximal possible theoretical MCC value [14]). Figure 4 presents training and testing percentage efficiency together with rule complexity reduction Δ (11).

5 Discussion and Conclusions

Elimination of inappropriate rule premise conditions can lead to an improvement of generalization quality of the rule set. For each method, we can notice that it is possible to choose ϵ within the interval $[0, 0.5]$ according to the training result that allow reducing the number of rule premise conditions. Results of testing show that the generalization efficiency stays on satisfying level or is even better than without the rule set refinement.

The aim of the study in not to evaluate different clustering methods as the basis of rule set generation. Usually clustering methods are the first step of the rule premise generation and have the greatest impact on the classification efficiency result. Testing efficiencies in Fig. 4 differ for various clustering methods. Changes in efficiency with no doubt indicate that experimenting clustering ideas in application for rule-based classifiers is important, but does not resolve the problem of the rule set complexity. This complexity depends not only on the number of rules but especially on the number of conditions in their premises which influence their interpretability. Our experimental results show that the evaluation of the cluster prototype-based rule premise can be helpful because

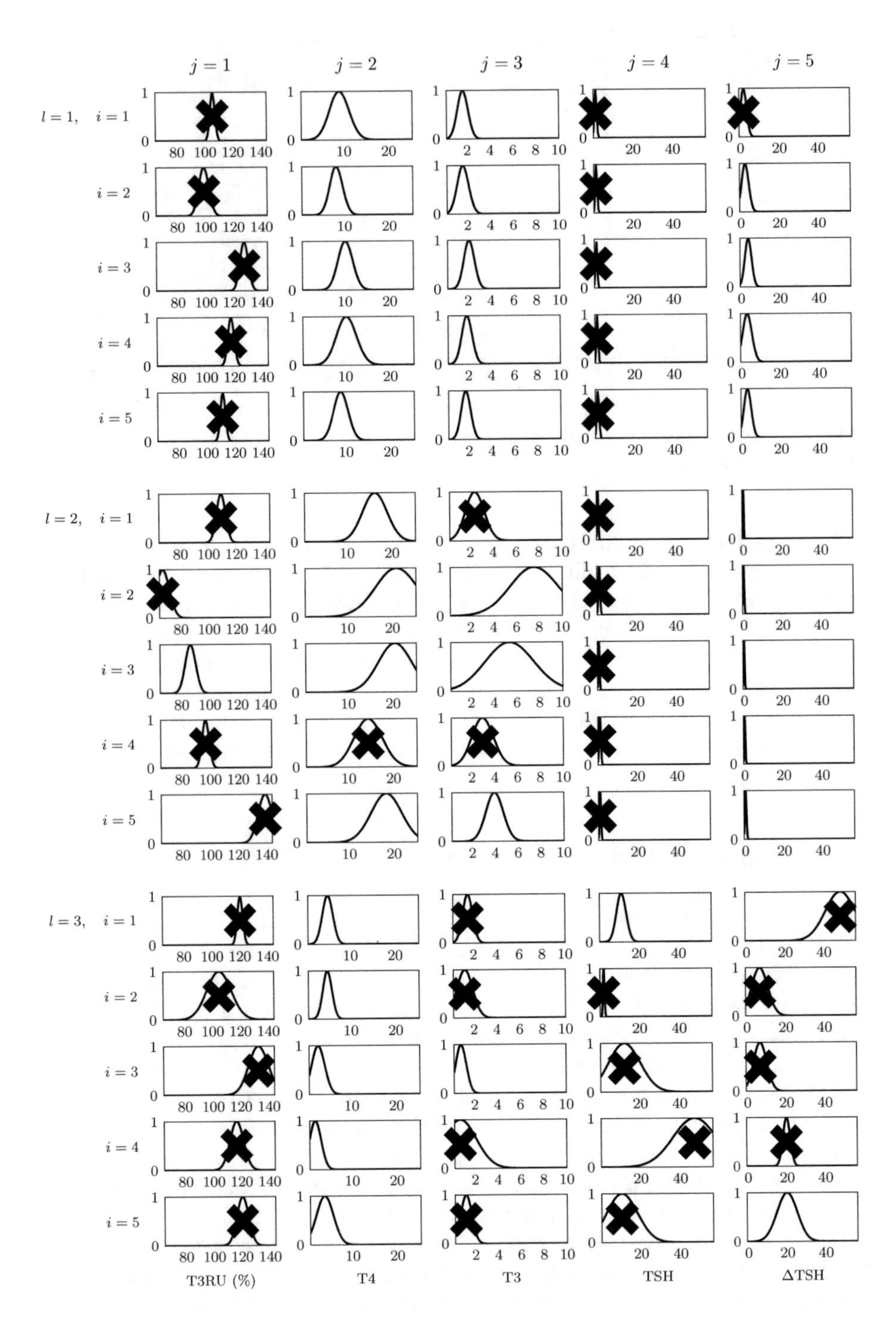

Fig. 3. Refined rule set after evaluation of the premise conditions with MCC

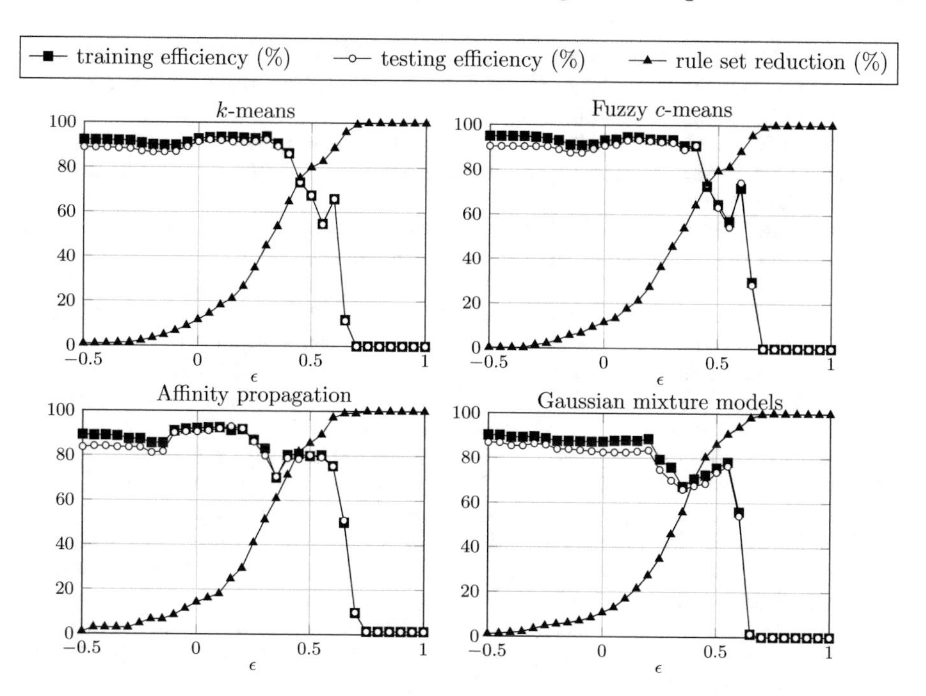

Fig. 4. Performance of the rule-based diagnosis in the 10-fold cross-validation procedure

it makes it possible to detect the ambiguous part and refine diagnostic rule set. This refinement leads to better interpretability together with satisfactory diagnosis efficiency. Presented exemplar results need further studies that concern: automatic rule set tuning without choosing evaluation value threshold ϵ or investigation of clustering ideas that provide cluster prototypes with already reduced dimension.

Acknowledgements. This research is financed from the statutory activities (BKM/510/Rau- 3/2018 & BK/206/Rau-3/2018) of the Institute of Electronics of the Silesian University of Technology.

References

1. Bezdek JC (1982) Pattern recognition with fuzzy objective function algorithms. Plenum Press, New York
2. Chen M-Y, Linkens DA (2004) Rule-base self-generation and simplification for data-driven fuzzy models. Fuzzy Sets Syst. 142:243–265. https://doi.org/10.1016/S0165-0114(03)00160-X
3. Frey BJ, Dueck D (2007) Clustering by passing messages between data points. Science 315(5714):972–976. https://doi.org/10.1126/science.1136800

4. Gacto MJ, Alcala R, Herrera F (2011) Interpretability of linguistic fuzzy rule-based systems: an overview of interpretability measures. Inf Sci 181:4340–4360. https://doi.org/10.1016/j.ins.2011.02.021
5. Guojun G, Ma C, Wu J (2007) Data Clustering: Theory, Algorithms, and Applications. ASA-SIAM Series on Statistics and Applied Probability. SIAM, Philadelphia, ASA, Alexandria
6. Gorzalczany MB, Rudzinski F (2016) A multi-objective genetic optimization for fast, fuzzy rule-based credit classification with balanced accuracy and interpretability. Appl Soft Comput 40:206–220. https://doi.org/10.1016/j.asoc.2015.11.037
7. Jezewski M, Czabanski R, Leski JM, Jezewski J (2018) Fuzzy classifier based on clustering with pairs of ϵ-hyperballs and its application to support fetal state assessment. Expert Syst Appl 118:109–126. https://doi.org/10.1016/j.eswa.2018.09.030
8. Moreno-Torres JG, Saez JA, Herrera F (2012) Study on the impact of partition-induced dataset shift on k-fold cross-validation. IEEE Trans Neural Netw Learn Syst 23(8):1304–1312. https://doi.org/10.1109/TNNLS.2012.2199516
9. McLachaln GJ, Basford KE (1988) Mixture models: inference and applications to clustering. Statistics: Textbooks and Monographs, Marcel Dekker
10. MacQueen JB (1967) Some methods for classification and analysis of multivariate observations. In: Proceedings of 5th Berkeley symposium on mathematical statistics and probability, vol 1. University of California Press, pp 281–297
11. Xingchen H, Pedrycz W, Xianmin W (2018) Fuzzy classifiers with information granules in feature space and logic-based computing. Pattern Recognit 80:156–167. https://doi.org/10.1016/j.patcog.2018.03.011
12. Porebski S, Porwik P, Straszecka E, Orczyk T (2018) Liver fibrosis diagnosis support using the Dempster–Shafer theory extended for fuzzy focal elements. Eng Appl Artif Intell 76:67–79. https://doi.org/10.1016/j.engappai.2018.09.004
13. Porebski S, Straszecka E (2018) Using fuzzy numbers for modeling series of medical measurements in a diagnosis support based on the Dempster–Shafer theory. In: Rutkowski L, Scherer R, Korytkowski M, Pedrycz W, Tadeusiewicz R, Zurada J (eds) Artificial Intelligence and Soft Computing, ICAISC 2018. Lecture Notes in Computer Science, vol 10842. Springer, Cham. https://doi.org/10.1007/978-3-319-91262-2_20
14. Powers DMW (2011) Evaluation from precision recall and F-measure to ROC informedness, markedness & correlation. J Mach Learn Technol 2(1):37–63
15. Shafer G (1976) The mathematical theory of evidence. Princeton University Press, New Jersey
16. Shin H, Roth HR, Gao M, Lu L, Xu Z, Nogues I, Yao J, Mollura D, Summers RM (2016) Deep convolutional neural networks for computer-aided detection: CNN architectures, dataset characteristics and transfer learning. IEEE Trans Med Imag 35(5):1285–1298. https://doi.org/10.1109/TMI.2016.2528162
17. Straszecka E (2010) Combining knowledge from different sources. Expert Syst 27(1):40–52. https://doi.org/10.1111/j.1468-0394.2009.00507.x
18. Tsekouras GE (2016) Fuzzy rule base simplification using multidimensional scaling and constrained optimization. Fuzzy Sets Syst 297:46–72. https://doi.org/10.1016/j.fss.2015.10.009
19. Wilk T, Wozniak M (2012) Soft computing methods applied to combination of one-class classifiers. Neurocomputing 75(1):185–193. https://doi.org/10.1016/j.neucom.2011.02.023
20. Zadeh LA (1999) Fuzzy logic = computing with words. In: Zadeh LA, Kacprzyk J (eds) Computing with Words in Information/Intelligent Systems 1: Foundations, vol 3. Physica-Verlag, Heidelberg

Exploiting Label Interdependencies in Multi-label Classification

Kinga Glinka(✉), Agnieszka Wosiak, and Danuta Zakrzewska

Institute of Information Technology, Lodz University of Technology,
Wólczańska 215, Lodz, Poland
kinga.glinka@edu.p.lodz.pl,
{agnieszka.wosiak,danuta.zakrzewska}@p.lodz.pl

Abstract. Multi-label classification problems very often concern multidimensional datasets, thus, performance of the method is problematic in many cases. Exploiting label dependencies may ameliorate classification results. In the paper, new effective problem transformation method which uses label interdependencies is introduced. Experiments conducted on several benchmarking datasets showed the good performance of the presented technique, regarding six evaluation metrics, including the most restricting Classification Accuracy and confirmed by statistical inference. The obtained results are compared with those obtained by the most popular problem transformation methods.

Keywords: Multi-label classification · Label interdependence · Problem transformation methods

1 Introduction

Classification is one of the most important tasks in machine learning area. It consists in automatic assignment of instances to predefined classes within the specified categories. In real life, in many cases single labels are not sufficient to indicate appropriate classes for data instances and multi-label classification, where more than one class is predicted, should be considered.

For multi-label classification problems, with X for the domain of observations and with L for the finite set of class labels, a training set T is defined as $T = \{(x_i, Y_i) : x_i \in X, Y_i \subseteq L, i = 1, 2, ..., n\}$ and $S = \{(x_i, Y_i) : x_i \in X, Y_i \subseteq L, i = 1, 2, ..., m\}$ denotes a test set. The goal of multi-label classification is to find a mapping of instance $x_i \in X$ to a subset of possible labels: $h : X \rightarrow 2^L$. The subset $K \in 2^L$ is considered as set of relevant labels, while $L \setminus K$ indicates the complement subset of irrelevant labels for given instance x_i.

From among possible application domains of multi-label classification one should mention classification of text, music, semantic scene, protein function as well as medical data [8]. Applications of multi-label classification often have to deal with multidimensional datasets with many attributes and relatively small number of instances at the same time. Moreover, labels often have correlations

R. Burduk et al. (Eds.): CORES 2019, AISC 977, pp. 57–66, 2020.
https://doi.org/10.1007/978-3-030-19738-4_7

with each other and if we do not consider their dependencies, we may loose important information. Researchers pointed out that including label dependencies into multi-label classification techniques can improve the performance of the classifiers (see [4] for example).

We have already considered label dependencies in the previous research [7], where multi-label classification method *Labels Chain (LC)* that uses a mapping of the relationship between labels, has been investigated. This technique requires a certain number of multi-class classifiers and it includes result labels as new attributes in the classification process. The investigations presented in [19] demonstrate that the group of methods that exploit dependencies outperforms the techniques without label dependencies. In the current research, we examine new alternative problem transformation method: *Hierarchy of Label Interdependencies (HoLI)* based on label pairs interdependence. We examine the performance of the proposed approach and compare its effectiveness with the state-of-art techniques via numerical experiments. We use eight multi-label datasets of different characteristics. The experimental results are evaluated via quality metrics and confirmed by statistical inference. The results demonstrate that the proposed approach outperforms the techniques in the most of the considered cases regarding the multi-label problems.

The remainder of the paper is organized as follows. In the next section, relevant work is summarized. Section three introduces the proposed method, presents evaluation metrics and the datasets that will be used during experiments. In section four, the experimental results are introduced and discussed. The final section presents the study's conclusions and delineates future research.

2 Related Work

Multi-label classification is intended to predict more than one class label for data instance. Thus, applications of multi-label classifications require specialized, dedicated methods in order to provide satisfactory accuracy and effectiveness. There exist three main categories of multi-label classification methods: (i) problem transformation methods, (ii) algorithm adaptation methods and (iii) ensemble methods [10]. In this research, we focus on the first category. Problem transformation methods, independently of the specific learning algorithms, transform multi-label classification problem into one or more traditional single-label tasks. Such approach allows to apply a variety of well-known classification algorithms to multi-label classification problems.

Including label correlations to multi-label classification has already been proposed in different approaches. Most of them have been examined on multimedia datasets. Huang et al. [9] proposed the framework called Group Sensitive Classifier Chains (GSCC). They assumed that similar objects share the same label correlations and tend to have similar labels. GSCC firstly expands the feature space of label space and then cluster them. Classifier chains are built on each cluster taking into account group specific label dependency graph.

Active learning approach was investigated for image classification in [20] and [21]. Ye et al. used cosine similarity to evaluate the correlations between the labels. Then, they develop an active learning selection strategy based on correlations to select label pairs for labeling [20]. Zhang et al., in turn, proposed a high-order correlation driven active learning approach. They indicated that additionally to pair-wise label correlations high-order correlation is also informative. Association rule mining was adopted to discover informative label correlation [21]. Both of the approaches have been evaluated by experiments done on several datasets.

There exist many multi-label classification methods with label correlation, which are not dedicated to multimedia datsets but can be used for classification of that kind of data. In this paper, we consider: Label Power-set (LP) [10,18], Classifier Chains (CC) [14] and Ensembles of Classifier Chains (ECC) [15].

3 Materials and Methods

3.1 Proposed Approach

The proposed novel method HoLI is based on LP approach [10,18], however, it has more complex transform phase. The algorithm uses the strength of the dependencies of the labels when building new classes. Firstly, it counts number of single labels c_{y_i} from L and possible label pairs occurrences $c_{y_i y_j}$ that are further used as strength of the labels correlation:

$$c_{y_i} = \sum_{n=1}^{N} p(x_n, Y_n), \quad p(x_n, Y_n) = \begin{cases} 1, & \{y_i\} \subseteq Y_n \\ 0, & \text{otherwise} \end{cases} \tag{1}$$

$$c_{y_i y_j} = \sum_{n=1}^{N} p(x_n, Y_n), \quad p(x_n, Y_n) = \begin{cases} 1, & \{y_i, y_j\} \subseteq Y_n \\ 0, & \text{otherwise} \end{cases} \tag{2}$$

Counted pairs $c_{y_i y_j}$ that have not even appeared are skipped in the next step. Both single labels y_i and pairs of them $y_i y_j$ become new labels and then, they are sorted by number of occurrences descending. Finally, according to the established order, new class of all set of labels is created, similarly to LP approach. The classification is conducted in relation to the newly created labels. The details of the classification process using our HoLI method is presented in Algorithm 1. It is divided into three main parts of classification process: *Transformation*, *Learning* and *Result*. The proposed approach relates mainly to the transformation stage.

The HoLI method relies on relationships among labels and overcome the label independence problem of many techniques. It takes into account not only pairs of labels, but also individual ones to deal with the detection of individual classes. Therefore, it can be successfully used for any type of data. The advantage of the proposed technique over traditional LP approach consists in more precise specification of newly created classes, which can result in more accurate classification. In HoLI technique, there are more label combinations, therefore, the classification is more precise and reduces the risk of confusing the classes. At the same time, additional transformations do not significantly change computational complexity of the basic algorithm.

Algorithm 1. Hierarchy of Label Interdependencies (HoLI)

```
Input  : T ∩ S = ∅,
         Training set T ≠ ∅, T = {(x_i, Y_i) : x_i ∈ X, Y_i ⊆ L, i = 1, 2, ..., N},
         Test set S ≠ ∅, S = {(x_i, Y_i) : x_i ∈ X, Y_i ⊆ L, i = 1, 2, ..., M},
Output: Single-label classifier h(·),
         S' = {(x_i, Y'_i) : x_i ∈ X, Y'_i ⊆ L, i = 1, 2, ..., M}, |S'| = |S|,
         Evaluation of classification results
   /* TRANSFORMATION                                                        */
 1 data(List of counters C_List = {c_y1, c_y1y2, ..., c_y1y|L|, ..., c_y|L-1|y|L|},
 2 List of new labels y'_List = {y'_1, y'_2, ..., y'_|L'|})
 3 X' = {}
 4 foreach (x, Y) ∈ T do
 5     foreach y_i in binary labels L do
 6         if y_i = 1 then
 7             c_yi ← c_yi + 1                           /* Count single labels */
 8         end
 9         foreach y_j in binary labels L, i ≠ j do
10             if y_i = 1 and y_j = 1 then
11                 c_yiyj ← c_yiyj + 1                    /* Count label pairs */
12             end
13         end
14     end
15 end
16 foreach c_yiyj in C_List do
17     if c_yiyj = 1 then
18         Add to list y'_List ← y_i y_j        /* Save only present label pairs */
19     end
20 end
21 foreach c_yi in C_List do
22     Add to list y'_List ← y_i y_j                /* Save all single labels */
23 end
24 SORT y'_List by values from C_List DESC                        /* Sort labels */
25 foreach (x, Y) ∈ T do
26     X' ← x ∪ y'_List                                /* Create new instances */
27 end
   /* LEARNING                                                              */
28 Y' = {}
29 foreach x ∈ X' do
30     Y' = h(x)                                  /* Build classification model */
31 end
   /* RESULT                                                                */
32 foreach (x, Y') ∈ S do
33     Evaluate in terms of evaluation measures        /* Check assigned labels */
34     if Y' = Y then
35         result ← correct
36     else
37         result ← incorrect
38     end
39 end
```

3.2 Evaluation

Performance of multi-label classification can be evaluated using two groups of metrics: example-based and label-based ones [8]. In this research, we limit our investigations to example-based metrics. However, different type of classification tasks may require different performance measures, as was stated by Sokolova and Lapalme, who investigated applicability of performance measures for text classification [16]. Thus, macro and micro-strategies of label-based metrics are also worth considering.

We applied six measures that base on the average differences between the true and predicted labels, and range in $[0, 1]$. All of the metrics are calculated separately for each instance and the average of all instances is determined. The measures are described as in [18]. *Classification Accuracy (CA)* is known as to be the basic and the strictest evaluation measure. It ignores partially correct predicted labels and requires them to be an exact match of the real label set. *Accuracy (Acc)* is the proportion between the number of correctly predicted labels and the total number of active labels in both sets – true and predicted. *Precision (Prec)* is defined as the ratio of the number of correctly predicted labels to the total number of predicted labels, while *Recall (Rec)* returns the ratio of correctly predicted labels to all truly relevant labels. *F1 mesasure (F1)*, also known as *F-measure*, is considered as the combination of Prec and Rec metrics, and it is computed as the harmonic mean of them. *Hamming Loss (HL)* is the most popular loss function, commonly used in evaluation of multi-label classification performance. It is the fraction of incorrectly classified labels over the total number of all possible labels.

To compare the new approach with state-of-art techniques from the literature, problem transformation methods are considered. HoLI belongs to this group of algorithms. We compare the performance of HoLI with seven popular problem transformation methods: Binary Relevance (BR) [18], Classifier Chains (CC) [14], Calibrated Label Ranking (CLR) [6], Label Power-set (LP) [10,18], Hierarchy Of Multi-label CkassifiERs (HOMER) [17], Pruned Problem Transformation (PPT) [12] and Pruned Sets (PS) [13]. Methods that allow to use various classifiers were tested with conjunction with eight base single-label classifiers: kNN (k-nearest neighbours), NB (naive Bayes), C4.5 decision tree, PART decision tree, SVM, rule-based JRip, rule-based MODLEM and fuzzy FURIA.

Regarding the fact that existence of label dependencies can improve the effectiveness of the classifiers, we investigated label dependencies in datasets chosen for evaluation. They can refer to: positive correlation, where both considered labels are present in the particular case, or negative correlation, where one label do not occur if the second label is present (and vice versa).

According to surveys on similarity of binary vectors (see [3], for example), different measures based on pair counting were considered. The pair counting approach is based on the number of occurrences of matches, differentiated as: "positive match", "negative match" and "absence mismatch". Most of formulas either increase in value with "negative matches" or depreciate values by "absence mismatches" [3]. As a result, we evaluate label dependencies by assessing the *p*-value of the Chi-square test [11].

Eight popular multi-label datasets from Mulan repository, representing different domains were considered for evaluation [1]. Their characteristics are presented in Table 1. The first column of Table 1 presents the name of the dataset and the second one its domain. Three subsequent columns contain the numbers of data instances, attributes and labels. Three last columns contain respectively values of cardinality, density and dimensionality expressed as proportion between numbers of attributes and records.

Table 1. Dataset characteristics

Dataset	Domain	Instances $N+M$	Attributes D	Label set L	Cardinality	Density	$D/(N+M)$
birds	Audio	645	260	19	1.014	0.053	0.403
emotions	Audio	593	72	6	1.868	0.311	0.121
flags	Images	194	19	7	3.392	0.485	0.098
scene	Images	2407	294	6	1.074	0.179	0.122
genbase	Biology	662	1186	27	1.252	0.046	1.791
yeast	Biology	2417	103	14	4.237	0.303	0.043
medical	Text	978	1449	45	1.245	0.028	1.482
enron	Text	1702	1001	53	3.378	0.064	0.588

4 Experiments and Discussion

The aim of the experiments was to examine the effectiveness of the proposed HoLI method and to compare its performance with state-of-art techniques from the literature. Moreover, we verified advantages of the HoLI method over the classic LP method. The experiments were carried out on the datasets presented in Table 1, considering the methods and classifiers enumerated in Sect. 3.2. All the examined techniques have been evaluated by using the metrics: CA, Acc, Prec, Rec, F1 measure and HL. The experiments started with the initial step, which aimed at confirming that label dependencies occur in the considered datasets. In the next steps, the procedure has been conducted twice for each dataset: with 60/40 split of instances (60% instances in the training set and 40% in the test set) and with 10-fold cross-validation. The experimental environment was based on WEKA Open Source [2] and Mulan library [1].

According to Sect. 3.2, the Chi-square test was used to assess associations between labels; results were recognized as statistically significant if p-value $<$ 0.05. The analysis revealed the diversity of label characteristics in datasets. There was one dataset with 100% ratio of statistically significant label associations to all possible pairs of labels (*scene*) and no dataset with no statistically significant dependencies. The average percentage of statistically significant associations between labels for all the datasets was 60%. Such results entitled to exploit label dependencies and to use the presented approach on the considered datasets.

For the most of the examined cases, the experiments showed good performance of the HoLI method in comparison to the other considered techniques. Regarding average values of the metrics, we stated that our approach obtained better average results considering 60/40 split as well as 10-fold cross validation cases. In further considerations we focus on CA, which is the most restrictive metrics and may give a good image of advantages of the proposed approach.

The best results of CA obtained for evaluated methods (among all classifiers) are presented in Tables 2 and 3. HoLI results are in the first rows of the tables. For each set, the best results from among all the methods are bold. The column *Avg* contains the average results for the method relatively to all the sets. The next column indicates the difference in the average result between the proposed HoLI approach and the other techniques – positive values mean the predominance of the proposed method. The *Avg rank* values in the last column refer to the Friedman test used for comparison of many classifiers [5]. It is also worth pointing out that in the both cases, the 60/40 split and 10-fold cross-validation, HoLI obtained better CA results than HOMER (for all datasets in the experiments with split and for 6/8 cases for cross-validations), indicated in the literature as one of the most effective multi-label classification methods [10].

Table 2. Results of classification accuracy [%]: 60/40 split

	birds	emotions	flags	scene	genbase	yeast	medical	enron	Avg	Difference with HoLI	Avg rank
HoLI	**51.17**	**31.12**	**28.13**	65.48	**95.41**	20.30	**72.45**	4.80	**46.11**		6.48
BR	49.77	29.59	26.56	62.97	94.50	19.55	67.80	4.45	44.40	1.71	3.21
CC	50.70	**31.12**	**28.13**	**66.25**	94.50	22.31	68.73	5.16	45.86	0.25	6.16
LP	48.83	30.61	26.56	64.48	**95.41**	21.93	**72.45**	**5.16**	45.68	0.43	6.11
HOMER	50.23	29.59	20.31	63.48	94.95	19.67	68.11	4.27	43.83	2.28	3.38
PPT	44.60	27.04	26.56	56.05	92.66	17.54	72.14	4.63	42.65	3.46	3.09
PS	49.30	29.59	**28.13**	65.49	92.66	**22.68**	73.07	5.16	45.76	0.35	5.05
CLR	50.23	28.57	**28.13**	62.34	94.50	19.55	20.74	4.63	38.59	7.52	2.52

Table 2 presents CA results of multi-label sets for 60/40 split. Most of the presented results concern application of kNN and C4.5 decision tree classifiers, but the best notes for particular datasets (bold values) were mostly obtained for MODLEM and FURIA classifiers. The method indicated as the best of all tested techniques is HoLI. The highest classification results of 95.41% were reported for *genbase* dataset, for which the HoLI and LP methods proved to be equally effective. The average values of HoLI in terms of all considered measures have been also compared with the average values from the best results obtained by other multi-label classification methods. For 60/40 split on datasets, the average results of all metrics for HoLI are as follows: CA 46,11%, Acc 61.91%, Prec 70.49%, Rec 68.03%, F1 measure 67.63% and HL 11.89%. For CA and Prec, proposed approach turned out to be more effective than average result from other techniques. The percentage advantages were 2.29% for CA and 0.28% for Prec (Table 4).

In case of 10-fold cross-validation (Table 3), most of the best results of CA over all classifiers were also attained for kNN and C4.5 classifiers, and the highest results for particular datasets gave kNN. The experiments again indicated HoLI as the most effective method based on average results from all collections. This time, HoLI overcame other methods, with differences ranging from 0.01% to 3.54%. For HoLI method, the considered metrics have reached average values: CA 45.33% Acc 61.23%, Prec 69.65%, Rec 67.74%, F1 measure 66.97% and HL 12.64%. According to the results from Table 4, the proposed approach obtained better average effects for the strictest evaluation measure – CA.

Table 3. Results of classification accuracy [%]: 10-fold cross-validation

	birds	emotions	flags	scene	genbase	yeast	medical	enron	Avg	Difference with HoLI	Avg rank
HoLI	51.50	27.66	25.32	56.55	96.67	21.18	**69.94**	**13.78**	**45.33**		5.44
BR	**54.43**	27.99	19.63	54.76	**96.82**	21.31	67.79	5.92	43.58	1.75	4.44
CC	53.20	28.83	24.79	57.30	**96.82**	**23.87**	69.63	8.10	45.32	0.01	6.19
LP	51.50	27.15	25.32	56.72	**96.82**	21.35	69.84	**13.78**	45.31	0.02	6.06
HOMER	51.02	**29.18**	15.97	55.51	**96.82**	20.73	67.08	6.33	42.83	2.50	3.81
PPT	49.94	25.29	22.74	51.68	94.40	19.15	70.66	13.37	43.40	1.93	2.81
PS	51.19	26.97	**25.87**	**56.92**	94.40	19.94	69.74	13.49	44.82	0.51	4.44
CLR	52.57	26.81	21.26	53.48	96.67	20.19	57.14	6.16	41.79	3.54	2.81

Table 4. Average results of all metrics [%]

	60/40 split						10-fold cross-validation					
	CA	Acc	Prec	Rec	F1	HL	CA	Acc	Prec	Rec	F1	HL
HoLI	46.11	61.91	70.49	68.03	67.63	11.89	45.33	61.23	69.65	67.74	66.97	12.64
Others	43.82	62.10	70.21	73.33	68.46	11.47	43.86	61.77	70.74	74.06	68.08	11.94

As part of experiments, we verified directly the advantage of HoLI approach over its basic LP algorithm. For 8 multi-label collections in 2 variants of classification (60/40 split and 10-fold cross-validation), the difference in favour of the HoLI method is positive and is equal to 3.55%.

Statistical analysis also showed the advantage of HoLI method. The overall significance level of differences for all mentioned classification techniques using the ANOVA Friedman test was below 0.01 for the both validation techniques (10-fold cross-validation and 60/40 split). The detailed post-hoc analysis was performed by comparing absolute differences between average ranks of classifiers over all the considered datasets and validation techniques. The results of the post-hoc tests are visualized with the diagram presented in the Fig. 1 [5]. The diagram compares all the algorithms against each other. The top line of the diagram is the axis on which we plot the average ranks of each method sorted by the values, in the descending order, i.e. the lowest and best ranks are to the right. The positions of average ranks for each classifier are marked with vertical lines and captioned with their names. Moreover, the groups of algorithms that are not significantly different in terms of classification accuracy are connected with horizontal lines. The HoLI method significantly outperformed most of the considered problem transformation techniques, including CLR, PPT, HOMER and BR. The statistical significance of differences between HoLI and the referencing LP method equaled to 0.31.

Summing up, the proposed HoLI approach proved to be effective for the considered multi-label datasets. Moreover, HoLI performed best for most of the considered cases and allowed to identify effectively multiple labels. Referring to average results for all the collections, HoLI turned out to be the best method in terms of the strictest evaluation measure, CA. Summary comparison with the

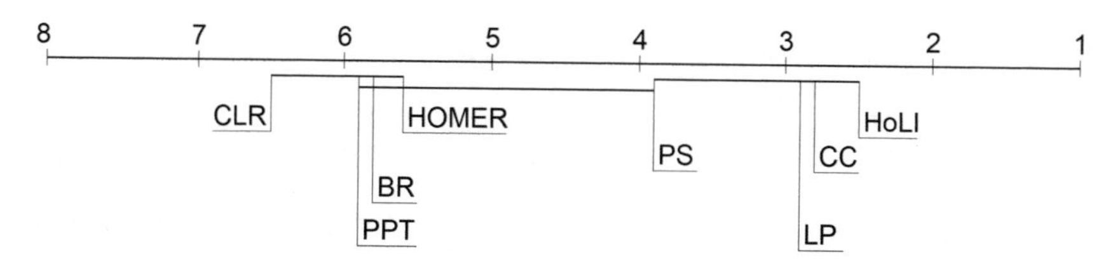

Fig. 1. Visualisation of post-hoc test for comparison of classifiers

classic LP method also showed the advantage of the proposed solution. Thus, we can conclude that the exploiting dependencies based on the strength of interdependence of label pairs is a promising and competitive solution for existing, state-of-art techniques.

5 Conclusions

In this paper, we present new effective multi-label classification method HoLI. The proposed problem transformation algorithm uses label pairs interdependence in the transformation phase. The performance of HoLI is examined and compared with seven problem transformation methods including baseline LP algorithm. The methods have been used with eight single-label classifiers. All the techniques we considered have been evaluated by experiments performed on eight datasets of different characteristics. Investigations presented the good performance of the HoLI technique on the considered datasets regarding six example-based evaluation metrics, especially the most restricting CA. Experiment results confirmed by statistical verification showed that HoLI algorithm outperformed the examined group of methods.

Future research consists of additional investigations of the proposed method. First, its performance should be compared with multi-label classification methods from other groups like algorithm adaptation techniques or ensemble of classifiers. Second, macro and micro-strategies of label-based metrics applied for each dataset should be the further step of investigations. Then, the results obtained for more datasets of different characteristics should be examined. Worth considering is also application of the proposed method to multi-perspective datasets.

References

1. MULAN: A Java Library for Multi-Label Learning. http://mulan.sourceforge.net/
2. Weka 3: Data mining software in Java. http://www.cs.waikato.ac.nz/ml/weka/index.html
3. Choi SS, Cha SH, Tappert CC (2010) A survey of binary similarity and distance measures. J Syst Cybern Inform 8(1):43–48
4. Dembczyński K, Cheng W, Hüllermeier E (2010) Bayes optimal multilabel classification via probabilistic classifier chains. In: Fürnkranz J, Joachims, T (eds) Proceedings of the 27th international conference on machine learning (ICML 2010). Omnipress, pp 279–286

5. Demsar J (2006) Statistical comparisons of classifiers over multiple data sets. J Mach Learn Res 7:1–30
6. Fürnkranz J, Hüllermeier E, Mencía EL, Brinker K (2008) Multilabel classification via calibrated label ranking. Mach Learn 73(2):133–153. https://doi.org/10.1007/s10994-008-5064-8
7. Glinka K, Zakrzewska D (2016) Effective multi-label classification method for multidimensional datasets. In: Andreasen T et al (eds) Flexible query answering systems 2015: Proceedings of the 11th international conference FQAS 2015, Cracow, Poland, October 2015. Springer International Publishing, pp 127–138. https://doi.org/10.1007/978-3-319-26154-6_10
8. Herrera F, Charte F, Rivera AJ, del Jesus MJ (2016) Multilabel classification: problem analysis, metrics and techniques, 1st edn. Springer Publishing Company Incorporated, Heidelberg
9. Huang J, Li G, Wang S, Zhang W, Huang Q (2015) Group sensitive classifier chains for multi-label classification. In: 2015 IEEE international conference on multimedia and expo (ICME), pp 1–6. https://doi.org/10.1109/ICME.2015.7177400
10. Madjarov G, Kocev D, Gjorgjevikj D, Džeroski S (2012) An extensive experimental comparison of methods for multi-label learning. Pattern Recogn 45(9):3084–3104. https://doi.org/10.1016/j.patcog.2012.03.004
11. McHugh ML (2013) The chi-square test of independence. Biochemia Medica 23(2):143–149. https://doi.org/10.11613/BM.2013.018
12. Read J (2008) A pruned problem transformation method for multi-label classification. In: Proceedings of the 2008 New Zealand computer science research student conference, NZCSRS, pp 143–150
13. Read J, Pfahringer B, Holmes G (2008) Multi-label classification using ensembles of pruned sets. In: 2008 eighth IEEE international conference on data mining, pp 995–1000. https://doi.org/10.1109/ICDM.2008.74
14. Read J, Pfahringer B, Holmes G, Frank E (2009) Classifier chains for multi-label classification. In: Buntine W, Grobelnik M, Mladenic D, Shawe-Taylor J (eds) Machine learning and knowledge discovery in databases. LNCS, vol 5782. Springer-Verlag, pp 254–269. https://doi.org/10.1007/978-3-642-04174-7_17
15. Read J, Pfahringer B, Holmes G, Frank E (2011) Classifier chains for multi-label classification. Mach Learn 85(3):333–359. https://doi.org/10.1007/s10994-011-5256-5
16. Sokolova M, Lapalme G (2009) A systematic analysis of performance measures for classification tasks. Inf Process Manag 45:427–437
17. Tsoumakas G, Katakis I, Vlahavas I (2008) Effective and efficient multilabel classification in domains with large number of labels. In: Proceedings of the ECML/PKDD 2008 workshop on mining multidimensional data (MMD 2008)
18. Tsoumakas G, Katakis I, Vlahavas I (2010) Mining multi-label data. In: Maimon O, Rokach, L (eds) Data mining and knowledge discovery handbook. Springer, pp 667–685. https://doi.org/10.1007/978-0-387-09823-4_34
19. Wosiak A, Glinka K, Zakrzewska D (2018) Multi-label classification methods for improving comorbidities identification. Comput Biol Med 100:279–288. https://doi.org/10.1016/j.compbiomed.2017.07.006
20. Ye C, Wu J, Sheng VS, Zhao P, Cui Z (2015) Multi-label active learning with label correlation for image classification. In: 2015 IEEE international conference on image processing (ICIP), pp 3437–3441. https://doi.org/10.1109/ICIP.2015.7351442
21. Zhang M, Zhou Z (2014) A review on multi-label learning algorithms. IEEE Trans Knowl Data Eng 26(8):1819–1837. https://doi.org/10.1109/TKDE.2013.39

Toward Shareable Multi-abstraction-level Feature Extractor Based on a Bayesian Network

Kaneharu Nishino(✉), Hiroshi Tezuka, and Mary Inaba

Department of Creative Informatics,
Graduate School of Information Science and Technology,
The University of Tokyo, Tokyo, Japan
{nishino.kaneharu,tezuka.hiroshi,mary}@ci.i.u-tokyo.ac.jp

Abstract. In this study, we propose a Multi-abstraction-level Feature extractor with a Bayesian network (MFB) that can output intermediate patterns as feature values and can be shared across different abstraction level classifiers. To leverage the patterns from intermediate layers, we implemented a bidirectional network based on a Bayesian network to accurately calculate posterior probabilities. Experimental testing confirmed that a MFB could be constructed successfully on an actual computer to achieve pattern extraction.

Keywords: Machine learning · Feature extraction · Bayesian network

1 Introduction

1.1 Motivation

Modern deep neural networks offer high performance for various tasks. Their hierarchical architecture can be divided into two parts: a feature extractor and a classifier (Fig. 1). The feature extractor is the first part of the network that identifies patterns in the input data as a feature values. The classifier, the latter part of the network, then categorizes these feature values into classes.

The feature extractor of a deep neural network is known to have an abstraction hierarchy [1]: nodes in lower layers are activated in response to small, simple patterns (such as edges) while in higher layers, nodes are activated in response to large, complex patterns. This pattern-abstraction hierarchy can be applied to various classifiers [3]. For example, unsupervised learning can be used to extract patterns with complex invariances [11], while semi-supervised learning has also been shown to offer high performance [9].

This feature extractor is considered to be similar to a hierarchy of the visual cortex (shown in Fig. 2) that is a famous example of this architecture. Locally oriented bars activate neurons in the V1 area of the visual cortex. Neurons in the V2 area then detect combinations of bars such as polygonal lines. In higher

R. Burduk et al. (Eds.): CORES 2019, AISC 977, pp. 67–78, 2020.
https://doi.org/10.1007/978-3-030-19738-4_8

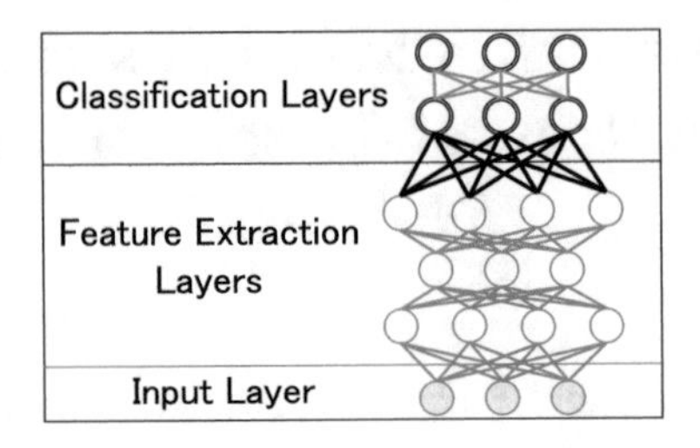

Fig. 1. Typical architecture of a deep neural network.

areas, such as the V4 area, anterior inferior temporal cortex, and posterior inferior temporal cortex, each neuron has a larger receptive field and can detect more complex patterns [2]. In this manner, highly complex patterns, such as human faces, can be recognized irrespective of position and direction. Many unsupervised deep-learning models that reproduce this property of neuronal networks in the visual pathway have been proposed. Feature extractors and their abstraction hierarchy show good performance in the recognition and classification of objects.

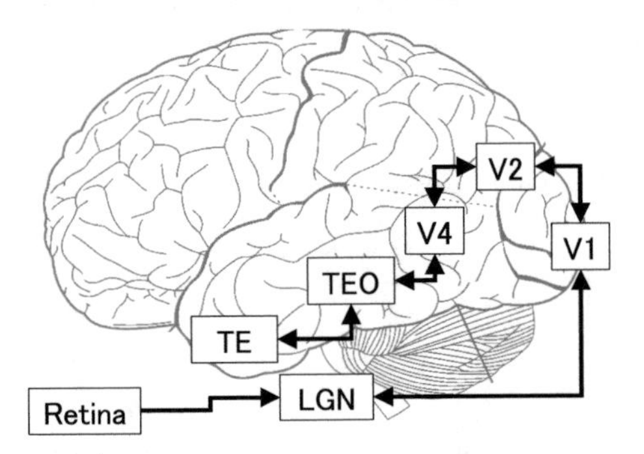

Fig. 2. Pathways between visual areas in the visual cortex. Based on [5] Fig. 728.

In this study, we attempt to construct a shareable feature extractor that can be used for various classes once trained. We consider the pattern recognition network in brains is shared across many classification tasks. Though deep neural networks are generally specialized in each task through fine-tuning, feature extractors in brains should not be specialized and not be limited to each task.

1.2 Shareable Feature Extractor

We attempt to construct a shareable feature extractor which can be used for various classes. Its network should not be limited to particular classes. We consider that it can only be constructed through unsupervised learning, without fine-tuning, or by the visual cortex acquiring its architecture. Particular classes should not change the network, in order to maintain its abstraction hierarchy.

A deep generative model is a type of unsupervised method in deep learning. In this model, extracted patterns are interpreted as basis elements of the input data. Deep networks abstract patterns hierarchically and eliminate small mutations such as noise, differences in position or direction, and transformation. They use these extracted patterns as abstracted factors for the feature values that are output to the classifier. However, in most conventional deep networks, only patterns of the top layer of feature extractors are used as feature values while those of intermediate layers are disregarded, as shown in Fig. 1. The abstraction level of the feature values depends on how many layers the feature extractor contains so that it is important to tune the number of layers according to the characteristics of the given labeled data.

Here, to use the same feature extractor across widely different classes, it is desirable to use features of all abstraction levels because the requirements of the abstract features classifiers need are unknown. Therefore, we consider that intermediate patterns in the feature extractor should be used as more than simply an intermediate representation to be output to the highest layer. Some studies employ multiple abstraction level features [6,18], and lower features may offer more benefits than higher features, especially in transfer learning [3,4]. We hypothesize that the classifier will be able to use multiple abstraction level features by including intermediate patterns as additional feature values (as shown in Fig. 3) and that the usage of them improves the classification ability. We surmise that the use of features of multiple abstraction levels can make a feature extractor shareable across layers without tuning the numbers of layers.

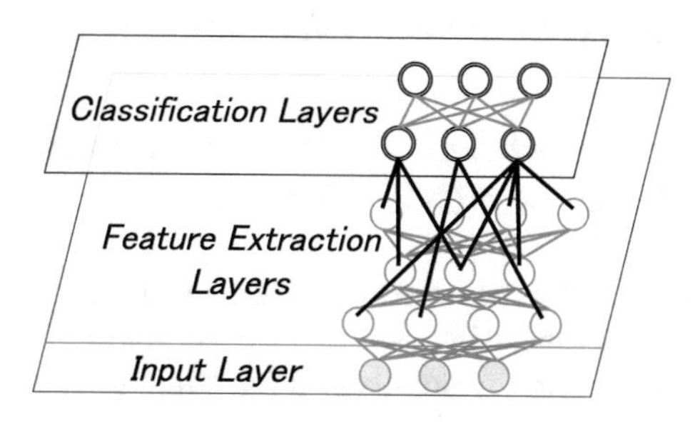

Fig. 3. The architecture proposed for the multiple abstraction level feature extractor.

Moreover, with respect to the use of intermediate patterns, we refer to some studies employing Bayesian networks. A Bayesian network has bidirectional information flow between layers, and such bidirectional paths are found in the cerebral cortex, such as the connection between V1 and V2, as shown in Fig. 2. In higher areas of the visual cortex, these paths are considered to code predictions and attention; in lower areas, they are considered to participate in the filling-in phenomenon [17]. Some studies have modeled these bidirections using a Bayesian network, and employ it as the framework of hierarchical networks for pattern recognition [7,8,14]. In these models, the networks can use information about higher abstract patterns to recognize lower patterns. Using higher information, they can recognize intermediate patterns more accurately [14].

Therefore, to construct a shareable feature extractor, we design a model MFB. It is a multi-layer network constructed in the framework of a Bayesian network and trained through unsupervised learning in order to output intermediate patterns as input feature values for classifiers.

1.3 Related Work

Regarding to related works, Ichisugi [8] and Hosoya [7] show their works that construct a multi-layer Bayesian network. These are similar to MFB in the point that they employ a Bayesian network as the framework of pattern recognition.

MFB differs to these works in two points: multiple abstraction level features and pooling by child.

We consider that multiple abstraction level features of intermediate patterns improves the classification ability and that the architecture to use the multiple abstraction level features shown in Fig. 3 is important. Such architecture is not described in the related work. MFB employ this architecture and recognize intermediate patterns more accurately using bidirectional inference.

In addition, the previous studies employ multivalued variables in contrast to MFB which employs binary variables. In their model, each value of variables corresponds to each pattern. Although the function such like pooling is implemented, each range of pooling is limited to states of one variable. MFB employ the pooling by child and it can pool patterns over different variables, not limited to one variable. MFB can obtain pooling function more flexibly.

2 Multi-abstraction-level Feature Extractor Using a Bayesian Network (MFB)

A Multi-abstraction-level Feature extractor using a Bayesian network (MFB) is a multi-layer feature extractor in the framework of a Bayesian network constructed by pattern extraction through unsupervised learning. It calculates the posterior probability of each intermediate pattern by inferences using belief propagation [16]. A simplified MFB architecture is shown in Fig. 4. The MFB has observable continuous variables as input layer to the lowest layer and contains hierarchical layers of binary variables as hidden layers above the input layer. Each input node has one hidden variable, called a child. On this child layer, the MFB has a layer of variables called parents, each corresponding to a child. The parent and child layers are initially fully connected but the links are subsequently tuned through unsupervised learning to extract patterns from the input data as hidden variables (i.e., binary variables representing either existent [True, T] or non-existent [False, F]) for each corresponding pattern.

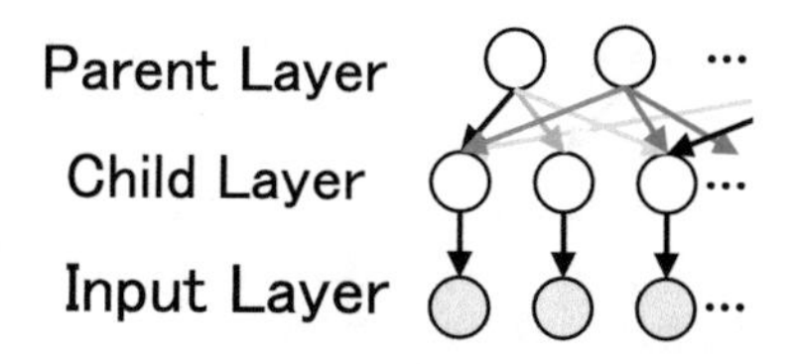

Fig. 4. MFB with one parent layer

A naive Bayesian network has problems in the number of parameters and the invariance to small mutation. Therefore, the MFB involves several assumptions and limitations that are described as follows.

2.1 Parameter Reduction

To construct a binary Bayesian network, the parameters for conditional probabilities must be set. However, the number of conditional probabilities increases exponentially as $O(2^n)$ with n parent variables. Therefore we introduced One-parent-T assumption and Link Strength to reduce the parameter.

One-parent-T Assumption. We introduce an assumption called the one-parent-T (1pT), which states that only one of the parents sharing a child can be in the T state. Under 1pT assumption, the conditional probabilities that a child is in the T or F state when two or more of its parents are in the T state are both 0. This reduces the number of parameters to $O(n)$.

This assumption also enforces sparsity in the recognition for parent layers as parents send messages via children to inhibit each other in the form of lateral inhibition. This effect results in sparsity of recognized patterns and high performance when mapping a feature space. Sparsity is an important property in many machine-learning models that is needed to suppress the representation ability of a model and avoid over-fitting.

Link Strength. Although the 1pT assumption reduces the number of parameters, it limits the conditional probabilities too strongly, such that they cannot express a situation in which a variable is independent. To solve this problem, we introduce parameters, called Link Strengths (LSs), that are used to formulate conditional probabilities.

An LS represents the strength of a link between variables. It takes on a continuous value between 0 (completely cut) and 1 (fully connected): when the LS is 0, the corresponding link is completely cut and the effect of the 1pT assumption is weakened, whereas when the LS is 1, the link is fully connected and the effect of the 1pT assumption is enhanced.

Conditional probabilities are formulated using LSs as follows. Herein, we refer to the probability that a child variable, v, takes the T state in the case where only a parent variable, u, takes the T state as $p(v|u)$:

$$p(v|u) = (1 - s_{uv})(\frac{\rho_{uv} s_{uv}}{1 - s_{uv}} + \rho_v) \quad (1)$$

where s_{uv} is the LS between u and v, ρ_{uv} is a parameter called the Coefficient of conditional Probability (CP), that represents the value of $p(v|u)$ when $s_{uv} = 1$, and ρ_v denotes the conditional probability of $v = T$ in the case that no parent variable takes the T state. Using the LSs and CPs, the conditional probability that only k parents $u_1, u_2, \ldots, u_k$ take the T state, $p(v|u_1, u_2, \ldots, u_k)$ is given by Eq. 2, where U denotes the set of parents u of a child v:

$$p(v|u_1, \ldots, u_k) = \prod_{j=1}^{k} (1 - s_{u_j v}) \left(\sum_{j=1}^{k} \frac{\rho_{u_j v} s_{u_j v}}{1 - s_{u_j v}} + \rho_v \right) \quad (2)$$

In addition, the conditional probability that $v = F$ under the same conditions, $p(\bar{v}|u_1, u_2, \ldots, u_k)$ is given by Eq. 3.

$$p(\bar{v}|u_1, \ldots, u_k) = \prod_{j=1}^{k} (1 - s_{u_j v}) \left(\sum_{j=1}^{k} \frac{1 - \rho_{u_j v} s_{u_j v}}{1 - s_{u_j v}} + 1 - \rho_v \right) \quad (3)$$

Learning Links of Parent Layers. LSs and CPs of links between higher parent nodes and lower child nodes are tuned by a stochastic maximum-likelihood method for input data. This method miximumizes $p(D) = \pi_k p(\boldsymbol{D}_k)$, likelihood of input data D where each input datum is $\boldsymbol{D}_k$. The update differences when a datum $\boldsymbol{D}_k$ is input are given by Eqs. 4 and 5 below.

$$\Delta\rho_{uv} = \alpha \frac{\partial \ln p(\boldsymbol{D}_k)}{\partial \rho_{uv}} = \alpha \frac{s_{uv}}{1 - s_{uv} + \frac{1}{m_{uv}}} \frac{(m_{vD} + 1)(m_{Dv} - 1)}{m_{Dv} m_{vD} + 1} C_v \quad (4)$$

$$\Delta s_{uv} = \beta \frac{\partial \ln p(\boldsymbol{D}_k)}{\partial s_{uv}} = \beta \frac{\rho_{uv} m_{Dv} + 1 - \rho_{uv}}{1 - s_{uv} + \frac{1}{m_{uv}}} \frac{m_{vD} + 1}{m_{Dv} m_{vD} + 1} \frac{m_{vu} - 1}{m_{vu} - 1 + s_{uv}} C_v \quad (5)$$

where α and β are the learning coefficients, and C_v is given as blow.

$$C_v = \prod_{u' \in U} \frac{1 - s_{u'v} + \frac{1}{m_{u'v}}}{1 + \frac{1}{m_{u'v}}} \quad (6)$$

In this training, parameters are trained in order to represent each input data and patterns of parents are learned to able to reproduce input data.

2.2 Invariance of Pooling by Child

To construct a multi-layered Bayesian network, it is natural to put a new parent layer at the top of a layer so that the old layer contains children of the new parents [7,8]. However, this simple architecture cannot be used for a binary Bayesian network. The parent variables are inferred based on a product of the messages from its children. Therefore, parent variables can only recognize patterns represented by the AND operations of child patterns. This architecture does not provide invariance to small mutations in the lower patterns.

To avoid this, original pooling layers, such as those used in convolutional neural networks [10], are implemented in the MFB.

Architecture of Pooling by Child. The architecture of a pooling layer is shown in Fig. 5: it comprises low-children connected to lower parents, a mid-grandchild connected to each low-child, a high-child connected to each mid-grandchild, and observable variables as children of the mid-grandchildren.

A message from a low-child to a mid-grandchild behaves like an average weighted by messages from lower parents, so that it can be handled as a pooling of inferences. The observable node is given a constant value and conditional probabilities are set such that a message from low-child equals a message to high-child. Thus, this layer can pool the inferences of a lower parents and send them to a higher parents. The MFB is constructed by alternately connecting parent layers and pooling layers (as shown in Fig. 6).

Learning Links of Pooling Layers. The links between a higher pooling layer and a lower parent layer are learned through unsupervised learning. We implemented a training algorithm based on the infomax principle [13] to update them. In this training, parents sharing one node of a pooling layer inhibit each other so that a set of parent nodes that are exclusively activated (have high probability for T state) remains connected to the same pooling node. For example, when a parent node is activated, higher links that propagate an inhibitory message is weakened and higher links that propagate an excitatory message is strengthened.

Through this training, a pooling node collects parent nodes that are activated exclusively, such as a set of patterns of the same position or minor different patterns. This results in pooling patterns with minor mutations such as noise, differences in position or direction, and transformation.

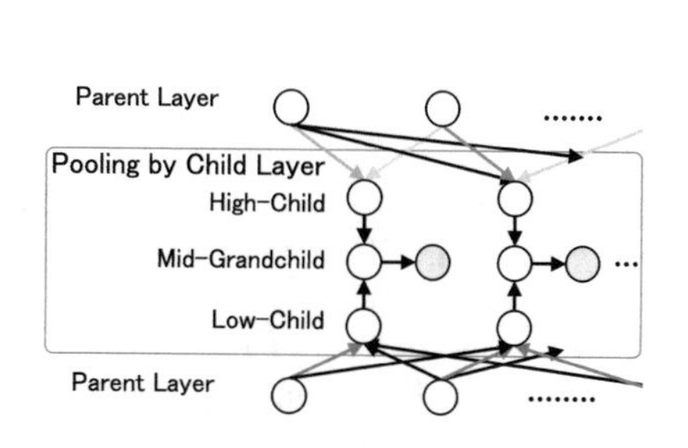

Fig. 5. Pooling layer in the MFB

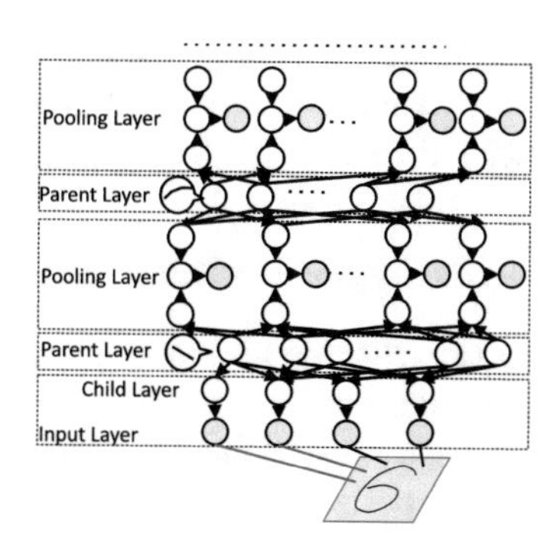

Fig. 6. Architecture of the MFB

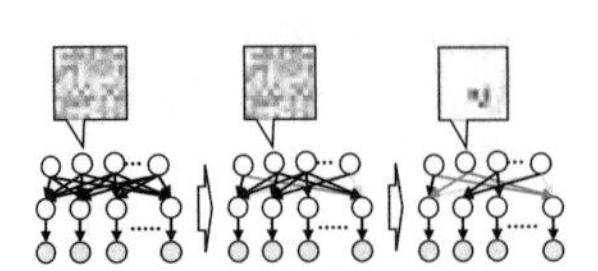

Fig. 7. Pattern extraction with the MFB

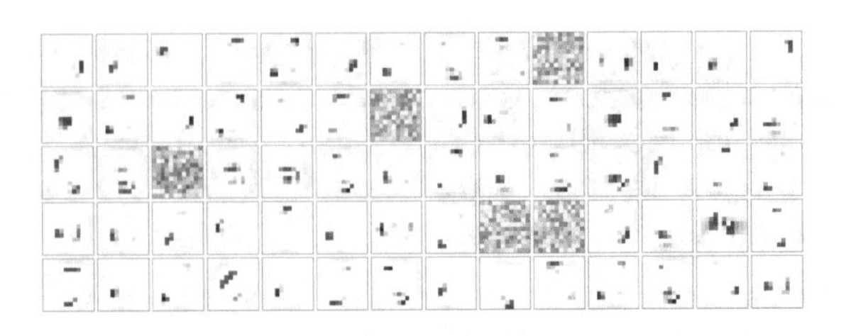

Fig. 8. Extracted patterns from MNIST. We show 98 patterns from 196 extracted patterns due to the space limitation.

3 Experimental Results

3.1 Pattern Extraction Experiments

In order to confirm the MFB can practically be constructed on computers, we performed experiments to extract patterns from an MNIST dataset [12].

Each input image was 12×12 pixels downsampled from 24×24 pixel area in the center of each MNIST image. The network had one parent layer of 196 parents and one child layer of 144 children. As the initial state, links between the parent and the children were randomly connected. Then pattern extraction proceeds as shown in Fig. 7.

The result of training is shown in Fig. 8. Each box corresponds to a parent variable. The pixels in the box represent the parameters of the links corresponding to each position. The pixel colors denote the CP or the excitation potential (red denotes a large CP, indicating an excitatory pixel, and blue denotes a small CP, indicating an inhibitory pixel). The pixel transparency denotes the LS or the strength of the connection (dark pixels denote large LSs or strong connection and light pixels denote low LSs or weak connections).

Based on Fig. 8, several patterns were identified as follows: Small lines of only excitatory pixels, excitatory line adjacent to inhibitory lines, only excitatory probs, and only inhibitory probs. These patterns are similar to the result of denoising autoencoders [19], which is a popular method of feature extraction. Thus it was concluded that the MFB could be constructed on an actual computer and extract patterns.

3.2 Pooling Layer Training Experiments

To confirm the learning ability of pooling layers, we have performed experiments [15]. It uses randomly-generated 1×10 belt-like input images in which no two successive pixels were white (as shown in Fig. 9). The structure of the network used in this experiment is as shown in Fig. 9a. Each of the input layer, child layer and parent layer has 10 nodes, and the pooling layer has 15 pooling units. Each child has a parent and their link is fixed. Each parent is activated when corresponding input pixel is white. Links between parents and low-child nodes of

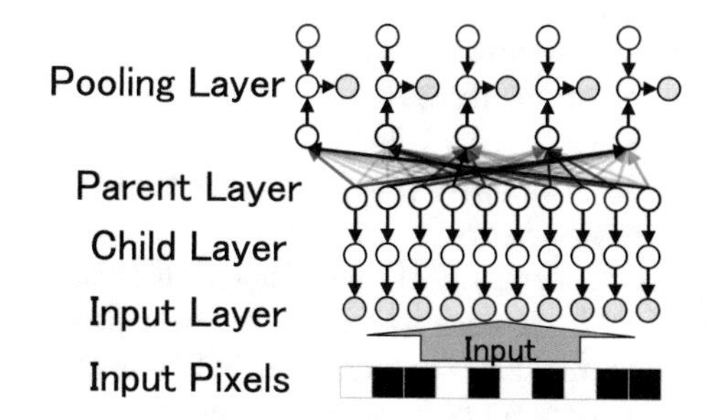

(a) Trained network structure

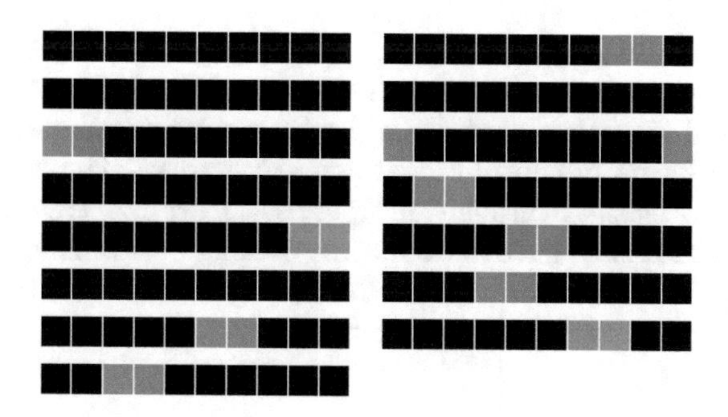

(b) LSs of links of each pooling node

Fig. 9. Pooling Experiments adopted from [15]. a: A trained network structure and a belt-like input. b: The LSs of trained links of the pooling layer. Each belt corresponds to a pooling unit and each panel represents a LS of a link to the parent corresponding to each position.

pooling nodes were initially connected randomly and trained. As no two successive pixels were white, adjacent parents are exclusively activated, and therefore they were expected to become connected to the same pooling node.

The result is as shown in Fig. 9b. Each belt represents links from each low-child node to all parent nodes. The belt contains 10 pixels that represent LSs of links to parents: each pixel corresponds to each parent. The colors of panels indicate values of the LS. Black pixels denote 0 and brighter pixels indicate larger LSs. We consider that adjacent parents have links to the same low-child as we expected and that MFB obtained pooling ability correctly.

4 Discussion

4.1 Bidirectional Recognition of Multiple Abstraction Features

MFB is based on Bayesian networks so that its inference is done by belief propagation. In belief propagation, information, called messages, is exchanged via links in either direction. In this manner, a Bayesian network gives the posterior probabilities of any variable upon observing any set of variables. Moreover, using belief propagation, it performs high-level inference such as in the explaining-away effect [16], by which the most suitable factor that explains an observation is selected from competing factors.

Through belief propagation, multi-layer Bayesian networks can use information about higher abstract patterns to recognize lower patterns and improve the recognition of each intermediate pattern [14]. MFB uses this characteristic to output intermediate patterns as features more accurately.

4.2 Pooling Layer Training

MFB trains also links of pooling layers. Its pooling node can collect output of parents that have similar patterns with minor mutations. The range of pooling is obtained based on activities of parent nodes, not simply based on its position such as pooling layers in convolutional neural networks. The connection is not hardcoded and is learned through unsupervised learning. We expect that this property enables MFB to have more flexible pooling performance: pooling small mutations such as rotations and transformations of patterns, instead of being limited to changes in positions of patterns.

4.3 Disadvantages of Bidirectional Recognition

MFB performs its recognition based on loopy belief propagation [16]. Therefore MFB has to keep updating its messages until the messages converge. This results in a problem of calculation cost in recognition; in our experiments, MFB needs to iterate the update about 10 to 20 times until convergence.

Moreover, the convergence of messages is not guaranteed especially at initial state of the training. This worsen efficiency of training MFB. Usually network architectures in which messages can converge tend to be learned stably more than the network in which messages cannot converge through training in our experiments. However, we have not yet guaranteed this property theoretically.

4.4 Conclusion

In this paper, we proposed an MFB, a feature extractor that can be shared among various classes. This characteristic is achieved by utilizing intermediate patterns extracted by unsupervised learning as feature values and by implementing a bidirectional network based on brain informatics models. This allows a more accurate calculation of posterior probabilities as feature values. To implement the MFB, we developed several techniques. Using the 1pT assumption and formulation of the conditional probabilities with CPs and LSs, the number of parameters was reduced and the representability of independency was maintained. Moreover, an original pooling layer was implemented to obtain invariant patterns. It was experimentally confirmed that the MFB could be constructed successfully on an actual computer and utilized for pattern extraction and learning pooling. As future works, we are going to expand the number of layers to extract more complex patterns, and evaluate the performance of intermediate patterns in semi-supervised learning and transfer learning.

Acknowledgements. This work was supported by JSPS Grant-in-Aid for JSPS Fellows, JP17J09110.

References

1. Bengio Y, Courville A, Vincent P (2013) Representation learning: a review and new perspectives. IEEE Trans Pattern Anal Mach Intell 35(8):1798–1828
2. Dehaene S, Cohen L, Sigman M, Vinckier F (2005) The neural code for written words: a proposal. Trends Cogn Sci 9(7):335–341
3. Donahue J, Jia Y, Vinyals O, Hoffman J, Zhang N, Tzeng E, Darrell T (2014) DeCAF: a deep convolutional activation feature for generic visual recognition. In: International conference on machine learning, pp 647–655
4. Du Y, Zhang R, Zargari A, Thai TC, Gunderson CC, Moxley KM, Liu H, Zheng B, Qiu Y (2018) A performance comparison of low-and high-level features learned by deep convolutional neural networks in epithelium and stroma classification. In: Medical imaging 2018: digital pathology, vol 10581. International Society for Optics and Photonics, p 1058116
5. Gray H, Goss CM (1973) Anatomy of the human body, by Henry Gray, 29th american ed., edited by charles mayo goss. with new drawings by don m. alvarado. edn. Lea & Febiger Philadelphia
6. He Y, Kavukcuoglu K, Wang Y, Szlam A, Qi Y (2014) Unsupervised feature learning by deep sparse coding. In: Proceedings of the 2014 SIAM international conference on data mining. SIAM, pp 902–910
7. Hosoya H (2012) Multinomial Bayesian learning for modeling classical and nonclassical receptive field properties. Neural Comput 24(8):2119–2150. https://doi.org/10.1162/NECO_a_00310
8. Ichisugi Y, Sano T (2016) Regularization methods for the restricted Bayesian network besom. In: Proceedings of the neural information processing: 23rd international conference, ICONIP 2016, Part I. Springer, pp 290–299. https://doi.org/10.1007/978-3-319-46687-3_32
9. Kingma DP, Mohamed S, Rezende DJ, Welling M (2014) Semi-supervised learning with deep generative models. In: Advances in neural information processing systems, pp 3581–3589
10. Krizhevsky A, Sutskever I, Hinton GE (2012) ImageNet classification with deep convolutional neural networks. In: Advances in neural information processing systems, vol 25. Curran Associates, Inc., pp 1097–1105
11. Le Q (2013) Building high-level features using large scale unsupervised learning. In: 2013 IEEE International conference acoustics, speech and signal processing (ICASSP), pp 8595–8598. https://doi.org/10.1109/ICASSP.2013.6639343
12. LeCun Y (1998) The MNIST database of handwritten digits. http://yann.lecun.com/exdb/mnist/
13. Linsker R (1998) Self-organization in a perceptual network. Computer 21(3):105–117. https://doi.org/10.1109/2.36

14. Nakada H, Ichisugi Y (2018) Context-dependent robust text recognition using large-scale restricted Bayesian network. Proc Comput Sci 123:314–320
15. Nishino K, Inaba M (2018) Constructing hierarchical Bayesian networks with pooling. In: Proceedings of the thirty-second AAAI conference on artificial intelligence (AAAI-18). AAAI Press, pp 8125–8126
16. Pearl J (1985) Bayesian networks: a model of self-activated memory for evidential reasoning. In: Cognitive science society, pp 329–334
17. Ramachandran VS, Gregory RL (1991) Perceptual filling in of artificially induced scotomas in human vision. Nature 350(6320):699–702
18. Soleymani S, Dabouei A, Kazemi H, Dawson J, Nasrabadi NM (2018) Multi-level feature abstraction from convolutional neural networks for multimodal biometric identification. arXiv preprint arXiv:1807.01332
19. Vincent P, Larochelle H, Bengio Y, Manzagol PA (2008) Extracting and composing robust features with denoising autoencoders. In: Proceedings of the 25th international conference on machine learning, ICML 2008. ACM, New York, pp 1096–1103. https://doi.org/10.1145/1390156.1390294

Factorization Machines for Blog Feedback Prediction

Krisztian Buza[1(✉)] and Tomaš Horváth[1,2]

[1] Telekom Innovation Laboratories,
Department of Data Science and Engineering, Faculty of Informatics,
ELTE - Eötvös Loránd University, Budapest, Hungary
{buza,horvathtamas}@inf.elte.hu

[2] Institute of Computer Science, Faculty of Science,
Pavol Jozef Šafárik University, Košice, Slovakia
http://t-labs.elte.hu

Abstract. Estimation of the attention that a blog post is expected to receive is an important text mining task with potential applications in various domains, such as online advertisement or early recognition of highly influential fake news. In the blog feedback prediction task, the number of comments is used as proxy for the attention. Although factorization machines are generally well-suited for sparse, high-dimensional data with correlated features, their performance has not been systematically examined in the context of the blog feedback prediction task yet. In this paper, we evaluate factorization machines on a publicly available blog feedback prediction dataset. Comparing the results with other results from the literature, we conclude that factorization machines are competitive with multilayer perceptron networks, linear regression and RBF network. Additionally, we analyze how parameters (feature weights and interaction weights) of factorization machine are learned.

Keywords: Blog feedback prediction · Factorization machine · Machine learning

1 Introduction

Early recognition of highly influential posts in social media, such as blogs or tweets, is an essential task with various applications. For example, the identification of most visited blogs may be useful in online advertisement scenarios or in order to identify highly influential fake news in advance.

In the blog feedback prediction problem [2], the number of comments serves as a proxy for the attention, i.e., the task is to predict the number of comments that a blog will receive. In order to predict the number of comments, various features are used that refer to the textual content, source (blog or website) or other conditions, e.g. on which day of the week the post was published.

Recently, the blog feedback prediction task received considerable attention, see e.g. [4,7,8,10,12]. Various models have been used, such as state-of-the-art

R. Burduk et al. (Eds.): CORES 2019, AISC 977, pp. 79–85, 2020.
https://doi.org/10.1007/978-3-030-19738-4_9

variants of nearest neighbour regression [4], fuzzy systems [7], ensemble techniques [12], neural networks and decision trees [10].

Motivated by movie recommendation tasks, factorization machines [9] have been developed for supervised machine learning on sparse and high-dimensional data containing correlated features. Factorization machines are especially well-suited to represent correlations efficiently, i.e., given n features, factorization machines are able to represent all the pairwise correlations with $\mathcal{O}(n)$ parameters. Additionally, factorization machines can estimate correlations between features even if the correlation is not expressed explicitly, but has to be inferred. Last, but not least, compared with deep learning, factorization machines require much less training data. For the aforementioned reasons, factorization machines have been recognized more and more in the machine learning community.

Due to the large number of textual features that correspond to the frequency of most predictive words, the data in the blog feedback prediction problem is high dimensional and sparse. Additionally, some of the features are expected to be correlated. Thus, factorization machines are promising candidates for the blog feedback prediction task.

Despite the aforementioned facts, factorization machines have not been systematically studied in context of the blog feedback prediction problem. For completeness, we note that an initial experiment with factorization machines on the blog feedback data have been documented in the first version of a manuscript available at arXiv [13], however, the results were not evaluated in terms of typical evaluation metrics of the blog feedback prediction task, such as AUC or the number of hits. More importantly, the manuscript does not focus on the blog feedback prediction tasks and the aforementioned experiment is not included in subsequent versions of the manuscript, including its current version.

For the above reasons, in this work, we aim to examine whether factorization machines are suitable blog feedback prediction.

The rest of this paper is organized as follows: in Sect. 2 we review factorization machines. In Sect. 3 we describe the blog feedback data and our experiments. Finally, we conclude in Sect. 4 and point out potential directions of future work.

2 Factorization Machines

Given an instance $\mathbf{x} = (x_1, \ldots, x_n)$, a factorization machine [9] of second degree with f factors predicts its label as follows:

$$\hat{y}(\mathbf{x}) = w_0 + \sum_{i=1}^{n} w_i x_i + \sum_{i=1}^{n} \sum_{j=i+1}^{n} \Big(\sum_{k=1}^{f} v_{i,k} v_{j,k} \Big) x_i x_j \quad (1)$$

where $w_0 \ldots w_n$ and $v_{1,1} \ldots v_{n,f}$ are parameters of the model. The later describe the interactions between features, while we refer to $w_1 \ldots w_n$ as *feature weights*.

Given a labelled training dataset, the parameters of the model can be learned with stochastic gradient descent so that the sum of squared errors (or another objective function) is optimized. For details, we refer to Algorithm 1 and [9].

Algorithm 1. Training the Factorization Machine

1 Training data D, number of epochs e, learning rate η, standard deviation σ Weights $w_0, w_1, \ldots w_k$ and $v_{1,1}, \ldots, v_{n,f}$ Initialize $w_0, w_1, \ldots w_k$ and $v_{1,1}, \ldots, v_{n,f}$ from standard normal distribution with zero mean and standard deviation σ
2 **for** *epoch in* $1 \ldots e$ **do**
3   **for** *each* $(x, y) \in D$ *in random order* **do**
4     $\hat{y} \leftarrow w_0 + \sum_{i=1}^{n} w_i x_i + \sum_{i=1}^{n} \sum_{j=i+1}^{n} \left(\sum_{k=1}^{f} v_{i,k} v_{j,k} \right) x_i x_j$ $w_0 \leftarrow w_0 - \eta\, 2(\hat{y} - y)$
5   **for** i *in* $1 \ldots k$ **do**
6       $w_i \leftarrow w_i - \eta\, 2(\hat{y} - y) x_i$
7   **end**
8   **for** i *in* $1 \ldots n$ **do**
9   **for** j *in* $1 \ldots f$ **do**
10         $v_{i,j} \leftarrow v_{i,j} - \eta\, 2(\hat{y} - y) \left(x_i \sum_{k=1}^{n} v_{k,j} x_k - v_{i,j} x_i^2 \right)$
11   **end**
12   **end**
13   **end**
14 **end**
15 **return** $w_0, w_1, \ldots, w_k$ and $v_{1,1}, \ldots, v_{n,f}$

3 Experimental Evaluation

Blog Feedback Data. We performed experiments on the publicly available Blog Feedback Data[1]. The data contains 60021 instances and 281 features, including the target. Each instance refers to a blog post which is described by various features referring to textual content, the source (blog) on which the post was published and other conditions, e.g. on which day of week the post was published. The target is the number of comments that the post received within the next 24 h. For a more detailed description of the data and how it was created, see [2].

Experimental Protocol. In order to assist reproducibility and comparability, we used the predefined training and test sets associated with the data. We calculated the evaluation metrics (AUC@10 and Hits@10) for each test set and aggregated the results. This is exactly the same protocol as in [2], therefore, the results are directly comparable.

Evaluation Metrics. In order to evaluate the predictions, we used AUC@10 and Hits@10 which are defined as follows.

For each test split, we consider 10 blog pages that were predicted to have the largest number of feedbacks. We count how many out of these pages are among

[1] https://archive.ics.uci.edu/ml/datasets/BlogFeedback

the 10 pages that received the largest number of comments in reality. We call this evaluation measure Hits@10.

For the AUC, i.e., area under the receiver-operator curve, we considered as positive the 10 blog pages receiving the highest number of comments in the reality. Then, we ranked the pages according to their predicted number of comments and calculated AUC. We call this evaluation measure AUC@10. For both evaluation metrics, higher values indicate better performance.

Hyperparameters of the Factorization Machine. In our experiments, by default, we set the number of factors $f = 3$. In order to find appropriate model parameters, we minimized the sum of squared errors on the training data with stochastic gradient descent. We initialize the parameters (i.e., feature weights W_i and interaction weights $v_{i,k}$) from a standard normal distribution with zero mean and $\sigma = 10^{-8}$. We set the learning rate of stochastic gradient descent to $\eta = 10^{-12}$ and iterated over the training instances 1000 times, in other words: we learned the model in 1000 epochs.

Experimental Results. As can be seen in Table 1, factorization machine achieved AUC@10 of 0.869, while the average number of Hits@10 was 5.117. We note that factorization machine is a generalization of linear regression: in particular, a factorization machine with $f = 0$ is equivalent to linear regression, therefore, we show the results for linear regression in Table 1.

We also performed experiments with $f = 5$ factors. As the results were very similar to the results with $f = 3$ factors, for simplicity, we only report results with $f = 3$ factors.

Table 1. The performance of factorization machine (FM) with $f = 0$ (linear regression) and $f = 3$ factors on the blog feedback prediction task

	Linear regression	FM, $f = 3$
AUC@10	0.864	0.869
Hits@10	4.733	5.117

Comparison with Results Reported in the Literature. According to the results reported in [2], factorization machine outperforms various multilayer perceptron networks, linear regression and RBF network. In particular, none of these models achieved Hits@10 greater than 5, while the AUC was between 0.8 and 0.85 for these models, see Fig. 1 in [2]. On the other hand, one of the regression trees, M5P, achieved AUC around 0.9, while its performance in terms of Hit@10 was comparable to that of factorization machine.

Learning the Parameters of the Factorization Machine. In order to understand how parameters of the factorization machine are learned during stochastic gradient descent, we calculated for each feature weight w_i and for each interaction weight $v_{i,k}$, how much that parameter changed in total during each epoch. Next, we calculated the absolute value of the change for each parameter. Subsequently, we calculated the mean of these absolute values separately for the feature weights and interaction weights. We refer to these mean values as the magnitude of change. We plot the magnitude of change for the first 250 epochs in Fig. 1.

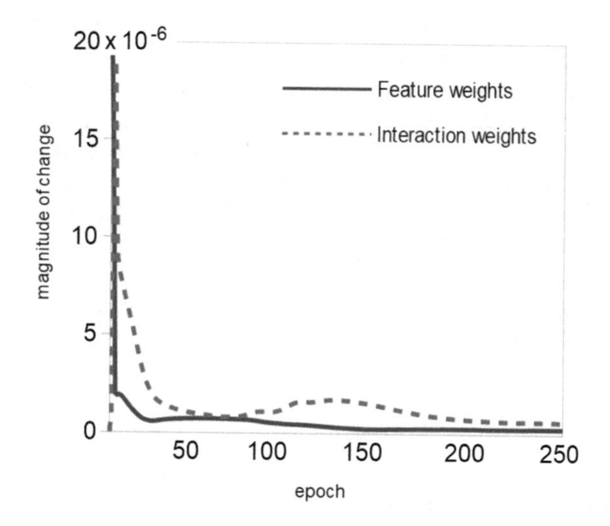

Fig. 1. Magnitude of change of feature weights and interaction weights as function of the number of epochs.

As one can see, at the beginning, the change of both feature weights and interaction weights is relatively large. As a global trend, the magnitude of change of feature weights is decreasing. In contrast, the magnitude of change of interaction weights is decreasing at the beginning, then it begins to increase again around the 70th epoch. Roughly from the 140th epoch on, the magnitude of change of interaction weights is decreasing again.

This analysis indicates that interaction weights may be more difficult to learn than feature weights. The relatively high magnitude of change of the interaction weights between the 100th and 160th epochs may be explained by the assumption that interaction weights can be learned effectively, once feature weights have relatively good values.

4 Conclusions and Outlook

In this paper, we considered the blog feedback prediction task and systematically examined factorization machines for this task. As expected, factorization machines are competitive models for blog feedback prediction. On the other hand, our analysis revealed that the development of efficient training algorithms may be challenging, because it seems to be the case that interaction weights can only be learned efficiently if feature weights already have reasonable values. Taking this observation into account, one may devise smart training algorithms in the future.

On the long term, it would also be interesting to adapt other techniques for blog feedback prediction, such as hybrid approaches, the incorporation of fuzzy-valued loss functions into factorization machines or model selection based on genetic algorithms, see e.g. [1,5,6]. Furthermore, we note that a special variant of the blog feedback prediction task, the so called *personalized blog feedback prediction* task has also been considered in the literature [3]. This motivates the idea of considering the situation as a *game* (in the sense of game theory) in which users have a limited budget of comments, i.e., each user is only able to comment on a limited number of blogs, while the users try to optimize the impact of their comments (payoff). Therefore, one could try approaches based on game theory, such as the one described in [11], for blog feedback prediction.

Acknowledgments. This work was supported by the project no. 20460-3/2018/FEKUTSTRAT within the Institutional Excellence Program in Higher Education of the Hungarian Ministry of Human Capacities.

References

1. Burduk R, Kurzyński M (2006) Two-stage binary classifier with fuzzy-valued loss function. Pattern Anal Appl 9(4):353–358
2. Buza K (2014) Feedback prediction for blogs. In: Data analysis, machine learning and knowledge discovery, pp 145–152. Springer
3. Buza K, Galambos I (2013) An application of link prediction in bipartite graphs: personalized blog feedback prediction. In: 8th Japanese-Hungarian symposium on discrete mathematics and its applications, June, pp 4–7. Citeseer
4. Buza K, Nanopoulos A, Nagy G (2015) Nearest neighbor regression in the presence of bad hubs. Knowl-Based Syst 86:250–260
5. Jackowski K, Krawczyk B, Woźniak M (2014) Improved adaptive splitting and selection: the hybrid training method of a classifier based on a feature space partitioning. Int J Neural Syst 24(03):1430007
6. Jackowski K, Wozniak M (2010) Method of classifier selection using the genetic approach. Expert Syst 27(2):114–128
7. Kaur H, Pannu HS (2018) Blog response volume prediction using adaptive neuro fuzzy inference system. In: 2018 9th international conference on computing, communication and networking technologies (ICCCNT), pp 1–6. IEEE
8. Kaur M, Verma P (2016) Comment volume prediction using regression. Int J Comput Appl 151(1): 1–9
9. Rendle S (2010) Factorization machines. In: 2010 IEEE 10th international conference on data mining (ICDM), pp 995–1000. IEEE
10. Singh K, Sandhu RK, Kumar D (2015) Comment volume prediction using neural networks and decision trees. In: IEEE UKSim-AMSS 17th international conference on computer modelling and simulation, UKSim 2015, Cambridge, United Kingdom

11. Suciu M, Lung RI, Gaskó N, Dumitrescu D (2013) Differential evolution for discrete-time large dynamic games. In: 2013 IEEE congress on evolutionary computation (CEC), pp 2108–2113. IEEE
12. Uddin MT (2015) Automated blog feedback prediction with Ada-Boost classifier. In: 2015 international conference on informatics, electronics & vision (ICIEV), pp 1–5. IEEE
13. Yamada M, Lian W, Goyal A, Chen J, Wimalawarne K, Khan SA, Kaski S, Mamitsuka H, Chang Y (2015) Convex factorization machine for regression. arXiv preprint arXiv:1507.01073v1

Vertical and Horizontal Data Partitioning for Classifier Ensemble Learning

Amgad M. Mohammed[1,2,3](✉), Enrique Onieva[1,2], and Michał Woźniak[4]

[1] Deusto University, Bilbao, Spain
amgad.elsayed@deusto.es
[2] Deusto Institute of Technology (DeustoTech), Bilbao, Spain
[3] Menoufia University, Menoufia, Egypt
[4] Wrocław University of Science and Technology, Wrocław, Poland

Abstract. Multiple classifier systems have proven superiority over individual ones to solve classification tasks. One of the main issues in those solution relies in data size, when the amount of data to be analyzed becomes huge. In this paper, the performance of ensemble system to succeed by using only portions of the available data is analyzed. For this, extensive experimentation with homogeneous ensemble systems trained with 50% of data and 50% of features is performed, using bagging sampling schema. Simple and weighted majority voting schemes are implemented to combine the classifier outputs. Experimental results including 25 datasets show the benefit of using multiple classifiers trained on limited data. The ensemble size and the accuracy obtained with individual model trained over the entire dataset is compared.

Keywords: Classifier ensemble · Pattern classification · Data complexity

1 Introduction

Data mining is defined as the science for discovering useful and unexpected new patterns from data. One of the most important barriers for data mining is the size of the data used to train predictors. During the last time, the amount of data available for exploration has exponentially increased, due to use of computer technology and information systems. Classification is the process to assign a given object to one of the predefined categories based on its observed feature values [14]. According to Wolpert's *no free lunch theorem* [21] there is no best classifier suitable for all problems, but each of the model has its own area of competence depends on the assumptions made during its design.

In order to increase the classification confidence, some methods focus on maximizing the margins between the instances in different classes, as Support Vector Machines (SVM) [8]. Other methods calculate the observed probability for each class output based on the evidence provided by feature value as in Naive Bayes.

R. Burduk et al. (Eds.): CORES 2019, AISC 977, pp. 86–97, 2020.
https://doi.org/10.1007/978-3-030-19738-4_10

Instead of using a single method, ensemble systems relies on a pool of classifiers that cooperate to categorize the same new pattern [14,22].

The overall objective of this contribution is to study the suitability of homogeneous ensemble systems for different machine learning algorithms. The motivation concentrates on the suitability of multiple classifiers to be combined by majority voting or weighted majority voting over 50% of data and 50% of features. Bagging and random feature selection are the source to achieve diversity between classifier outputs.

The rest the article remains as follows. Section 2 explains concepts and components of ensemble systems, where it will be discussed methods to build and combine this kind of systems. In Sect. 3, the results of experiments carried out on 25 benchmark datasets are presented and discussed. The last section concludes the paper and presents the future works.

2 Ensemble Classification

In order to build an ensemble system, four aspects must be taken into account in its design:

1. The building phase, which denotes how the individual models are generated,
2. The prediction and output phase, which refers to how to get predictions from those models.
3. The ensemble pruning, which is responsible for choosing the most valuable individuals to the ensemble.
4. The combination schema, specify how to aggregate responses from models in a single decision.

Each one of the aspects are detailed in subsequent subsections.

2.1 Generating the Base Experts

The diversity of individual models is considered the corner stone for building ensemble systems. Uncorrelated and high diverse models are more promising to increase and enhance the general performance of the system. According to this, several techniques can be applied to improve diversity among individual models [14], such as data sampling methods (Bagging [1], Boosting [17], AdaBoost [7]), using different training parameters for the same learning algorithm, using different feature subsets to train each individual model [2] or using different base learning algorithms.

- *Data Sampling (bagging)*: Bagging based methods use different parts of the training set to build each individual model [1]. Bagging methodology can be described as parallel data sampling, since all individual models can be built and even implemented in parallel way. Each individual model receives a part of samples from the original dataset.

- *Data Sampling (boosting)*: Boosting aims to classify difficult and hard data samples [17]. The resampling is strategically geared to focus on the most informative data for each consecutive classifier. On the other side boosting is more sensitive to noise and outliers [5,6].
- *Attribute Sampling (vertical partitioning)*: There are several propositions based on this concept, such as *Random Subspace* [10], *Attribute Bagging* [3], or *Random Forest* [9] which employs the first mentioned approach for random decision trees which also use randomness during choosing "the best" attribute in decision tree induction.

2.2 Predictions of Experts

Once all the individual experts are trained, all these models are tested against unseen data in order to get estimations of the generalization capability. When all models provide predictions as specific class label not as ranked probability, the combination strategy can be directly applied. The prediction accuracy for each individual classifier depends on how it learns from the training data. Examples of individual experts are, among others: (1) *Naive-Bayes classifier*, which computes *posterior* probabilities of a categorical class variable depending on Bayes rule. (2) *Conditional Inference Tree*, that makes recursive partitioning for continuous, ordered, nominal and multivariate response variables in a conditional inference framework. (3) *Logistic regression model*, which measures the relationship between the categorical dependent variable and independent variables by estimating probabilities using a logistic function.

2.3 Ensemble Pruning

To some extend, combining few well-selected classifiers may result better results than considering all. The main challenge here is to choose complementary classifiers with high diversity among them and high accuracy. Selection of classifiers may be offline (static selection [13]) or online (dynamic selection [4]). In offline selection, the subset of classifiers is chosen once in order to classify all unseen instances; on the other hand, for online selection a different subset of classifiers can be chosen for each one of the unseen instances. Evolutionary algorithms and heuristic search methods are usually applied to solve this problem of selecting near optimal subset of models [15].

2.4 Combination Strategy

This feature denotes the methodology followed in order to aggregate the results of several models in a single one. The combiner balances the deviation between the diversity and the bias, also eases the mistakes raised by some models [18]. As clarified in [16]: The combination strategy depends on the output of the base classifiers whether class label vs. class-specific continuous outputs, as well as the combination mechanism may be trainable vs. non-trainable.

In trainable mechanisms, further training and learning is needed to be implemented, as in stacked generalization [20], mixture of experts [12] and behavior knowledge space [11]. Where in non-trainable ones, no extra training beyond the generation of the ensemble is defined.

Let's present the formal model of classification. $\mathcal{X}$ denotes feature space and $x \in \mathcal{X}$ is the example, i.e., x is the so-called feature vector which informs about attribute values. We will assume that we have d attributes at our disposal

$$x = \left\{x^{(1)}, x^{(2)}, ..., x^{(d)}\right\}, \text{ and } x \in \mathcal{X} = \mathcal{X}^{(1)} \times \mathcal{X}^{(2)} \times ... \times \mathcal{X}^{(d)} \tag{1}$$

The classification goal is to assign a given object described by its features x into one of the predefined categories, also called labels. Let's $\mathcal{M} = \{1, ..., M\}$ stands for the set of labels. The classification algorithm is any function Ψ with domain $\mathcal{X}$ and codomain $\mathcal{M}$

$$\Psi : \mathcal{X} \rightarrow \mathcal{M}. \tag{2}$$

Then if we define a set of individual classifiers Let's assume that we have a pool Π of n individual classifiers $\Pi = \{\Psi_1, \Psi_2, ..., \Psi_n\}$. The majority voting combine the binary outputs of base classifiers, the highest number of votes reserved as output of ensemble as in Eq. 3

$$\Psi(x) = \arg\max_{i \in \mathcal{M}} \sum_{k=1}^{n} [\Psi_k(x) = i] \tag{3}$$

Figure 1 shows an ensemble of three experts that cooperate to classify eight objects into a two-class problem (black or white classes). In order to take the decision (binary), at least $\lfloor K/2 \rfloor + 1$ classifiers must agree to choose a particular class label. Weighted majority voting models the situation in which some expert classifiers perform better than others, then weighting the decisions of those qualified experts may improve the general accuracy of the system. Each classifier Ψ_k receives weight w_k proportional to its estimated performance. The future performance of each classifier can be estimated over separate validation dataset. In its basic form, classifiers with higher accuracy receive higher weight [14], as shown in Eq. (4).

$$w_k = \log\left(\frac{\hat{P}_k}{1 - \hat{P}_k}\right), \quad 0 < \hat{P}_k < 1, \quad k = 1, 2, ..., n \tag{4}$$

where $\hat{\mathrm{P}}_k$ represents the estimation of accuracy for the expert Ψ_k. The final class label decided upon Eq. (5).

$$\Psi(x) = \arg\max_{j \in \mathcal{M}} \sum_{k=1}^{n} [\Psi_k(x) = j]\, w_k \tag{5}$$

Figure 2 shows an example of applying the weighted majority voting mechanism to aggregate three experts together, where the final ensemble output in this case will be the same as the best classifier from the pool.

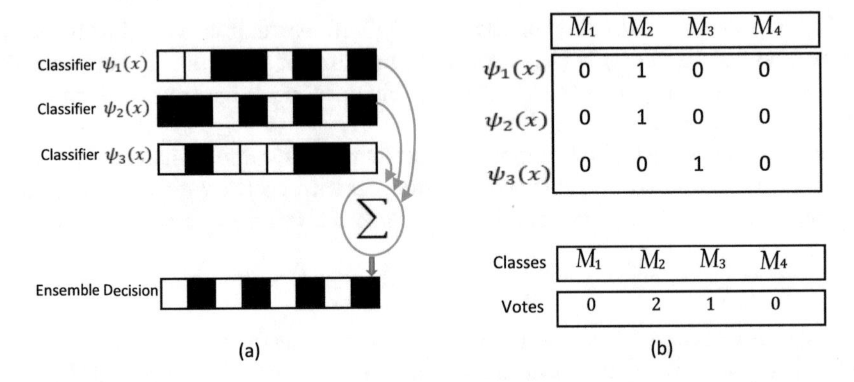

Fig. 1. (a) Majority voting of 3 experts classify 8 instances, (b) 3 experts solve 4-class problem

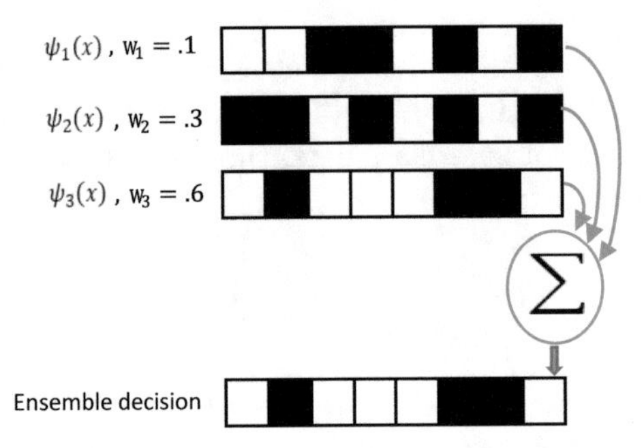

Fig. 2. Weighted Majority Voting for 3 experts classifying 8 instances

3 Experimental Study

3.1 Experimental Setup

The general architecture of the proposed ensemble system implemented for validation, is presented in Algorithm 1. The experimental setup consider the following parameters:

- The Pool Size is denoted by n. It represents the total number of individual classifiers to build the ensemble. For this work, configurations with $n = \{3, 10, 20, 30, 40, 50\}$ are considered. A particular individual in the ensemble is denoted as $\Psi_k,\ k \in \{1, 2, \ldots, n\}$.
- Each one of the models will be built using a percentage of the provided training set. This percentage is denoted by the parameter p_{tra}. In this work, p_{tra} is set to 0.5. In addition, each model is trained using a percentage of the attributes of the training data, denoted by p_{att}, which is set also to 0.5.
- In this work, homogeneous ensemble systems are tested, which means that the pool of classifiers are composed by models obtained by the same learning

algorithm. In this case, *Naive-Bayes* (NB), *Decision Trees* (DT) and *Multinomial Logistic Regression* (ML) are tested as *Base Classifiers* (*Base*()).
- Once the ensemble is built, a particular combination strategy *comb*() is used to aggregate predictions over the test dataset. In this article, single *majority voting* and *weighted voting* will be tested.

For the implementation, the R programming language is used. For Base Classifier models, packages *e1071*[1] (NB), *partykit*[2] (DT) and *nnet*[3] (ML) are used.

To make the results more reliable, 5 fold cross validation is applied 10 times over each one of the datasets.

A total number of 25 classification datasets taken from UCI Machine Learning Database Repository[4] will be used in the experimentation. The diversity of datasets ranges from binary to multi-class problems (up to 11), while the number of features ranges from 3 to 62. A description of datasets used in this work with their main characteristics can be found in Table 1.

3.2 Experimental Results

In order to test the performance of the proposed schema, this section presents results of the experimentation. The section is divided into an analysis of the effectiveness of different homogeneous ensemble systems and the study of the effect of the ensemble size over the general accuracy. In both cases, the average accuracy of the ensemble under each combination scheme is compared with the result of the model trained using the entire dataset, denoted as Complete (COM).

Performance Analysis. This experiment aims to study the performance of ensemble techniques against training model over larger data portion (COM). Each one of the ensembles are composed by $n = 50$ individual expert as pool size. Table 2 shows the average accuracy of the 50 runs for each dataset. Table 2 is divided into three groups, depending on the *Base* model considered in this work ($\{DT, NB, ML\}$), as well as each one of the groups show the average accuracy for the complete model (the *Base* model trained with all the data), and the accuracy for the ensemble using majority voting and weighted majority voting, denoted as ENS_M and ENS_W, respectively. Row named *count*, remark the number of times that the ensemble outperforms the *complete* model. Analyzing results in Table 2, the results show that weighted majority voting wins 13 dataset from the total of 25 to score higher aggregation accuracy than simple majority. Homogeneous ensembles of DT often guarantee higher performance over the complete model, with 24 higher scores for ENS_W over COM and ENS_M wins COM in 23 from the total 25 datasets.

[1] Package e1071: http://cran.r-project.org/web/packages/e1071/index.html.
[2] Package partykit: http://cran.r-project.org/package=partykit.
[3] Package nnet: http://cran.r-project.org/web/packages/nnet/index.html.
[4] UCI MachineLearning Database Repository: http://archive.ics.uci.edu/ml/.

Algorithm 1. The schema followed in the experimentation

Data: Train, Test, Pool Size n, Percent of rows to train p_{tra}, Percent of attributes p_{att} and Base model {NB ∨ DT ∨ LM}
Result: COM, ENS_M, ENS_W
1 $EnsembleTrain \leftarrow$ Stratified Sampling with p_{tra} from Train
2 **for** $k \leftarrow 1$ **to** n **do**
3   R← Bagging sample from $EnsembleTrain$
4   C← Sample p_{att} features from $EnsembleTrain$
5   $Train_k \leftarrow$ R rows and C columns from $EnsembleTrain$
6   Model $\Psi_k \leftarrow$ Base($Train_k$)
7   $Pred_k \leftarrow$ predict responses of Ψ_k over Test
8 **end**
9 complete ← Base(Train)
10 COM ← Accuracy of complete over Test
11 ENS_M ← Accuracy of combination of all $pred_k$ using majority voting
12 ENS_W ← Accuracy of combination of all $pred_k$ using weighted voting

Table 1. Characteristics of the selected datasets.

	DataSet	Rows	Features	Classes		DataSet	Rows	Features	Classes
1	haberman	306	3	2	14	iris	150	4	3
2	mammographic	830	5	2	15	balance	625	4	3
3	breast	277	9	2	16	newthyroid	215	5	3
4	wisconsin	683	9	2	17	wine	178	13	3
5	heart	270	13	2	18	lymphography	148	18	4
6	crx	653	15	2	19	vehicle	846	18	4
7	bands	365	19	2	20	cleveland	297	13	5
8	ring	7400	20	2	21	dermatology	358	34	6
9	twonorm	7400	20	2	22	satimage	6435	36	6
10	ionosphere	351	33	2	23	penbased	10992	16	10
11	spectfheart	267	44	2	24	optdigits	5620	62	10
12	spambase	4597	57	2	25	texture	5500	40	11
13	sonar	208	60	2					

We use Friedman's procedure to compute the average ranks that represent the effectiveness associated with each algorithm. For this and other statistical procedures check the web link[5]. The last line of Table 2 shows the average rankings related to Friedman test (*AR_Friedman*).

Also we apply the Wilcoxon's signed ranks statistical test [19], this is a non-parametric pairwise test that aims to detect significant differences between two sample means. The results of the Wilcoxon test are shown in Table 3. ✓sign means that the algorithm in the row outperforms the algorithm in the column at the 10% significance level with respect to accuracy over test. An empty table

[5] Statistical inference procedures: http://sci2s.ugr.es/sicidm/.

Table 2. Average accuracy over 50 run for each dataset, (*) represents that the ensemble performs better than the complete model.

	Dataset	*DT*			*NB*			*ML*		
		COM	ENS_M	ENS_W	COM	ENS_M	ENS_W	COM	ENS_M	ENS_W
1	haberman	0.716	0.735*	0.735*	**0.747**	0.736	0.736	0.743	0.735	0.735
2	mammographic	0.818	0.826*	0.827*	0.810	0.816*	0.817*	0.826	0.827*	**0.830***
3	breast	0.717	0.734*	0.734*	**0.735**	0.727	0.727	0.696	0.719*	0.719*
4	wisconsin	0.953	0.968*	0.968*	0.961	0.970*	**0.971***	0.967	0.966	0.966
5	heart	0.758	0.825*	0.820*	**0.844**	0.823	0.826	0.840	0.841*	0.839
6	crx	0.861	0.855	**0.862***	0.771	0.781*	0.804*	0.851	0.843	0.858*
7	bands	0.628	0.630*	0.630*	0.465	0.635*	0.649*	0.665	0.658	**0.667***
8	ring	0.860	0.942*	0.939*	**0.980**	0.976	0.976	0.759	0.754	0.756
9	twonorm	0.829	0.961*	0.962*	**0.979**	0.976	0.976	0.978	0.976	0.976
10	ionosphere	0.899	0.904*	**0.908***	0.894	0.877	0.892	0.874	0.878*	0.881*
11	spectfheart	0.770	0.794*	**0.795***	0.685	0.737*	0.740*	0.791	0.784	0.784
12	spambase	0.906	0.928*	**0.929***	0.714	0.681	0.696	0.925	0.917	0.919
13	sonar	0.719	0.736*	0.739*	0.683	0.692*	0.698*	0.755	0.781*	**0.784***
14	iris	0.941	0.935	0.940	0.953	0.944	0.948	**0.959**	0.951	0.955
15	balance	0.785	0.829*	0.828*	**0.901**	0.851	0.850	0.892	0.847	0.842
16	newthyroid	0.914	0.946*	0.946*	**0.966**	0.949	0.959	0.962	0.947	0.954
17	wine	0.894	0.944*	0.942*	0.974	0.971	0.969	0.969	**0.975***	0.974*
18	lymphography	0.695	0.770*	0.775*	0.814	0.803	0.812	0.763	0.839*	**0.841***
19	vehicle	0.658	0.693*	0.699*	0.454	0.469*	0.383	**0.800**	0.746	0.751
20	cleveland	0.549	0.572*	0.564*	0.538	0.528	0.224	**0.587**	0.584	0.580
21	dermatology	0.939	0.955*	0.960*	0.853	0.897*	0.913*	0.951	0.970*	**0.972***
22	satimage	0.852	0.883*	**0.884***	0.796	0.798*	0.798*	0.859	0.855	0.856
23	penbased	0.941	**0.977***	0.976*	0.858	0.852	0.853	0.954	0.911	0.913
24	optdigits	0.853	0.957*	0.958*	0.875	0.882*	0.888*	0.942	0.966*	**0.967***
25	texture	0.902	0.948*	0.948*	0.775	0.775	0.777*	0.993	0.994*	**0.995***
Count		1	23	24	14	10	10	13	10	11
AR_Friedman		6.92	4.76	4.32	5.44	6.22	5.58	3.88	4.28	3.6

cell indicates that the algorithm in the row and column do not significantly differ from each other with P-value higher than 0.10.

From Table 3, the ensemble system with NB as base classifier does not outperform any complete model. While using DT as base classifier guarantee better accuracy over the complete model by using only small data portions. Also it is interesting to see that ensemble systems with both ML and NB aggregated by weighted majority outperforms the simple majority aggregation method, each in its group.

Since we care relative performance instead of absolute performance, the accuracy of the ensemble systems should be normalized according to the accuracy of complete model in each group separately. The accuracy of complete model is regarded as 1.0. The reported accuracy of other ensembles calculated as the

Table 3. Wilcoxon Test, Comparison between 9 methods with respect to the accuracy over test, ✓means that method in the row outperforms the method in the column

Method	DT_COM	DT_M	DT_W	NB_COM	NB_M	NB_W	ML_COM	ML_M	ML_W
DT_COM									
DT_M	✓			✓	✓	✓			
DT_W	✓			✓	✓	✓			
NB_COM									
NB_M									
NB_W					✓				
ML_COM	✓			✓	✓	✓			
ML_M	✓			✓	✓	✓			
ML_W	✓			✓	✓	✓		✓	

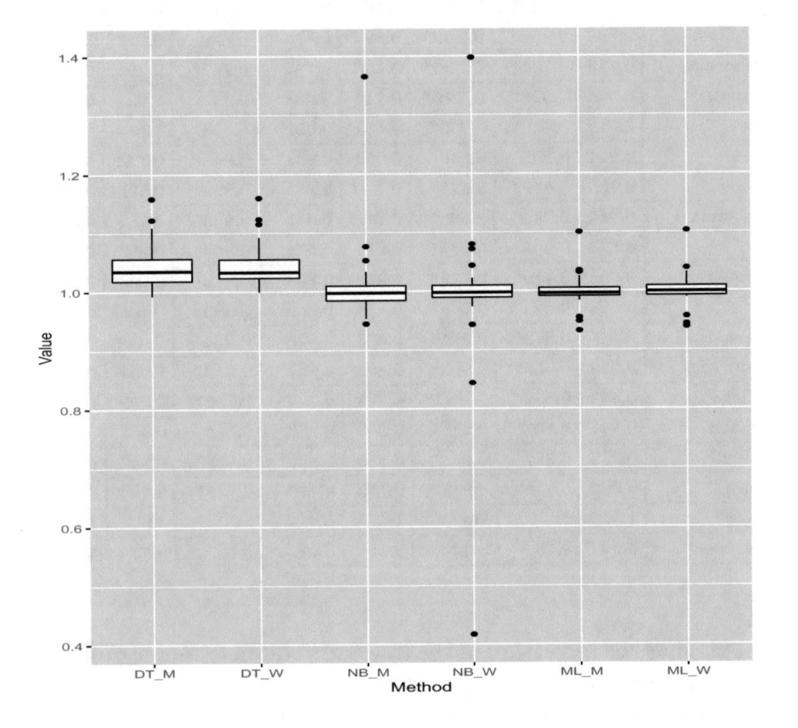

Fig. 3. Relative performance of ensemble systems over COM.

ratio against COM. Figure 3 shows the box plots of the relative accuracy over COM, where DT preferred as base classifier with no values less than 1.

Ensemble Size Analysis. This second part of the experiments aims to analyze the effect of ensemble size on the accuracy of the model. In Table 4, we find that the ensemble accuracy increases as the number of individual experts increase. The sixth column of Table 4, shows the percentage by which the ensemble

Table 4. Analysis of the Ensemble size, (*) represents that ensemble better than complete model

Expert type	Pool_Size	COM	ENS_M	ENS_W	ENS_M|COM	ENS_W|COM
DT	3	**0.8149**	0.8029	0.8051	0.4752	0.4832
	10	0.8149	0.8349*	**0.8373***	0.7704	0.7976
	20	0.8152	0.8427*	**0.8443***	0.8032	0.8208
	30	0.8151	0.8477*	**0.8488***	0.8432	0.8600
	40	0.8147	0.8494*	**0.8499***	0.8584	0.8696
	50	0.8142	0.8499*	**0.8507***	0.8632	0.8784
NB	3	**0.8002**	0.7786	0.7826	0.3520	0.3704
	10	**0.7996**	0.7938	0.7927	0.4376	0.4672
	20	**0.7996**	0.7989	0.7939	0.4888	0.5024
	30	0.7993	**0.8044***	0.7949	0.5328	0.5416
	40	0.7999	**0.8038***	0.7939	0.5368	0.5488
	50	0.8010	**0.8058***	0.7952	0.5480	0.5592
ML	3	**0.8514**	0.8154	0.8152	0.2512	0.2488
	10	**0.8527**	0.8398	0.8421	0.4104	0.4456
	20	**0.8525**	0.8451	0.8479	0.4496	0.4880
	30	**0.8523**	0.8482	0.8505	0.4832	0.5176
	40	0.8529	0.8511	**0.8531***	0.5112	0.5384
	50	0.8505	0.8505	**0.8525***	0.5160	0.5584

majority is higher than the complete model. This means that when the individual expert is DT and the ensemble pool size 50 we get 86% of the general runs where the ensemble by majority voting is better than the complete model. The same concept apply to the seventh column of Table 4, when the individual is ML and pool size is 50 we get 55% from the general experiments where ensemble by weighted majority voting is better than the complete model, which seems more interesting as this value increases also when the number of individual experts is increased.

4 Conclusions and Future Work

This article reviewed two common combination schemes to aggregate several individual classifiers under an ensemble model. The main focus of this paper is to study the performance changing of ensemble systems when built over small portion of the data, in comparison with accuracy obtained by training complete models using big portion of the data. Benefits derive from this approach is the reduction of time needed to train individual experts, since they are using a portion of the data, as well as the possibility of parallelization of the construction of the ensemble. Characteristics that are quite desirable, in particular, when dealing with costly to train classification models or high amounts of data.

In order to perform the study, three base models, as well as two combination schemes are tested over 25 datasets, and their results have been compared with those coming from the complete model. The number of experts needed in the ensemble system needed to outperforms the base classifier have been also studied. In general, weighted majority voting performs better than the simple majority voting. As well as homogeneous ensemble systems composed by decision trees often guarantee better results. The ensemble system discussed in this article are homogeneous, where all the individual experts are of the same classifier type.

Future work will be focused on heterogeneous classifier systems, also, the inclusion of mechanisms for the selection of classifiers from the available pool will be studied. The concept of selecting the most informative samples for individual classifier learning will also be studied.

Acknowledgements. This project has received funding from the European Union's Horizon 2020 research and innovation programme under the Marie Skłodowska-Curie grant agreement No 665959. In addition, Michał Woźniak was supported by the statutory funds of the Department of Systems and Computer Networks, Faculty of Electronics, Wrocław University of Science and Technology.

References

1. Breiman L (1996) Bagging predictors. Mach Learn 24(2):123–140
2. Breiman L (2001) Random forests. Mach Learn 45:5–32
3. Bryll RK, Gutierrez-Osuna R, Quek FKH (2003) Attribute bagging: improving accuracy of classifier ensembles by using random feature subsets. Pattern Recogn 36(6):1291–1302
4. Cruz RM, Sabourin R, Cavalcanti GD (2018) Dynamic classifier selection: recent advances and perspectives. Inf Fusion 41:195–216
5. Dietterich TG (2000) Ensemble methods in machine learning. LNCS, vol 1857, pp 1–15
6. Drucker H, Cortes C, Jackel LD, LeCun Y, Vapnik V (1994) Boosting and other ensemble methods. Neural Comput 6(6):1289–1301
7. Freund Y, Schapire RE (1997) A decision-theoretic generalization of on-line learning and an application to boosting. J Comput Syst Sci 55(1):119–139
8. Hearst MA, Dumais ST, Osman E, Platt J, Scholkopf B (1998) Support vector machines. IEEE Intell Syst 13(4):18–28
9. Ho TK (1995) Random decision forests. In: Proceedings of the third international conference on document analysis and recognition, ICDAR 1995, vol 1. IEEE Computer Society, Washington, p 278
10. Ho TK (1998) The random subspace method for constructing decision forests. IEEE Trans Pattern Anal Mach Intell 20(8):832–844
11. Huang Y, Suen C (1993) Behavior-knowledge space method for combination of multiple classifiers. In: IEEE conference on computer vision and pattern recognition, pp 79–87. https://doi.org/10.1109/CVPR.1993.1626170
12. Jacobs RA, Jordan MI, Nowlan SJ, Hinton GE (1991) Adaptive mixtures of local experts. Neural Comput 3(1):79–87
13. Krawczyk B, Woźniak M, Schaefer G (2014) Cost-sensitive decision tree ensembles for effective imbalanced classification. Appl Soft Comput J 14(PART C):554–562
14. Kuncheva LI (2014) Combining pattern classifiers: methods and algorithms. Wiley, Hoboken

15. Martinez-Muñoz G, Hernández-Lobato D, Suarez A (2009) An analysis of ensemble pruning techniques based on ordered aggregation. IEEE Trans Pattern Anal Mach Intell 31(2):245–259
16. Polikar R (2006) Ensemble based systems in decision making. IEEE Circ Syst Mag 6(3):21–45
17. Schapire RE (1990) The strength of weak learnability. Mach Learn 5(2):197–227
18. Soares S, Antunes CH, Araújo R (2013) Comparison of a genetic algorithm and simulated annealing for automatic neural network ensemble development. Neurocomputing 121:498–511
19. Wilcoxon F (1945) Individual comparisons by ranking methods. Biometrics Bull 1(6):80–83
20. Wolpert DH (1992) Stacked generalization. Neural Netw 5(2):241–259
21. Wolpert DH (2001) The supervised learning no-free-lunch theorems. In: Proceedings of the 6th online world conference on soft computing in industrial applications, pp 25–42
22. Woźniak M, Graña M (2014) A survey of multiple classifier systems as hybrid systems. Inf Fusion 16:3–17

Segmentation of Subretinal Hyperreflective Material and Pigment Epithelial Detachment Using Kernel Graph Cut

Melinda Katona[1], Attila Kovács[2], Rózsa Dégi[2], and László G. Nyúl[1](✉)

[1] Interdisciplinary Excellence Centre, Department of Image Processing and Computer Graphics, University of Szeged, Árpád tér 2, Szeged 6720, Hungary
{mkatona,nyul}@inf.u-szeged.hu

[2] Department of Ophthalmology, University of Szeged, Korányi fasor 10-11, Szeged 6720, Hungary
{kovacs.attila,degi.rozsa}@med.u-szeged.hu

Abstract. Optical Coherence Tomography (OCT) is one of the most advanced, non-invasive method of eye examination. Age-related macular degeneration (AMD) is one of the most frequent reasons of acquired blindness and it has two forms. Our aim is to develop automatic methods that can accurately identify and characterize biomarkers in SD-OCT images, related to wet AMD. Detection of biomarkers can be challenging because of their variable shape, size, location and reflectivity. In this paper, we present an automatic method to localize subretinal hyperreflective material (SHRM) and pigment epithelial detachment (PED) via kernel graph cut. The proposed method is evaluated using an annotated dataset by ophthalmologists. The Dice coefficient was 0.81 (±0.11) in the case of PED and 0.77 (±0.11) for SHRM. In many cases, the ophthalmologist cannot clearly determine the exact location and extent of the biomarkers, so our achieved results are promising.

Keywords: Age-related macular degeneration · Subretinal hyperreflective material · Pigment epithelial detachment · Optical coherence tomography

1 Introduction

Age-related Macular Degeneration (AMD) is a common eye condition in people above 50 years worldwide, currently affecting 170 million people globally. AMD means degeneration of the central part of the retina, responsible for fine vision. Untreated patients lose their shape- and face recognition, reading ability, and central vision [3]. Two subtypes of AMD are known; the dry (non-exudative) and the wet (exudative, neovascular) form. From these two subtypes the wet form

R. Burduk et al. (Eds.): CORES 2019, AISC 977, pp. 98–105, 2020.
https://doi.org/10.1007/978-3-030-19738-4_11

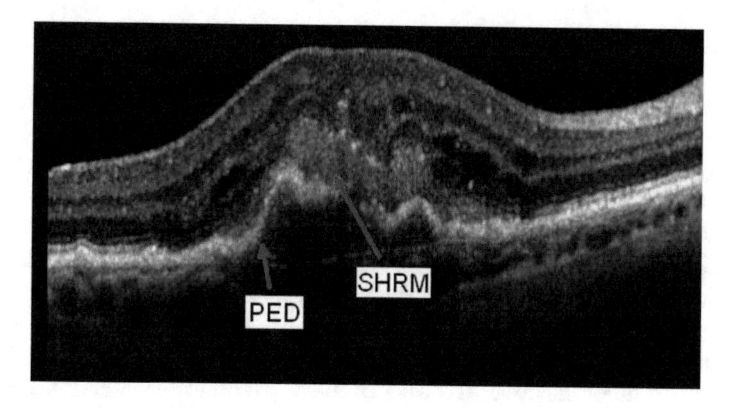

Fig. 1. Optical Coherence Tomography (SD-OCT) image of the retina with PED and SHRM biomarkers. Our medical experts use it in inverted display in daily routine.

causes rapid and serious visual impairment accounting for 10% of the cases. In this neovascular type of the disease, fluid and blood leakage occurs from abnormal angiogenesis from the choroid, leading to photoreceptor damage. Vascular endothelial growth factor (VEGF) plays crucial role in the pathogenesis of the wet form [12], leading to a treatment option by blocking it with periodically repeated anti-VEGF intravitreal injections, the first choice of treatment in neovascular AMD [8]. Nowadays, image processing tasks are solved by training different type of neural networks. Lee et al. [6] concerned with segmentation of PED and SHRM simultaneously. In this approach, CNN-based deep neural network was trained.

In the last decade, optical coherence tomography (OCT) has become a basic imaging technique in diagnosing and following neovascular AMD and its response to treatment, since it provides high-resolution cross-sectional images of the retina. With OCT we can detect not only the layers of the retina, but we are also capable of identifying the detailed morphology and the effects of the disease, the so-called OCT biomarkers, such as subretinal/intraretinal fluid accumulation (SRF/IRF), pigment epithelial detachment (PED), subretinal hyperreflective material (SHRM), outer retinal tubulation (ORT) or hyperreflective foci (HRF) (see, e.g., Fig. 1). These markers help the clinical decision-making process for treating/re-treating/observing a patient [1]. In our previous studies [4,5], we have already discussed IRF, SRF, ORT, and HRF, the present analysis focussed on PED and SHRM.

More reports have dealt with PED, which is an elevated lesion below the retinal pigment epithelium (RPE). Reports showed that PED should be monitored precisely, and any growth in size urgently treated with anti-VEGF injections, thus preventing sustained vision loss. SHRM can be detected on OCT as a medium- to hyperreflective mass between the neurosensory retinal layers and RPE.

To improve the treatment procedure, there is a need for more precise measurements, hence our aim was to create an algorithm which automatically identifies and characterizes two of the above-mentioned biomarkers, namely the PED and the SHRM.

A large number of publications in the scientific literature deal with the analysis of retinal images. Nowadays, most papers use machine learning methods to solve various image processing problems. Using these approaches, it is possible to classify the retinal disease by using some biomarkers. Srinivasan et al. [10] developed a method which can determine RPE layer after some preprocessing steps to improve image quality. To classify the individual eye diseases, they were calculated HOG descriptors and they used these features to train an SVM. In addition, to differentiate AMD and Diabetic Macular Edema (DME), Sugmk et al. [11] have applied a binary classification based on different threshold results. Some of the procedures work with color fundus images. Mohaimin et al. [7] localized the disease regions via color and boundary based segmentation method. In these cases, the determination of the RPE layer and/or the localization of PED was the target for classification.

In many cases, it is not sufficient for a medical doctor to decide what illness may affect the patient's eyes, but also the extent, numeracy, etc. of the biomarkers. We have already worked on the topic of localization of outer retinal layers [4] and detection of PED [5]. In this paper, we deal with segregation of subretinal hyperreflective material and pigment epithelial detachment. For this, we introduce a graph cut based approach to determine biomarkers. The achieved results were compared with annotated data by ophthalmologists.

2 Materials and Methods

The OCT images were acquired using Heidelberg Spectralis OCT (Spectralis, Heidelberg Engineering, Heidelberg, Germany) on wet age-related macular degeneration. Patients were either treated naively or with anti-VEGF intravitreal injections. The annotated images contained several biomarkers, such as subretinal fluid, PED, Cyst, SHRM. In this study, we used only PED and SHRM annotation and 18 image sequences were available. These are recordings of 12 different patients. These sequences consisted of 49 or 25 slices taken with a 6×6 mm pattern size and 251 μm slice distance. Slices had a resolution of 512×496 pixels with pixel sizes 11.74 and 3.87 μm and a quality score above 23 dB.

2.1 Graph Cut Based ILM and RPE Layer Segmentation

We worked with 2D images because we could not exploit 3D information directly to segment the retina layers, since there are some anomalies among slices of the OCT volume, due to the image acquisition and registration process (within the device's software). In the literature, a number of procedures are based on graph cut [2] since it is a robust method against other techniques. As a result, no preprocessing of the input data has been performed, although in many cases the image is very noisy. Graph cut is a semi-automatic method that requires seed points. There are several existing approaches to automatically assigning them. We used kernel k-means to automatically calculate these points. We map data points in the input space onto a high-dimensional feature space using Gaussian radial basis kernel function:

$$k(x, y) = \exp\left(-\frac{||x - y||^2}{2\sigma^2}\right),$$

where $k(x, y)$ is a point in higher dimensional space. The number of clusters is determined by how noisy the picture is. So, we calculated no-reference image quality score for the image using the Naturalness Image Quality Evaluator (NIQE). This model based on a set of local features extraction from an image then fits the feature vectors to a multivariate Gaussian. In the higher quality images, 5 clusters were isolated empirically, while in the other cases this number increased. In cluster image, value 0 represents the darkest region and $k - 1$ is the brightest, where k is the number of clusters.

We used graph cut [9] to optimize partitioning. This gives a better classification. PED and SHRM are located near the RPE layer, so its detection helps later classification. The ILM elevation can be determined after a simple Otsu thresholding because the foreground and background can be clearly distinguished from that part of the retina in the clustered image. RPE layer has higher reflectivity in the original image, so it has also higher cluster number in the clustered image. Using a method in our previous publication [4], we calculated vertical projection in every 10th column to determine boundary and we chose local minimum from the projected data.

To determine all pixels which can produce the layer, we fitted a curve to the resampled points. Our data probably include outliers, because Choriocapillaris and Choroidal vessels are located under the RPE layer, so The intensity of these regions vary. These salient points may mislead the fitting, so eliminating them is important. The distribution of the points is not a normal, due to the distortions, therefore outliers are defined as elements more than 1.5 interquartile ranges above the upper quartile (75%) or below the lower quartile (25%). We fitted shape-preserving piecewise cubic spline interpolation to determine RPE layer.

2.2 Classification of PED and SHRM

Subretinal Hyperreflective Material, as its name suggests, is likely composed of many components, including fluid, fibrin, blood, etc., and its composition changes over time. So, the reflectivity of SHRM is heterogeneous. By contrast, Pigment Epithelial Detachment has lower intensity, so they can be separated from each other. Nevertheless, in many cases, their location and presence are unclear.

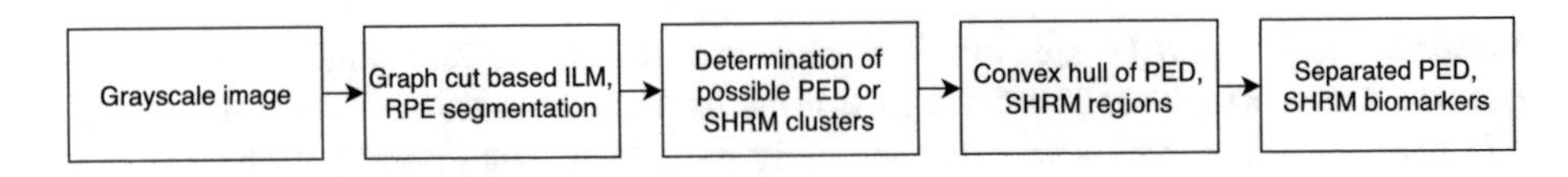

Fig. 2. Flowchart of the proposed method.

The pixels of the PED and the SHRM regions are roughly in the same intensity range, and therefore the results of a proper clustering can form a separate region. These biomarkers are usually located around the RPE layer, along or near the distortion, so clusters were sampled from these areas only. Patients with other lesions may also be observed, so we have also taken into account whether the patient has a cyst or fluid area. We also used our work referred to in Subsect. 2.1 to determine the location of distortion and to detect other biomarkers (cysts, fluid areas). We only investigated in a specific range over the RPE layer. For this, the threshold was determined as follows:

$$T = \begin{cases} \max(CF_{(x,\max(y))}), & \text{if } \max(CF_{(x,\max(y))}) > 0 \\ RT_x * 0.2, & \text{otherwise} \end{cases},$$

where CF represents the binary image with detected cyst and subretinal fluid, x is the actual x position, $\max(y)$ is the actual maximum y position and RT denotes the calculated retinal thickness (from ILM and RPE layer).

The collected clusters produce SHRM regions. Those cluster(s) will be PED area which defined as a non-SHRM segment below the RPE layer. They may also have amorphous shapes because they are made up of different sub-areas, so we have defined their convex shapes to accomplish the final result. The key stages of the procedure are summarized in Fig. 2.

3 Results

We evaluated segmentation ability in terms of overlap between each segmented area by graders. We considered data obtained from medical doctors as ground truth.

3.1 Evaluation Metrics

We used Dice coefficients to measure the ratio of overlap.

$$\text{Dice coefficient} = \frac{2 \cdot |\text{X} \cap \text{Y}|}{|\text{X}| + |\text{Y}|},$$

where $|X|$ and $|Y|$ are the numbers of pixels representing the segmented regions, while $|X \cap Y|$ are the overlapping regions in pixels.

In addition, for the further characterization of the procedure, the sensitivity was also calculated.

$$\text{Sensitivity} = \frac{\text{TP}}{\text{TP} + \text{FP}},$$

where TP denotes the numbers of overlapping pixels, FP represents the numbers of segmented, but not overlapping pixels.

3.2 Localization of PED and SHRM

We implemented our proposed method in MATLAB, using the Image Processing Toolbox, and compared the results with annotated data by ophthalmologists. The location of the possible SHRM has played an important role in determining PED. In many cases, SHRM is not clearly detectable. In spite of all this, as you can see in Fig. 3, in both cases, the average Dice coefficients are above 0.75 and the sensitivity is 0.93 for PED and 0.77 for SHRM. The detection of SHRM is not yet widely investigated in the literature. In contrast, several articles deal with the delimitation of PED. The CNN-based method mentioned in Sect. 1 also determines the biomarkers with similar efficiency. This means that it is not necessary to have a complex system with large image database to segment the affected regions. Most errors are in parts where the ophthalmologist cannot clearly define the area of the given biomarker.

The kernel k-means more computationally than the basic k-means which separates linearly, so the running time is also increasing. This means that the running time of the process is 30 s on average. In the Fig. 4, and you can see some examples of automatic results and also the manual annotation for a visual comparison.

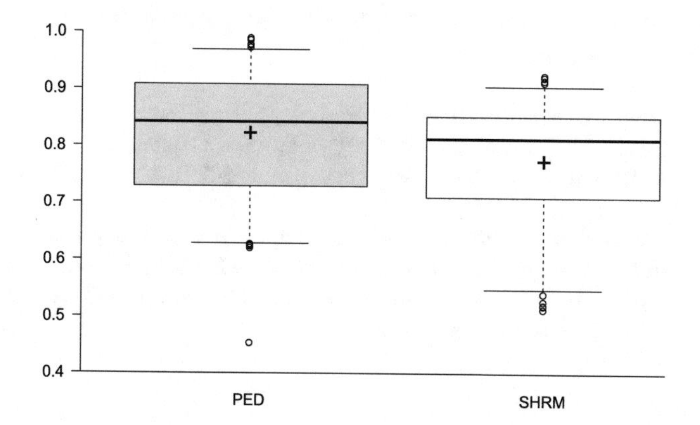

Fig. 3. Box plots representing the Dice coefficients of PED and SHRM generated by proposed method. In both cases, the mean Dice coefficients were greater than 0.7. Black square = mean Dice coefficient; empty circles = outliers; SHRM = subretinal hyperreflective material; PED = pigment epithelial detachment.

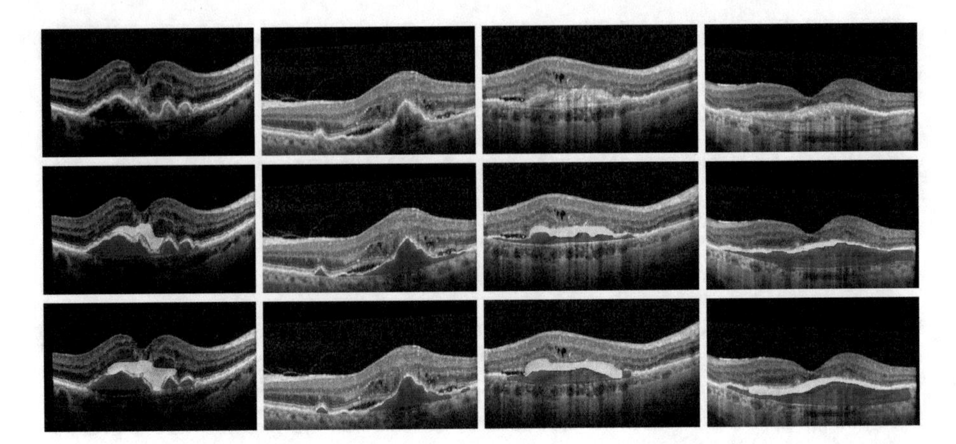

Fig. 4. Illustration of annotated (in middle row) and detected (bottom row) PED and SHRM. Red regions show PED yellow represents SHRM.

4 Conclusion

We introduced a graph cut based approach to detect PED and SHRM regions on SD-OCT B-scans. The examined biomarkers have different reflectivity, but in a significant part of the cases, their extent cannot be clearly defined. In the process, ILM and RPE layers were also determined, which served as the basis for our main goal. The results were compared with data annotated by medical doctors. We calculated measures to quantify features of OCT in point of wet AMD patients. This automated localization method can help the quantitative assessment of the OCT biomarkers by providing automatic tools to detect abnormalities and to describe by objective metrics the current state and longitudinal changes during disease evolution and treatment.

Acknowledgements. Melinda Katona was supported by the UNKP-18-3 New National Excellence Program of the Ministry of Human Capacities.

Ministry of Human Capacities, Hungary grant 20391-3/2018/FEKUSTRAT is acknowledged.

Melinda Katona and László G. Nyúl were supported by the project "Integrated program for training new generation of scientists in the fields of computer science", No. EFOP-3.6.3-VEKOP-16-2017-0002. The project has been supported by the European Union and co-funded by the European Social Fund.

References

1. Coscas G et al (2009) Clinical features and natural history of AMD on OCT, pp 195–274. Berlin, Heidelberg
2. Dodo BI, Li Y, Eltayef K, Liu X (2018) Graph-cut segmentation of retinal layers from OCT images. In: Proceedings of the 11th international joint conference on biomedical engineering systems and technologies, BIOIMAGING, vol 2. INSTICC, SciTePress, pp 35–42. https://doi.org/10.5220/0006580600350042

3. Hee MR et al (1996) Optical coherence tomography of age-related macular degeneration and choroidal neovascularization. Ophthalmology 103(8):1260–1270
4. Katona M, Kovács A, Dégi R, Nyúl LG (2017) Automatic detection of subretinal fluid and cyst in retinal images. In: Image analysis and processing - ICIAP 2017, Proceedings, Part I, pp 606–616
5. Katona M, Kovács A, Varga L, Grósz T, Dombi J, Dégi R, Nyúl LG (2018) Automatic detection and characterization of biomarkers in OCT images. In: Image analysis and recognition, pp 706–714
6. Lee H, Kang KE, Chung H, Kim HC (2018) Automated segmentation of lesions including subretinal hyperreflective material in neovascular age-related macular degeneration. Am. J. Ophthalmol. 191:64–75. https://doi.org/10.1016/j.ajo.2018.04.007. http://www.sciencedirect.com/science/article/pii/S0002939418301673
7. Mohaimin SM, Saha SK, Khan AM, Arif ASM, Kanagasingam Y (2018) Automated method for the detection and segmentation of drusen in colour fundus image for the diagnosis of age-related macular degeneration. IET Image Process. 12(6):919–927
8. Rosenfeld PJ, Brown DM, Heier JS, Boyer DS, Kaiser PK, Chung CY, Kim RY (2006) Ranibizumab for neovascular age-related macular degeneration. New Engl. J. Med. 355(14):1419–1431
9. Salah MB, Mitiche A, Ayed IB (2011) Multiregion image segmentation by parametric kernel graph cuts. IEEE Trans. Image Process. 20(2):545–557
10. Srinivasan PP, Kim LA, Mettu PS, Cousins SW, Comer GM, Izatt JA, Farsiu S (2014) Fully automated detection of diabetic macular edema and dry age-related macular degeneration from optical coherence tomography images. Biomed. Opt. Express 5(10):3568–3577
11. Sugmk J, Kiattisin S, Leelasantitham A (2014) Automated classification between age-related macular degeneration and diabetic macular edema in OCT image using image segmentation. In: The 7th 2014 biomedical engineering international conference, pp 1–4
12. Velez-Montoya R et al (2013) Current knowledge and trends in age-related macular degeneration: today's and future treatments. Retina 334:1487–1502

Road Tracking Using Deep Reinforcement Learning for Self-driving Car Applications

Raid Rafi Omar Al-Nima[1,2], Tingting Han[1(✉)], and Taolue Chen[1]

[1] DCSIS, Birkbeck, University of London, London, UK
tingting@dcs.bbk.ac.uk

[2] Technical Engineering College of Mosul, Northern Technical University, Mosul, Iraq

Abstract. Deep reinforcement learning has received wide attentions recently. It combines deep learning with reinforcement learning and shows to be able to solve unprecedented challenging tasks. This paper proposes an efficient approach based on deep reinforcement learning to tackle the road tracking problem arisen from self-driving car applications. We propose a new neural network which collects input states from forward car facing views and produces suitable road tracking actions. The actions are derived from encoding the tracking directions and movements. We perform extensive experiments and demonstrate the efficacy of our approach. In particular, our approach has achieved 93.94% driving accuracy, outperforming the state-of-the-art approaches in literature.

Keywords: Road tracking · Deep reinforcement learning · Self-driving

1 Introduction

Road tracking is one of the most challenging tasks emerged from key applications such as autonomous driving. It aims to automatically guide a car through the correct track without crashing other cars or objects. Reinforcement learning (RL), in a nutshell, is concerned with an agent interacting with the environment, learning an optimal policy, by trial and error, for sequential decision making problems. The combination of deep neural networks and reinforcement learning gives rise to deep reinforcement learning (deep RL, DRL). We hypothesise that DRL can be one of the most promising methods to address the road tracking challenge, and this paper reports our approaches and results to apply DRL to accomplish some typical tasks in road tracking.

One of the primary tasks in road tracking is to track objects, e.g., walking people, buildings, cars, etc. [3,5,10,17]. Other tracking tasks include tracking and controlling velocity [7], tracking periodic gene repressilator [16], tracking

T. Chen is supported by EPSRC grant (EP/P00430X/1), ARC Discovery Project (DP160101652, DP180100691), and NSFC grant (No. 61662035). R. Al-Nima and T. Han are supported by EPSRC grant (EP/P015387/1).

R. Burduk et al. (Eds.): CORES 2019, AISC 977, pp. 106–116, 2020.
https://doi.org/10.1007/978-3-030-19738-4_12

single and double pendulum [6], and tracking maximum powers for wind speed conversion systems [18]. Applying DRL techniques to road tracking has been reported, but has not been explored thoroughly. For instance, Perot *et al.* [14] investigated tracking roads of a driving car where DRL was used. The main problem thereof is that the driving car oscillates around the main road track. This is due to the selected reward in the study, where it utilised the oriented angle of the road with the car speed. More generally, Yun *et al.* [19,20] proposed an action-decision network (ADNet) to track objects (not necessarily for road tracking purposes), where a pre-trained Convolutional Neural Network (CNN) was firstly employed followed by the reinforcement learning.

Our work contributes to this area by suggesting an effective road tracking procedure based on the deep reinforcement learning and by employing notions from Markov Decision Processes, where states are used as inputs; rewards are utilised to evaluate the tracking policy and actions are predicted to provide new states. Effective coding for various road tracking possibilities of actions has been considered such as turning to the left or right and recognising crossing object(s). The established codes have been used as regression information in the proposed DRL to train the neural network and maintain the actions.

Related Work. Due to space restriction, we can only discuss related work briefly, focusing on those pertinent to applying reinforcement learning for tracking objects purpose. For tracking objects, Grigore [5] proposed a tracking system controller by using a recurrent neural network combined with the reinforcement learning. Cohen and Pavlovic [3] suggested an effective and vigorous real-time tracker, which utilised reinforcement learning and was applied to tracking personal faces. Liu and Su proposed an object tracking method to determine the features of the tracking object [10]. Supančič and Ramanan constructed a learning policy for tracking objects. The reinforcement learning was applied to video streams to provide on-line decision [17].

2 Modelling

In this section, we address various essential issues to model the road tracking, such as how to identify road crossing object(s), how to encode different tracking directions, and how to design the deep reinforcement learning network to determine the next action, etc.

The Database. Our research is based on the SYNTHIA-SEQS-05 database [15]. Video sequences were recorded and saved, from which one extracted front, rear, left and right view road images around the car for both right and left steering cars. Each view covers a range of angle up to 100°. As a result, a large number of simulated image frames were provided, each has a resolution of $760 \times 1,280 \times 3$ pixel.

The images were taken in four different environments: (1) clear environment in spring, (2) fog, (3) rain and (4) heavy-rain environment. Examples from the four databases are given in Table 1. In our study, we look at *forward* view road frames for a *right* steering car in all *four* environments.

Table 1. Examples of the four employed environments

Environment	Examples
Spring	
Fog	
Rain	
Heavy-rain	

2.1 Road Tracking

To differentiate the main objects on the road, the databases provide specific colours for certain objects in the images, e.g., sky is grey, buildings are brown, roads are purple, side walks are blue and road markings are green. Overall, the purple and green colours refer to the allowed driving regions.

Based on the images, we further divide each view (image) into a safety zone and a danger zone. If objects are spotted in the safety zone the car can keep moving, as the zone is further away, allowing ample time to stop or slow down if necessary. The danger zone is the area that a car must stop if any objects are recognised. Such zones can be found in Fig. 1.

To determine a safety zone or a danger zone, a virtual triangle and trapezoid are drawn. The base of the triangle and trapezoid moves in the road direction, and the crossing lines slope toward the tracking direction. The height of the triangle (and the trapezoid) is two thirds of the image. The top one third of the trapezoid is the safety zone and the bottom two thirds is the danger zone. The shared point of the triangle and the trapezoid is called an *anchor point*. It denotes the road tracking direction by following the track centre.

By drawing the virtual triangle in the trapezoid area we can divide the region into left, right and centre region, with which the direction of any crossing objects can be specified. The car can, for instance, avoid an object crossing from the left side by moving to the right, if the space is empty there. If objects are identified from both sides in the danger zone, then the car has to stop.

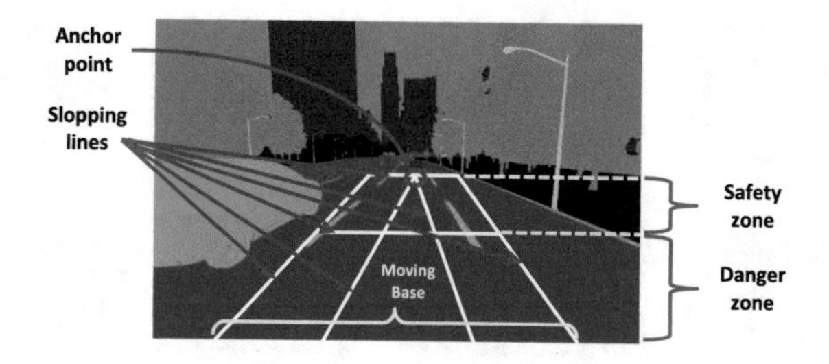

Fig. 1. The suggested front view road tracking zones, lines and anchor point

2.2 Actions Codes

At each decision point, a car can take one of the following actions: straight on, turn left or right, reverse and stop. In our work, we use a 5-digit binary number to encode each action, see Table 2.

Table 2. The road tracking actions with their suggested codes and descriptions

Action sign	Binary code	Equivalent decimal code	Description
Straightforward	01110	14	Follow the straight track
Turn left	11100	28	Follow the track to the left
Turn right	00111	7	Follow the track to the right
Stop	00000	0	Object(s) identified from both sides
Backwards	11111	−31	Reverse

When a crossing object is detected, it will change the code from that side with a "0". For instance, a car is moving forward with action 01110, then an object appears in the danger zone from left, the action code becomes 00111, indicating that the car should turn right to avoid the object. If another object appears from the right, the code will be 00110 and the car will have to stop. A car has to stop as long as there are no three consecutive 1's. We model all such cases as 00000 to reduce the number of coding values. Figure 2 shows various coding cases. The cases in the red dashed box are all coded as 00000.

The binary codes can be converted to desired equivalent decimal codes in the standard way. The only exception is the backward direction, where a negative sign (−) is added to refer to the reverse movement. The reverse action will only be considered if the car is out of the track. The decimal codes are used in the regression layer of the proposed network as will be explained later.

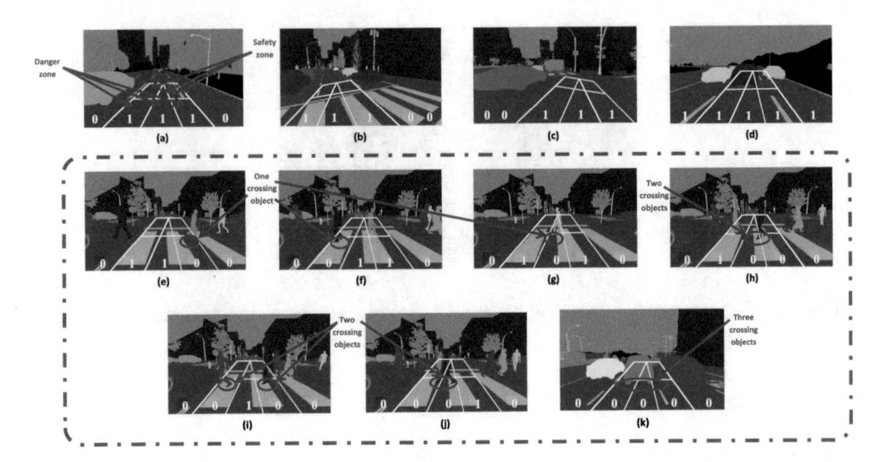

Fig. 2. Segmented images for road tracking: (a) straightforward, (b) turning left, (c) turning right, (d) reverse or backward, (e–g) stopping action because of a single crossing object, (h–j) stopping action because of two crossing objects and (k) stopping action because of three crossing objects

2.3 Proposed DRL-RT

In this section, we present the neural network (NN) used in the Deep Reinforcement Learning framework for Road Tracking (DRL-RT), which consists of eight layers: two convolution layers, two Rectified Linear Unit (ReLU) layers, a pooling layer, a fully connected layer, a regression layer and a classification layer. Theoretical explanations of the main deep NN layers can be found in [13]. Figure 3 depicts the proposed DRL-RT network.

We now elaborate the NN architecture as follows: (a) The input of the NN is an image of a car facing view which is considered as the current state. The dimension of the input image is reduced to $254 \times 427 \times 3$ pixels to speed up the training. (b) The first layer is a convolution layer which consists of 5 filters, each of which has a filter size of 10×10 pixels. This layer is to extract the main features of the input image. This is followed by a ReLU layer which removes the negative values and maintains the positive values of the previous layer. (c) The third layer is again a convolution layer, consisting of 5 filters, each of which has a filter size of 5×5 pixels. This layer extracts more features from the input images. A ReLU layer is employed, which rectifies the negative values. We note that it has empirically been established that using two convolution layers with two ReLU layers can well analyse the information before being compressed by applying the next layer. (d) A pooling layer of a maximum type is applied as the fifth layer. The filter size here is 3×3 pixels with a stride of 3 pixels. (e) The sixth layer is a fully connected layer. It collects the outputs (2 dimensional) from the previous layer and produces a series (1 dimensional) of values for the next layer. (f) In the seventh regression layer, a series of directional road tracking

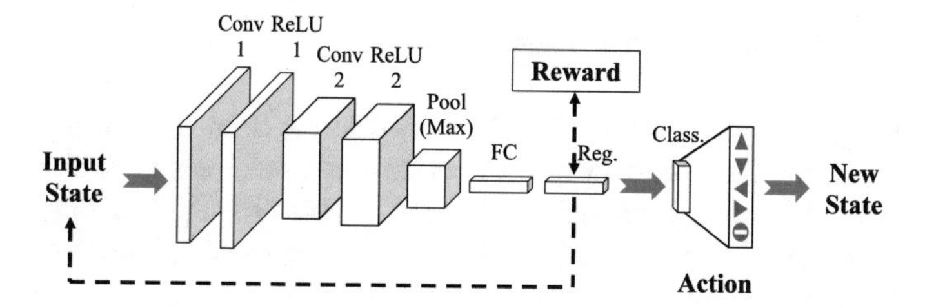

Fig. 3. The main design of the proposed DRL-RT. It consists of two convolution layers, two ReLU layers, a pooling layer, a fully connected layer, a regression layer and a classification layer

codes are generated. The successful tracking codes in this layer produce positive rewards, whereas, unsuccessful tracking codes generate negative rewards.

The network should propagate the information forward and backward to update the network weights during the training stage to obtain as many positive rewards as possible. Given the codes in the regression layer, it is the classification layer's task to generate a new action—one of the five as in Table 2.

For the theory underpinning the DRL-RT network, the underlying model of the network is essentially cast into a Markov Decision Process (MDP) framework with the following instantiations: (i) *States.* The states in the MDP are the views (images). (ii) *Rewards.* The reward R takes a simple form, i.e., the correct tracking is considered as (+1), whereas, the reward R of the incorrect tracking is considered as (−1). The correct and incorrect tracking are specified by comparing the regression layer outputs with the desired tracking codes. Measuring the successful process of the road tracking will be based on obtaining as many positive rewards as possible. (iii) *Policy search.* The DRL-RT network is based on the policy search. DRL-RT collects input images as current states S_t (or current views) and gets advantages from rewards R to generate actions A. The actions then predict new states S_{t+1} by following the track of the road. The process will be repeated in a multi-episodic manner.

3 Results

All implementations were performed on a PC with 8 GB RAM and Intel Core i5 processor (3.2 GHz). Only the microprocessor was used for training and testing. The databases have 264, 284, 268 and 248 frames for spring, fog, rain and heavy-rain environments, respectively. The following experiments were carried out:

(1) Training 2/3 of all the frames and testing the remaining 1/3 frames. The frames are randomly selected. Consequently, the testing stage is repeated several times after adding different types of noises to the testing frames.
(2) Training and testing each database individually. Here, the frames are equally divided between the training and testing stages (50% each), where the odd-indexed frames are used in the training stage and the even-indexed frames are used in the testing stage.

3.1 Training and Testing Stages

Training Stage. The suggested DRL-RT network has been separately trained for each environment. The following parameters have been assigned for the trainings: Adaptive Moment Estimation (ADAM) optimizer [9], learning rate equal to 0.0003, gradient decay factor (β_1) equal to 0.9, squared gradient decay factor (β_2) equal to 0.99 and mini batch size equal to 128.

Testing Stage. In the testing stage we use (driving) accuracy to measure how well our model performs. The driving accuracy is defined as the percentage of appropriate tracking actions by the driving car.

Group (1) Implementation: In this group, 709 out of 1064 (i.e., 2/3) frames are randomly chosen for training and the remaining 355 (i.e., 1/3) frames are used for testing. We then repeat the experiment by adding various types of noises to the testing frames.

Table 3. The driving accuracies of Group (1) Implementation

Testing specification	Parameters	Driving accuracy
No noises	—	95.49%
+ Gaussian noise	Mean = 0.1, Var = 0.001	89.86%
+ Poisson noise	—	94.37%
+ Salt & Pepper noise	Density = 0.001	94.93%
+ Salt & Pepper noise	Density = 0.005	93.80%
+ Speckle noise	Mean = 0, Var = 0.01	94.65%

It can be seen from Table 3 that a remarkable driving accuracy of 95.49% is achieved with no additional noises. The accuracy decreases after applying different types of noises, but is still reasonably acceptable. Gaussian noise has the worst effect, where the driving accuracy attained 89.86% (for Mean = 0.1 and Var = 0.001). This type of noise distributes the noise over to all the pixels of the input image, so it significantly affects the DRL-RT inputs. Nevertheless, the driving accuracy is still high. Other types of noises (Poisson, Salt & Pepper and Speckle) have reported high and comparable results. The driving accuracy achieved 94.37% after adding the Poisson noise, which was generated from the input data instead of adding artificial noise to the data [11]. By adding Salt & Pepper noise, the accuracies obtained 94.93% and 93.80% for the densities of 0.001 and 0.005, respectively. This type of noise influences some pixels, depending on the density. Therefore, the performance is still high after adding this noise. Finally, the speckle method multiplies a uniform noise with the original data and adds the noise back on [11]. We still attain a high accuracy (94.65%) in our models.

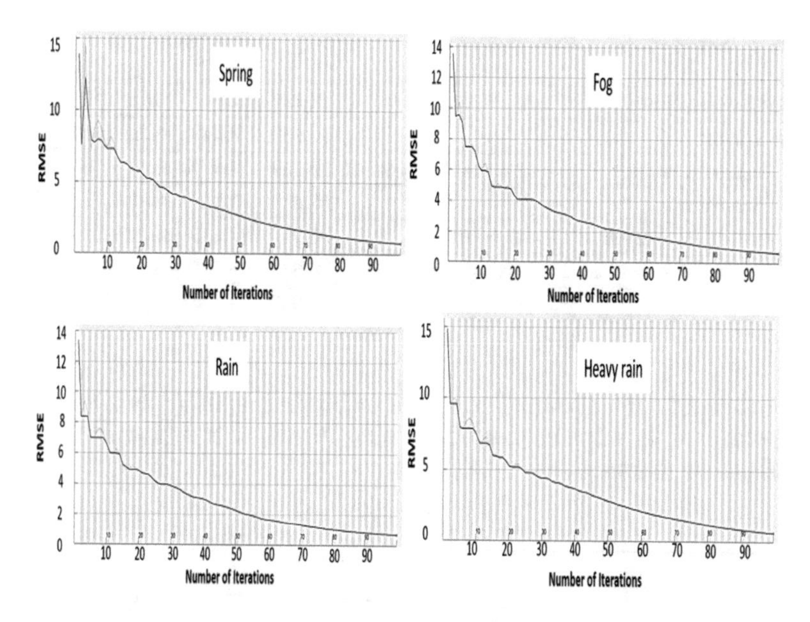

Fig. 4. The training RMSE vs #iterations for the four environments

The above experiments confirm that our proposed approach can deal with different types of noises and produce acceptable outcomes.

Group (2) Implementation: The training performance of the DRL-RT for the four databases is given in Fig. 4. This figure demonstrates the relationships between the Root Mean Square Error (RMSE) and the training iterations during the NN training stages. The RMSE values are usually exploited to demonstrate the differences between desired values and output values. These differences are usually reduced along with the training iterations.

The first row of Table 4 shows that the driving accuracy attained its highest value of 93.94% by using the spring environment database. This is because that the DRL-RT has analysed very clear provided images. The fog environment database obtained a high driving accuracy of 93.66%. Here, the overall views are blurred (or in a low quality) but all the input information can still be distinguished, so the accuracy is only slightly reduced with respect to the spring views. The driving accuracy of the rain environment database achieved 89.55% and this is due to the noise effects of rain drops on image views. Finally, the inferior driving accuracy of 84.68% was recorded for the heavy-rain environment database as the amount of rain drops (or noise) is increased there.

In Group (2) experiments, we also explored the effects of adding noises to the test cases, see Row 2–6 in Table 4. Noticeably, the results here are lower than the experiment of the Group (1) counterparts (Table 3). This is because (i) the number of frames in Table 4 (individual database) is smaller than that in Table 3 (all four databases); (ii) Group (1) used 2/3 of the frames for training, and Group (2) only used 1/2. A larger and more diverse training set would usually result in

Table 4. The driving accuracies of Group (2) Implementation

Testing specification	Parameters	Spring	Fog	Rain	H-Rain
No noises	—	93.94%	93.66%	89.55%	84.68%
+ Gaussian noise	Mean = 0.1, Var = 0.001	79.55%	65.49%	59.70%	66.94%
+ Poisson noise	—	88.64%	90.14%	91.04%	83.87%
+ Salt & Pepper noise	Density = 0.001	90.91%	91.55%	89.55%	84.68%
+ Salt & Pepper noise	Density = 0.005	90.91%	86.62%	82.84%	84.68%
+ Speckle noise	Mean = 0, Var = 0.01	88.64%	90.14%	92.54%	83.87%

a higher accuracy. If we compare different environments in Table 4, spring has a high or comparable value because there were already noises (fog or rain) in the other three environments.

3.2 Comparisons

We have investigated and simulated various deep learning approaches to establish a fair comparison between our DRL-RT method and other work. Table 5 shows the accuracies of different deep learning networks by applying the SYNTHIA-SEQS-05-SPRING database. (Some parameters were adapted to allow acceptable comparisons.) The reason of selecting this database is that it has the clear environment, which is suitable to eliminate undesired effects and ensure a fair judgement.

Table 5. A comparison between the DRL-RT method and other suggested networks

Reference	Neural networks	Accuracy
Karaduman and Eren [8]	CNN	67.42%
Bojarski *et al.* [2]	CNN	74.24%
George and Routray [4]	CNN	83.33%
Yun *et al.* [19,20]	ADNet	83.33%
Mnih *et al.* [12]	DQN	88.64%
This work	DRL-RT	93.94%

The suggested CNN in [8] obtained an inferior tracking accuracy of 67.42%. This is due to the architecture of this network, where it was constructed to classify directions of traffic signs. More specifically, a pooling layer was applied after each convolution layer and no ReLU layers were used. This caused compressing and wasting useful extracted features after each convolution layer. The CNN used in [2] attained a low accuracy of 74.24%. The main drawback of this network is that it considers steering angles to be tracked, which increases the errors of obtaining precise outputs. The CNN used in [4] achieved 83.33%. This is also due to the architecture of this network, which was designed for classifying eye gaze directions. The ADNet used in [19,20] attained the same accuracy of 83.33%. The essential problem lies in the rewards used there, which were basically designed for recognising moved objects, as the rewards are updated in the stop action. In addition, the ADnet architecture is not entirely appropriate for road tracking tasks.

The work which is closest to ours is the influential Deep Q-Network (DQN) used in [12] and illustrated in [1], which achieved a reasonable accuracy of 88.64%. We believe that this is due to the introduction of DRL (i.e., the deep network and the Q-learning). Our proposed method has shown superior performance by attaining the accuracy of 93.94%. This may be due to the overall structure of our road tracking method including the network architecture, tracking policy and designed codes.

4 Conclusion

In this paper, a deep reinforcement neural network DRL-RT has been proposed for road tracking. This network was trained and tested to guide driving cars under different weather environments. Different tracking instances were coded to represent the appropriate road tracking. The MDP concept is used here, where the network accepted states and produced actions by taking advantages from rewards. This study has been compared with other work and showed superior performance. In particular, we have achieved an accuracy of 93.94% in a clear environment, and accuracies 93.66%, 89.55% and 84.68% under unclear environments of fog, rain and heavy-rain respectively. These would set up a new baseline for further studies.

References

1. Arulkumaran K, Deisenroth MP, Brundage M, Bharath AA (2017) A brief survey of deep reinforcement learning. arXiv preprint arXiv:1708.05866
2. Bojarski M, Del Testa D, Dworakowski D, Firner B, Flepp B, Goyal P, Jackel LD, Monfort M, Muller U, Zhang J et al (2016) End to end learning for self-driving cars. arXiv preprint arXiv:1604.07316
3. Cohen A, Pavlovic V (2010) Reinforcement learning for robust and efficient real-world tracking. In: 20th international conference on pattern recognition, pp 2989–2992
4. George A, Routray A (2016) Real-time eye gaze direction classification using convolutional neural network. In: International conference on signal processing and communications (SPCOM), pp 1–5
5. Grigore O (2000) Reinforcement learning neural network used in a tracking system controller. In: Proceedings of the 9th IEEE international workshop on robot and human interactive communication, pp 69–73
6. Hall J, Rasmussen CE, Maciejowski J (2011) Reinforcement learning with reference tracking control in continuous state spaces. In: 50th IEEE conference on decision and control and european control conference, pp 6019–6024

7. Jinlin X, Weigong Z, Zongyang G (2009) Neurofuzzy velocity tracking control with reinforcement learning. In: 9th international conference on electronic measurement instruments, pp 3:465–3:468
8. Karaduman M, Eren H (2017) Deep learning based traffic direction sign detection and determining driving style. In: International conference on computer science and engineering (UBMK), pp 1046–1050
9. Kingma DP, Ba J (2014) Adam: a method for stochastic optimization. arXiv preprint arXiv:1412.6980
10. Liu F, Su J (2004) Reinforcement learning-based feature learning for object tracking. In: Proceedings of the 17th international conference on pattern recognition, ICPR 2004, vol 2, pp 748–751
11. MATLAB: image processing toolbox, for use with MATLAB®, computation, visualization, programming. version 3. The MathWorks Inc., Natick, MA (2001)
12. Mnih V, Kavukcuoglu K, Silver D, Rusu A, Veness J, Bellemare M, Graves A, Riedmiller M, Fidjeland AK, Ostrovski G et al (2015) Human-level control through deep reinforcement learning. Nature 518(7540):529
13. Omar RR, Han T, Al-Sumaidaee SAM, Chen T (2019) Deep finger texture learning for verifying people. IET Biom 8:40–48. https://doi.org/10.1049/iet-bmt.2018.5066
14. Perot E, Jaritz M, Toromanoff M, de Charette R (2017) End-to-end driving in a realistic racing game with deep reinforcement learning. In: IEEE conference on computer vision and pattern recognition workshops, pp 474–475
15. Ros G, Sellart L, Materzynska J, Vazquez D, Lopez AM (2016) The synthia dataset: a large collection of synthetic images for semantic segmentation of urban scenes. In: IEEE conference on computer vision and pattern recognition (CVPR), pp 3234–3243
16. Sootla A, Strelkowa N, Ernst D, Barahona M, Stan G (2013) On periodic reference tracking using batch-mode reinforcement learning with application to gene regulatory network control. In: 52nd IEEE conference on decision and control, pp 4086–4091
17. Supančič J, Ramanan D (2017) Tracking as online decision-making: learning a policy from streaming videos with reinforcement learning. In: 2017 IEEE international conference on computer vision (ICCV), pp 322–331
18. Wei C, Zhang Z, Qiao W, Qu L (2015) Reinforcement-learning-based intelligent maximum power point tracking control for wind energy conversion systems. IEEE Trans Ind Electron 62(10):6360–6370
19. Yun S, Choi J, Yoo Y, Yun K, Choi JY (2017) Action-decision networks for visual tracking with deep reinforcement learning. In: IEEE conference on computer vision and pattern recognition (CVPR), pp 1349–1358
20. Yun S, Choi J, Yoo Y, Yun K, Choi JY (2018) Action-driven visual object tracking with deep reinforcement learning. IEEE Trans Neural Netw Learn Syst 29(6):2239–2252

New Heuristics for TCP Retransmission Timers

Robert Janowski[1(✉)], Michał Grabowski[1], and Piotr Arabas[2]

[1] Warsaw School of Computer Science, M. Edelmana 17, 00-169 Warsaw, Poland
{rjanowski,mgrabowski}@poczta.wwsi.edu.pl
[2] Research and Academic Computer Network NASK,
Kolska 12, 01-045 Warsaw, Poland
parabas@nask.pl

Abstract. We propose and analyze experimentally two new heuristics for TCP retransmission timeout algorithms. The first one refers to Recursive Weighted Median retransmission algorithm. We propose to use autocorrelations of RTT samples in tuning of parameters of this algorithm. It appears that the Recursive Weighted Median algorithm equipped with tuning along this heuristics outperforms its original version with recommended weights of RTT samples (at least on analyzed here sequences of RTT samples of well correlated real TCP transmissions). The second heuristics is related to "impulsiveness" of TCP transmissions. We propose to use 1 - the entropy of RTT samples as a quantitative measure of the impulsiveness. We use the impulsiveness of a dynamically sliding window of a few most recent RTT samples as a dynamic factor modifying the weight of RTT sample in the classic Jacobson's algorithm. The proposed version of Jacobson's algorithm outperforms its classic version on analyzed here sequences of RTT samples. The original version of Recursive Weighted Median algorithm is outperformed as well.

Keywords: TCP · Retransmission timers · Jacobson · Weighted Median · Entropy

1 Introduction

Most of today Internet traffic is transferred using Transmission Control Protocol (TCP) [2]. TCP protocol ensures an end-to-end reliable data transfer that's why it is used by any service that requires total data integrity. Any deterioration of TCP performance also impacts the performance of a service that is using it as a mean of network transport. From this point of view the adjustment of TCP parameter values is very crucial as it decides about the effectiveness of TCP transport and the upper layer services. The retransmission is the one of TCP mechanisms that is strictly related to its performance. The key element of this mechanism is an algorithm that is used to calculate the retransmission timeout based on the previous round trip time (RTT) samples collected at the TCP

R. Burduk et al. (Eds.): CORES 2019, AISC 977, pp. 117–129, 2020.
https://doi.org/10.1007/978-3-030-19738-4_13

sender when it receives a TCP acknowledgement. Today TCP implementations use Jacobson's algorithm [1] but in the literature we can find some other proposals e.g. Recursive Weighted Median algorithm (RWM) [11]. Since this is so crucial issue with broad impact on Internet services we also decided to investigate it. We propose modifications of two RTO algorithms: the classical Jacobson's algorithm [1] and the Recursive Weighted Median algorithm (RWM) [11]. These algorithms are based on information offered by RTT samples only. There are other algorithms with this feature, for instance algorithms analyzing fractal structure of time series of RTT samples [5]. On the other hand, some other RTO algorithms make use of the additional information hidden in TCP congestion control mechanism, for instance Eifel timer [4]. Our proposed modifications obey the paradigm of Jacobson's and RWM algorithms: the information used is that offered by RTT samples only. The paper is organized as follows. In Sect. 2 we define the accepted here measures of performance of RTO algorithms. In Sect. 3 the analyzed algorithms are defined in details. In Sect. 4 the results of experiments are presented altogether with a description of analyzed files with RTT samples. Section 5 concludes the paper.

2 Basic Performance Metrics for RTO Algorithms

TCP protocol assures end-to-end reliable data transfer by using the retransmission mechanism that relies on explicit positive acknowledgements. It means that a TCP sender awaits an acknowledgement from a TCP receiver for each TCP data chunk, that is referred to as 'segment', that has been sent. The arrival of an acknowledgement informs a TCP sender about a reliable transfer of this particular segment it has received an acknowledgement for to the destination i.e. the TCP receiver. However the lack of an acknowledgement within the specified time, so called timeout, lets a TCP sender assume an unreliable transfer and the segment loss before reaching its destination i.e. the TCP receiver. In this case the TCP segment is retransmitted and the procedure is repeated until the reception of a TCP acknowledgement. This briefly described TCP retransmission mechanism is strictly related to the performance of TCP protocol because a TCP sender infers about network conditions based on the reception of an acknowledgement. Thus, when a TCP sender receives expected acknowledgement it not only knows about the reliable transfer of a segment to the destination but it also assumes that the network is underloaded so it can increase the rate it sends data to the destination. The sending rate is controlled by a variable called congestion window (cwnd) that indicates the amount of bytes that can be sent by a TCP sender at once. Depending on the phase of TCP operation the reception of an acknowledgement causes the cwnd to be raised by a different value. In case of slow start phase the value of cwnd is increased by the value of one Message Segment Size (MSS) while in congestion avoidance phase the value of cwnd is increased by 1/MSS. On the other side the lack of expected acknowledgement within the specified timeout lets a TCP sender assume the segment loss due to a network congestion. In a consequence a TCP sender tries to relief the network congestion by lowering its sending rate that is achieved by lowering the cwnd value. If a TCP sender receives triple duplicate acknowledgements it can follow the Fast Retransmit and Fast Recovery algorithms [2]. Not going into deep details of these algorithms we may summarize that a TCP sender sets the cwnd value to the half of its last value and continues to transfer TCP segments in congestion avoidance phase. Otherwise a TCP sender sets cwnd value to 1 and continues to transfer TCP segments in a slow start phase. If a segment is retransmitted more than once then the timeout value is doubled each time. This results in a very poor TCP performance not only because of the waste of time for retransmitting some TCP segments but primarily due to the restrictive lowering of cwnd value and the extension of the timeout value. Thus the process of TCP performance improvement should focus on the retransmission mechanism and the calculation of the timeout values in particular. The longer the timeout is the more time is wasted waiting for an acknowledgement what obviously leads to a poorer performance of TCP. On the other hand, the shorter the timeout is the greater probability of a spurious TCP retransmission i.e. deciding the acknowledgement hasn't come while it has been just a little bit delayed. This also leads to the deterioration of TCP performance. Thus a perfect retransmission algorithm would keep the retransmission timeout as low as possible but without spurious retransmissions. The number of retransmissions and the mean RTO are commonly accepted key performance indicators for TCP retransmission algorithm [11]. In the remaining part of this paper we assume the product of the two above metrics as a primary performance indicator of TCP retransmission algorithm. However neither of these two metrics used separately is sufficient to quantify the loss of performance. That's why instead of reporting the values of these two metrics in our opinion it is reasonable to take the product of the number of retransmissions and the mean RTO because of the practical aspects. In this way we are able to quantify the performance of a TCP retransmission algorithm using only one well defined metric. It is worth to mention that these two metrics: the number of TCP retransmissions and a mean RTO are exchangeable which means that a reduction of the value of one metric leads to a growth of the value of the other metric. Then some balance state must always exist that is an optimum for a given algorithm.

3 The Algorithms

In today's computer systems TCP/IP protocol stack is implemented in an operating system and a TCP retransmission timeout value is calculated dynamically according to some chosen algorithm with a set of predefined parameter values. The input data for a TCP retransmission algorithm is a series of round trip time (RTT) values that represent the time it takes a TCP segment to travel from a TCP sender to a TCP receiver and a relevant TCP acknowledgement in a backward direction. The RTT values let a TCP sender to infer current network conditions and approximate the timeout it should await a TCP acknowledgement for any TCP segment sent. In fact the only widely implemented TCP retransmission timeout (RTO) algorithm is the one defined by Jacobson [1]. According to this algorithm RTO value is calculated as a sum of smoothed round trip time (SRTT) and a component representing a variability - see (1).

$$RTO\,(n) = SRTT\,(n) + 4\sigma\,(n) \tag{1}$$

A smoothed RTT (SRTT) after collecting an n-th RTT sample is computed as an exponentially weighted moving average (EWMA) according to (2):

$$SRTT\,(n) = (1-\alpha)SRTT\,(n-1) + \alpha RTT\,(n) \tag{2}$$

where RTT(n) denotes the last measured RTT sample i.e. the n-th one, SRTT(n) and SRTT(n − 1) denote the SRTT value for the last (current) and the previous RTT measurements, respectively. Parameter α is an averaging factor with default value of 0.125 [2]. The component representing the variability of the measured RTT samples is calculated in a very similar way - see (3):

$$\sigma\,(n) = (1-\beta)\,\sigma\,(n-1) + \beta|SRTT\,(n-1) - RTT\,(n)\,| \tag{3}$$

where σ(n) and σ(n − 1) denote the variability value for the last (current) and the previous RTT measurements, respectively. β is an averaging factor with default value of 0.25 [2].

Jacobson's algorithm tries to follow the current trend of RTT values while not being so sensitive to the sudden changes of RTT values. That's why it relies also on the previous measurements either for smoothed RTT or variability calculations. The responsiveness of this algorithm to sudden changes of RTT values is hidden in the value of parameters β or α for σ or SRTT, respectively. The higher value of β (α) the algorithm is more responsive to sudden changes in RTT values by assigning a higher weight to the last measurement than to the previous ones. Linear recursion of Jacobson's algorithm (1), (2), (3) is a variation of a classic double exponential smoothing algorithm and SRTT converges to the average value of RTT samples if RTT samples exhibit Gaussian behavior [1]. Thus Jacobson's algorithm provides a good performance for RTT samples with Gaussian statistics. However the real TCP transmissions are often impulsive and do not obey Gaussian statistics [6,12]. The Internet traffic is often highly nonstationary with respect to probability distributions of RTT samples [6,12]. So linear filters are relevant to some extent only. The RWM algorithm is a certain remedy. The use of weighted median of a few recent RTT samples and recursively, previous weighted median yields RTOs that are less susceptible to impulsiveness of RTT samples and the variability of the ratio $\frac{RTT}{RTO}$ is decreased in this way.

The Recursive Weighted Median algorithm [11] uses a general concept of a median - a statistical measure of a random data set. In fact, the basic concept of a median is extended with weights assigned to each sample in a considered data set. The weighted median definition remains the same as in case of a "standard" median which is a threshold value of an ordered (sorted) data set up to which the 50% of total probability is accumulated. However the way it is calculated is different since instead of equally treating the samples as in a "standard" median definition here each sample 'd_i' is assigned a weight 'w_i'. This results in calculating a weighted median as a value 'd_j' in an ascending ordered data set $D = \{d_i, i = 1, ..., n\}$ that fulfills the following condition (4):

$$\text{Weighted median} = d_j \iff \sum_{i=1}^{j} w_i = \sum_{i=j+1}^{n} w_i \tag{4}$$

The assignment of weights lets to differentiate the samples 'd_j' that makes an RTO algorithm to be more or less sensitive to most recent RTT samples. This is the advantage emphasized by the authors of [11] in comparison to classical Jacobson's RTO algorithm. In general a data set $D = \{d_i, i = 1, \ldots, n\}$ is composed of RTT samples and previously calculated SRTT values. In practice, the authors of [11] based on extensive tests performed on RTT samples from real networks, recommended to limit this data set to M = 5 of previous RTT measurements and only one last SRTT value (N = 1). Then Recursive Weighted Median algorithm is executed in the following way:

1. For the n-th RTT sample a new value of SRTT is calculated as a weighted median with $W = \{w_i, i = 0, \ldots, N+M-1\}$ vector of weights over a data set composed of previous (M) RTT samples (starting from the n-th RTT sample backward) and the last (N) SRTT values:

$$SRTT\,(n) = MEDIAN(w_0 \diamond SRTT\,(n-1)\,, \ldots, w_{N-1} \diamond SRTT\,(n-N)\,,\\ w_N \diamond RTT\,(n)\,, \ldots, w_{N+M-1} \diamond RTT\,(n-M+1)) \tag{5}$$

   where $\diamond$ operator denotes replication of a second operand a number of times indicated by the first operand.
2. A component representing the variability of the RTT samples is calculated based on the accumulated statistics (until the n-th RTT sample):

$$\zeta\,(n) = \frac{\frac{1}{n}\sum_{k=1}^{n} |RTT(k) - \frac{1}{n}\sum_{i=1}^{n} RTT(i)|}{\frac{1}{n}\sum_{i=1}^{n} RTT(i)} \tag{6}$$

   It is defined as the mean absolute deviation of RTT samples to their mean.
3. The RTO value is calculated as a scaled SRTT value according to (7):

$$RTO(n) = SRTT(n)(1 + \mu\zeta(n)) \tag{7}$$

   where μ is a scale parameter that can be determined empirically as explained in [11]. The determination of μ parameter value is based on the demanded equality of the mean RTO value calculated according to Recursive Weighted Median and Jacobson's algorithms. The conclusion is that typical value of μ is in range [4.3–4.7] - see [11].

We propose a heuristics for choosing weights of RTT samples used in the Recursive Weighted Median computation. Since the weights might be interpreted as an indication of the importance and impact of previous RTT measurements on current SRTT approximation we propose to use weights equal to the autocorrelation coefficients. Since an autocorrelation coefficient determines the dependence of the current value on the previous one this seems a natural choice. We propose the weight w_i related to the i-th previous RTT sample to be assigned a value of the autocorrelation coefficient of lag 'i'. This rule is formally expressed in Eq. (8). In case of independent samples i.e. the autocorrelation coefficients are 0 at any lag, we propose to assign the weights the same value equal 1.

$$w_i = \rho_{RTT,RTT} = \frac{E\{(RTT\,(n) - RTT_{avg})\,(RTT\,(n-i) - RTT_{avg})\}}{\sigma_{RTT}\sigma_{RTT}} \quad \text{for } i = 1, ..., M \tag{8}$$

where RTT denotes the random variable representing the values of RTT samples and RTT_{avg} and σ_{RTT} denote their statistical measures: the average and the standard deviation, respectively.

Our second heuristics is concerned with divergence of recent RTT samples. We think that the impact of current RTT sample on SRTT value should be more meaningful when recent RTT samples exhibit high divergence (or high impulsiveness). This is imprecise and very rough idea and we propose a formalization of it. We agree with the thesis: a bigger divergence corresponds to a smaller entropy. Thus 1 - entropy of recent RTT samples is a natural candidate for quantitative measure of divergence of recent RTT samples and we simply set $\alpha = 1-$ entropy of recent RTT samples in all consecutive computations of RTO forecast. Surprisingly, this nonlinear idea of computing of RTOs yields a considerable diminution of the number of retransmissions while mean RTO remains, more or less, at the level of that yielded by the classic Jacobson's algorithm (at least for the sets of RTT samples analyzed here).

Our heuristics for the RTO calculation is based on the entropy determination for a set of RTT samples collected over some time horizon. The range of this time horizon is regulated by a user selectable 'k' parameter that defines an "entropy window". The appearance of a new RTT sample makes this "entropy window" move forward in this sense that a new RTT sample is included in the window while the last RTT sample is abandoned so the "entropy window" size remains constant and equal to 'k'. As a consequence the consecutive "entropy windows" are overlapping unless 'k' value equals 1. The calculated entropy is used for estimating smoothed RTT (SRTT) that is determined according to a general formula (9):

$$SRTT\,(n) = (1 - \alpha_e)SRTT\,(n-1) + \alpha_e RTT\,(n) \tag{9}$$

Formula (9) is very similar to (2) given for Jacobson's algorithm except the value of α_e parameter is calculated dynamically using entropy concept instead of being a constant. It is intended for an adaptive accommodation to changing RTT values taking more or less from the most recent RTT measurement by automatically adjusting the value of α_e parameter. This feature lets to achieve a good responsiveness of the algorithm especially in a case of impulsive traffic (repeatable and high RTT values) that results in lesser probability of overestimated RTO values. The algorithm for calculating α_e parameter value is outlined below:

1. For a given set D of RTT measurements collected until the n-th sample: $D = \{RTT(n-i), i = 0, \ldots, k-1\}$ where 'k' is an entropy window size, calculate the set ND of normalized RTT samples:

$$ND = \{p_i = \frac{RTT\,(n-i)}{\sum\limits_{j=0}^{k-1} RTT\,(n-j)} \text{ for } i = 0, \ldots, k-1\} \tag{10}$$

2. For the samples in ND that fulfill the condition (11)

$$p_i > 0.0000001 \tag{11}$$

   calculate the entropy (12):

$$E\,(n,k) = -\sum_{i=0}^{k-1} p_i log_k p_i \text{ where } p_i \in ND \tag{12}$$

   The main purpose to skip component not fulfilling the condition (11) is to avoid the impact of rounding errors introduced by small value components.
3. Calculate the value of α_e parameter:

$$\alpha_e = 1 - E\,(n,k) \tag{13}$$

   The RTT variability $\sigma(n)$ and RTO are calculated according to (14) and (15):

$$\sigma\,(n) = (1-\beta)\sigma\,(n-1) + \beta|SRTT\,(n-1) - RTT\,(n)| \tag{14}$$

$$RTO\,(n) = SRTT\,(n) + 4\sigma\,(n) \tag{15}$$

4 Results of Experiments

All three RTO algorithms introduced in Sect. 3 were evaluated against chosen performance metrics (see Sect. 2) over a wide range of data sets. These data sets can be classified into 3 groups:

1. RTT samples collected from a backbone of a real network [3]
2. RTT samples collected from a real mobile/MAN connection
3. RTT samples generated according to some chosen probability distributions

The intention of using different data sets was to evaluate the performance of considered RTO algorithms in a wide range of network characteristics they need to cope when implemented in real computer systems. The RTT samples in group 1 represent RTT delay measured in a backbone of a real network [3]. A backbone network traffic is usually highly aggregated and a correlation that is immanent feature of a single TCP flow due to a closed control loop might be not visible (measurable) because of the mix of UDP and TCP traffic as well as because of a very huge number of TCP flows that are not mutually correlated [9, 10].

The samples in the second group were collected during experiments performed in a real network environment involving downloading data from a remote location. Three scenarios were prepared as follows:

- 2a downloading data via mobile connection - tcp trace analysis (file "gla_tra"),
- 2b downloading data via connection between two institutions (MAN network) - tcp trace analysis (files "we_part1_tra" and "we_part1rtt"),
- 2c downloading data via connection between two institutions (MAN network) - tcp parameters sampling (file "we_reno-01").

In all scenarios the data to download (large video file) was placed on a server in one of institutes involved. The data was downloaded using **scp** run on a client node. The TCP versions set on both server and client node were: Cubic TCP in scenarios 2a and 2b and NewReno TCP in the scenario 2c. It must be noted that, although the algorithms behave differently in the congestion avoidance phase, they defaults to the same RTO timer mechanism in the case of a subsequent packet loss. The client node in the scenario 2a was connected via a mobile network of relatively poor quality and bandwidth (500 kb/s) resulting in a low download speed (approximately 100 kb/s). The connection crossed several networks including mobile access network, Warsaw Metropolitan Network and LAN inside the institute. In scenario 2b the client was located inside LAN network of another institute involved, resulting in a better quality as the path crossed only three networks: both institutes LANs and Warsaw Metropolitan Network. The data gathered in scenarios 2a and 2b were RTT samples of all sent and received TCP segments derived from full TCP trace files (pcap files) collected using **tcpdump** sniffer [7]. The traces were then processed using Perl scripts to correlate tcp sequence numbers and record a TCP segment RTT and distinct regular and retransmitted segments. The experiment in the scenario 2c was analogous to 2b but NewReno TCP was used and data was collected differently: instead of saving data traces TCP parameters were sampled every 0.1 s on the client using Linux **ss** command [8]. Although sampling lowers the resolution of collected data (previously every segment was analyzed), the application of ss command allows insights into TCP algorithm operation and collection of the following variables: congestion window (**cwnd**), slow start threshold (**ssthreshold**), RTO and RTT.

RTT samples in group 3 were artificially generated to create particular conditions to conduct our experiments. They were prepared in a way to emulate traffic with very different statistical characteristics e.g. a value of index of variation defined by (6) or autocorrelation coefficients. In this context the RTT samples generated according to "impulsive distribution" are of our special interest since they were prepared to emulate repeatable (in a stochastic manner) RTT spikes. The purpose of such a traffic was to create a test conditions to evaluate the responsiveness of the proposed RTO algorithms to sudden changes in RTT values. As stated in [11] it was the main advantage of the Recursive Weighted Median algorithm over Jacobson's.

Below we present the results of quantitative evaluation with respect to the number of retransmissions, the mean RTO and the product of them. The conditions of conducting the experiments were as follows:

- For Jacobson's algorithm we have set the default values of control parameters i.e. $\alpha = 0.125$ and $\beta = 0.25$ [1],
- For RWM algorithm we have set the number of previous RTT samples (M) and SRTT values (N) to M = 5 and N = 1, respectively. The values of weights were set according to recommendation given in [11] i.e. $w_0 = 0.5$ and $w_i = 0.875^{(i-1)}$ for i = 1, ..., M,
- For our Recursive Weighted Median heuristics we have set the number of previous RTT samples and SRTT values to M = 5 and N = 1, respectively. The values of weights were set according to autocorrelation coefficients as given in Eq. (8),
- For Entropy algorithm we have set entropy window size k = 5 and $\beta = 0.05$.

Before we present the evaluation of RTO algorithms we shortly summarize the statistical characteristics of RTT samples used in these experiments (Table 1). These characteristics include the ratio of mean absolute deviation of RTT samples to the their mean - (6) that is used in the Recursive Weighted Median algorithm. The average entropy measure depends on the choice of the value of 'k' parameter. We set this to 5, the same as for the entropy algorithm executed on the data sets. The main reason for setting k = 5 is to provide a similar decision delay as in the Recursive Weighted Median algorithm where the number of historical RTT samples accounted during weighted median calculation is just M = 5.

The analysis of the results from Table 1, especially the ones obtained for "impulsive distribution" could raise some suspicions that entropy might be a good measure of impulsiveness of RTT samples. This discovery was our motivation to make the α parameter (exploited in the calculation of SRTT) dependent on the value of the entropy calculated for some recent RTT samples. The number of historical RTT samples should be small enough (small entropy window size 'k') to avoid overlooking of single spikes that could go unnoticed with a higher entropy window size.

Table 1. Statistical characteristics of RTT samples used in experiments

Group of RTT samples	Type of RTT samples	Total number of RTT samples (N)	Autocorrelation coefficient at lag 1	Index of variation as defined in (6)	Average entropy $\sum_{i=k}^{N} \frac{E(i,k)}{N-k+1}$; for k = 5
Collected from a backbone of a real network	Samples1	25527	0	0.624	0.798
Generated according to some chosen probability distributions	Poisson distribution	20000	0	0.544	0.870
	Normal distribution	20000	0	0.143	0.991
	Nonstationary distribution	20000	0	0.659	0.915
	Uniform distribution	20000	0	0.495	0.890
	Bimodal distribution	20000	0	0.303	0.968
	Impulsive distribution	20000	−0.197	1.523	0.300
Collected from a mobile/MAN connection	Gla_tra	24542	0.678	0.134	0.997
	we_part1_tra	72014	0.710	1.351	0.963
	we_part1rtt	74236	0.680	1.514	0.937
	we_reno-01	6940	0.485	0.793	0.933

In Table 2 we have summarized the results of the evaluation of considered RTO algorithms. As a main comparison criterion we assumed a product of a number of retransmissions and a mean RTO as explained in Sect. 2. To obtained the results we have developed our own C++ application that reads samples from a given data set, calculates an RTO value for each RTT sample, then compares it with the next RTT sample in a data set and finally updates the value of a retransmission counter (#retrans) accordingly.

Table 2. Performance results of four RTO algorithms expressed in terms of the product of two main performance metrics: a number of retransmissions and a mean RTO value (#retrans x $\overline{RTO}$)

Type of RTT samples	Jacobson's algorithm (default values: $\alpha=0.125$, $\beta=0.25$)	RWM algorithm with weights set according to [11]	RWM algorithm with proposed setting of weights (our heuristics)	Entropy algorithm ($k=5$, $\beta=0.05$)
Samples1	271.37	700.27	700.27	108.79
Poisson distribution	164.96	399.46	399.46	115.09
Normal distribution	1090.88	485.34	485.34	212.07
Nonstationary distribution	207.35	680.46	680.46	126.92
Uniform distribution	173.91	9541.29	9541.29	0
Bimodal distribution	4.45	177.50	177.50	0
Impulsive distribution	3812.19	1495.60	1466.33	53.16
Gla_tra	189.11	22.35	22.16	18.15
we_part1_tra	57.88	74.74	58.19	59.25
we_part1rtt	137.089	237.65	198.90	148.76
we_reno-01	7847.69	3833.91	3833.66	4486.64

The analysis of the results from Table 2 reveals the following observations

- RWM algorithm outperforms Jacobson's one in case of "impulsive" RTT values. It was already noted in [11] and we just confirmed this conclusion. It is also better in 2 out of 4 cases with correlated traffic (RTT samples collected from a real Local Area Network) and artificial RTT samples following normal distribution. However in the remaining cases its performance expressed by the accepted metric is worse than for Jacobson's algorithm. This indicates that RWM algorithm needs weight adjustment per case and the rule for setting weights advised in [11] is not a universal one,
- Our heuristics for setting the weights for Recursive Weighted Median algorithm outperforms the version of weight setting proposed in [11] in case of RTT samples exhibiting positive correlations. At the same time it gives the same results for uncorrelated RTT samples. It appears that relating the values of weights to autocorrelation coefficients lets to determine the appropriate proportions between the importance of the current and past RTT samples,
- Our RTO algorithm based on the entropy concept outperforms the remaining algorithms in the sense of the accepted performance metric. It gives better results than Jacobson's algorithm in all cases except for two when RTT samples were highly correlated (sample sets: "we_part1_tra" and "we_part1rtt").

However these differences are in the range of 1–10% and the complexity of controlling the algorithms is similar since both algorithms require to set the values of two parameters (α and β for Jacobson's algorithm and β and 'k' for entropy algorithm). When compared to Recursive Weighted Median algorithm an entropy RTO algorithm gives better results except one case ("we_reno-01" RTT sample set). However the complexity of Recursive Weighted Median algorithm in the sense of a number of parameters one has to set the values of is too much higher than the complexity of the entropy RTO algorithm. The entropy algorithm doesn't require careful adjustment of many parameters as in case of the Recursive Weighted Median algorithm.

5 Conclusions and Final Remarks

Our proposals seem to be complementary. Modification of the RWM algorithm exhibits good performance on well auto-correlated sets of RTT samples while the algorithm with entropy has a substantial advantage on uncorrelated sets. It is possible, that the algorithm with entropy is an adequate proposal for backbone networks, since the traffic in such nets is uncorrelated rather by its nature [9, 10]. Gini index is a polynomial approximation of entropy and the use of this index instead of entropy makes arithmetical operations complexity of the proposed algorithm comparable with that of the original RWM algorithm. Thus these algorithms have similar potential of real application in operating systems. We have checked that the use of Gini index yields results very similar to that yielded by the algorithm with entropy. We can say that the quality of the results is the same. We have found a nonstationary stochastic process (alternating probability distributions Poisson - bimodal) for which the algorithm with entropy of RTT samples yields better performance (with respect to metrics defined in Sect. 2) than Jacobson's algorithm, for any (!) constant $\alpha \in (0, 1)$. That is why we think that the idea of making parameter α dependent on accumulated information hidden in some recent RTT samples is worth a deep analytical explanation.

References

1. Jacobson V (1988) Congestion avoidance and control. In: Proceedings of SIGCOMM 1988, Stanford, CA, pp 314–329
2. Postel J (1981) Transmission control protocol. IETF RFC-793
3. Mills D Network time synchronization research project. https://www.eecis.edel.edu/~ntp/ntp_spool/ntpstats/pogo/primary/rawstats.20010417. Accessed Jan 2019
4. Ludwig R, Sklower K (2000) The Eifel retransmission timer. ACMSIGCOMM Comput Commun Rev 30:17–27
5. Li Q, Mills DL (2001) Jitter-based delay-boundary prediction of wide-area networks. IEEE/ACM Trans Netw. 9:578–590
6. Li Q (2000) Delay characterization and performance control of wide-area networks. Ph.D. dissertation, University of Delaware, Newark (2000)

7. Jacobson V, Leres C, McCanne S Tcpdump. http://www.tcpdump.org. Accessed Jan 2019
8. Kerrisk M The Linux man-pages project. http://man7.org/linux/man-pages/man8/ss.8.html. Accessed Jan 2019
9. Qiu L, Zhang Y, Keshav S (2001) Understanding the performance of many TCP flows. Comput Netw 37(3–4):277–306
10. Iannaccone G, May M, Diot C (2001) Aggregate traffic performance with active queue management and drop from tail. SIGCOMM Comput Commun Rev 31(3):4–13
11. Ma L, Arce G, Barner K (2004) TCP retransmission timeout algorithm using weighted medians. IEEE Sig Process Lett 11(6):569–572
12. Paxson V, Floyd S (1997) Why we don't know how to simulate the internet. In: Proceedings of Winter Simulation Conference, pp 1037–1044

Fault-Prone Software Classes Recognition via Artificial Neural Network with Granular Dataset Balancing

Marek Pawlicki[1(✉)], Agata Giełczyk[1], Rafał Kozik[1,2], and Michał Choraś[1,2]

[1] UTP University of Science and Technology, Bydgoszcz, Bydgoszcz, Poland
marek.pawlicki@utp.edu.pl
[2] ITTI Sp. z o.o., Poznań, Poland

Abstract. In this paper we have investigated the fault proneness of the software source code using artificial intelligence methods. The main contribution lies on improving the data pre-processing step. Before we put the data into an Artificial Neural Network, are implementing PCA (Principal Component Analysis) and k-means clustering. The data-clustering step improves the quality of the whole dataset. Using the presented approach we were able to obtain 10% increase of accuracy of the fault detection. In order to ensure the most reliable results, we implement 10-fold cross-validation methodology during experiments.

Keywords: Pattern recognition · Faults detection · ANN · Data clustering

1 Introduction

The development of a reliable software system, especially at a low cost, can be a significant challenge. The product also has to be market-ready in a reasonable time. Failure detection and defect proneness prediction become crucial tools for reliable software creation, helping with decision making and resource allocation. Various metrics, such as code complexity, or number of revisions can help spot classes with high probability of bugs. Bug prediction, therefore, is a classification problem. Numerous classification methods have been employed to deal with this challenge, along with Artificial Neural Networks (ANN). While some researchers are reluctant to employ ANNs for their lack of transparency, however their prowess in modeling nonlinear functional relationships seem to make them well suited for the problem of defect prediction [12].

2 State of the Art

Software development is at times an amazingly complex and iterative process. However research indicates that there are ways to augment the process with

R. Burduk et al. (Eds.): CORES 2019, AISC 977, pp. 130–140, 2020.
https://doi.org/10.1007/978-3-030-19738-4_14

knowledge inferred with known data mining procedures. For example, Yanguang Shen and Jie Liu conduct a research into the application of Data Mining in Software Testing and Defects Analysis. The authors bring up the following data mining methods for defect testing:

- Association Analysis - an outline of the relations between data sets.
- Cluster Analysis - aggregates data points to form different clusters of features; can be used to clarify the patterns of error types, error stages etc.
- Classification of defects - based on an input set and a set of causes of defects, used to ensure the accuracy of quantitative analysis of software defect.
- Sequence Analysis - used to spot defective software by analyzing the trends of a time-ordered transaction dataset, allowing for detection of potential defects in early stages of development.

Some papers warn of the over-reliance on existing tools geared towards software assessment. In Assessing the Precision of FindBugs by mining Java Projects developed at a University by Vetro, Torchiano and Morisio [1] the authors undertake an assessment of a bug finding tool called FindBugs. Historically, bug finding tools experience a myriad of problems, including a high number of false positives, detecting bugs only partially, prioritizing the bugs in a manner not suited for the project at hand and raising the question of general economic feasibility. The authors hone down on the precision question of those tools. After a series of tests on data gathered from university Java coding assignments, the authors conclude that a proper way to use a bug finding tool is to prioritize the issues report so that the bugs that are most likely to lead to a defect are on top of the list. Almost 30% of the reported issues are bad predictors and should most likely be best left untampered with [1]. In a fairly old paper (2010), Applying Data Mining Techniques in Software Development authors survey the feasibility of using Data Mining principles to augment the software creation process. The identified process consists of Understanding and Analysis of users needs, Interpretation of data and noise removal, preprocessing of data and choosing a proper algorithm, creating a mining model, evaluating and interpreting the model. The paper briefly delves into Frequent Itemsets Mining, Timing Mining, and Classification. Frequent Itemsets Mining is used to uncover defect detection rules, and then association rules create a process model with Apriori and FP2 growth algorithms. The execution timing is briefly touched upon in Timing Mining part as the authors conclude that the technology was then in its infancy. Classification methods are used to predict unknown class label object class [7]. The most recent evaluations of the subject delve into the applications of the Data Mining principles to a far greater extent. For example, in [4] the authors propose a mining approach centred on figuring out the software vulnerability characteristics based on open source vulnerability datasets - Common Vulnerability and Exposure (CVE), Common Weakness Enumeration (CWE) and National Vulnerability Database (NVD). Vulnerability in this article is understood as any weakness in software or hardware logic that results in a negative effect on the confidentiality, integrity or availability of the system when exploited. The illustrated procedure performs the extraction, identification, and mining of the crucial characteristics of software vulnerabilities. The authors relate the probability of a vulnerability occurrence with the chance of a distinct vulnerability category or source code containing the root of the problem explained in the report. This allows to create a knowledge database and to form a procedure to mine the features of software vulnerabilities. After the extraction and identification of features, the approach proceeds to execute the mining by specifying the mining rules. Association rules are used as a determinant of all rules that are present in the database, which fulfil a certain confidence condition. Classification rules are used to designate a minute set of rules in the database which allows forming an accurate classifier. The classifiers performance is evaluated by taking inventory of two metrics - precision and recall, where precision is the rate of true positives to all positives, and recall is the rate of true positives to true positives augmented with the number of false negatives. After data preprocessing a dictionary of vulnerabilities is created, and textual indicators are listed. The method uses Frequency-Inverse Document Frequency (tf-idf) to assign weights to the textual indicators. Those are categorized as essential and non-essential vulnerabilities. The study demonstrates the effectiveness of the approach, noting vast improvement of the Data Mining approach over the manual method, with reported recall around 70% and precision around 60% [4]. In [17] authors target detecting fault-proneness in software classes. Fault proneness is a quality attribute in software development that lends itself to the prediction of potential software faults via machine learning methods. The authors aim to establish a link between object-oriented indicators and fault proneness at the class level. An adaptive neuro-fuzzy inference system (ANFIS) is used, as it amalgamates the benefits of the Fuzzy Inference System and the Artificial Neural Network approaches, along with a Levison-Marquardt updating. Six metrics serve as input, Weighted Methods/Class, Number Of Children, Depth of Inheritance, Response For a Class, Coupling Between Objects and Lack of Cohesion in Methods. The system returns a prediction rate of the class fault-proneness. The authors claim the effectiveness of predicting fault-prone classes at a level of 80% correct classifications and the ability to discover 90% of faulty classes in the best case scenario [17]. One more method fusing fuzzy approaches with data mining is considered in [2]. Multivariate analysis of variance MANOVA is a method of figuring out the effect multiple independent variables have on one dependent variable which uses covariance. The Gini Algorithm is used to create classification and decision trees, and fuzzy logic handles modelling uncertainty by employing partial memberships. The proposed approach uses a mix of those methods on the NASA dataset to detect possible software defects. MANOVA identifies the attributes that have the most impact on defects, and the remaining ones are disregarded. The method develops a fuzzy model for this concentrated dataset. The range values of membership functions are calculated with the Gini Algorithm. The resulting model is successfully tested and achieves an accuracy of 86% in identifying a non-defective software [2].

3 Bug Prediction Dataset

The dataset [8] used constitutes an accumulation of software development metrics. As the authors explain themselves, the ambition of the dataset is to establish a benchmark to test various bug prediction procedures and to judge weather a new routine is an enhancement over the preceding ones. The set provides features derived from source code metrics along with historical and process data. A number of bugs and severity of them is also provided. The dataset allows for defect prediction at the class level, therefore the data could be aggregated into package or subsystem level by aggregating class metrics.

The dataset accommodates information about 5 projects, these are: Eclipse JDT Core, Eclipse PDE UI, Equinox Framework, Lucene and Mylyn. Every project comes with a number of associated metrics, for the use of this work the change log data in the form of comma separated files was used, with features suggested by [14].

4 Principal Component Analysis

The process of finding a feature vector which contains the essence of the data, but with a reduced set of features is called dimensionality reduction. In other words, it is the challenge of building an n-dimensional projection that constitutes the representation of the data in a k-dimensional space. Besides the obvious computational benefits, dimensionality reduction techniques help prevent high dimensionality of input, which induces the phenomenon called 'curse of dimensionality'. The phenomenon causes various machine learning classifiers to underperform in when the number of dimensions inflates. [13] This comes with an exponential increase of samples needed for the classifiers to be accurate.

Principal Component Analysis (PCA) is an established method of dimensionality reduction. Essentially PCA finds the projection of the data in which the data's variance is maximised. In a classic example given in [13], if the data is distributed over a line, performing PCA would quickly inform that the variance over all the other directions is 0, therefore the features responsible for those distributions can be discarded. In this case despite the data-gathering process providing a strong signal in one direction, the data usually contains noise in numerous features. If the strength of the signal overbears the noise to a sufficient extent, extracting the projection which explains maximum variance has a significant chance of containing the essence of the data. The process can be repeated in order to find the next dimension with the largest variance.

5 Data Imbalance

A set is considered imbalanced when the classes are not uniformly represented. This seemingly trivial matter can cause machine learning algorithms to fall short of the expected performance. An example brought to attention in [5] considers a situation where a mammography dataset consists of no more than 2% of abnormalities. In that situation a classification of all the samples to the majority class would yield an accuracy of 98%, completely missing the point of constructing a machine learning algorithm detecting the minority class. With the prevalence of dataset imbalance in most of real-world research problems, two mainstream approaches to dealing with the challenge have emerged. These are different ways of data resampling, either by subsampling the majority class or by oversampling the minority class, and attaching a specific cost function to the training samples [5].

Additionally, recognising the aforementioned issue with accuracy as a performance metric a range of other methods of determining the efficiency of machine learning algorithms were adopted.

6 Artificial Neural Network

Artificial Neural Networks (ANN) are a versatile tool for modeling. Applied in a wide array of uses, they are a standard utility for data mining, providing classification, regression, clustering and time series analysis abilities. The premise of an ANN is that it mimics the learning capabilities of a biological neural network, with emphasis on the properties of neural networks found in human brains, however abstractly simplified [13].

The astonishing modeling ability of ANN as applied to pattern recognition stems from its high adaptability to data. This universal approximation ability is immensely valuable when dealing with real-world data, when the data is abundant, but the patterns concealed in the data are yet to be discovered.

The weights of the ANN are revised by the algorithm with the survey of consecutive data instances, enabling gaining knowledge from experience. Not only can the network attain the relations between the variables, but it can generalize to a sufficient extent so as to allow adequate performance on unforeseen data [16]. An Artificial Neural Network is essentially like fitting a line, plane, or hyper-plane though a dataset, defining the relationships that perhaps exist among the features [15].

An ANN with only one computational layer is frequently called a perceptron. This simplest form of ANN contains an input and an output (computational) layer. The input layer provides the data to the output layer, where computations are performed. Said input layer consists of d nodes that represent d features $X = [x_1 \dots x_d]$ and edges of weight $W = [w_1 \dots w_d]$. The output neuron computes $W \cdot X = \sum_{i=1}^{d}(w_i x_i)$. The binary prediction of either -1 or 1 is a mapping based on the sign of the real value of the result of that computation. Adding bias b helps the model perform in environments with high class distribution imbalance. Thus, the prediction of $\hat{y}$ is the result of the Eq. 1.

$$\hat{y} = sign\{W \cdot X + b\} = sign\{\sum_{i=1}^{d} w_i x_i + b\} \quad (1)$$

In this example, the *sign* plays the role of the activation funciton $\Phi(v)$. Different activation functions will be used in ANNs with multiple hidden layers, usually either the Rectified Linear Unit (ReLU) or Hard Tanh is utilised for the ease in training multilayered networks. The error of the prediction can be expressed as the difference between the real-life test value and the predicted value, namely $E(X) = y - \hat{y}$. If the error is different than 0 the weights should be revised. It becomes the aim of the perceptron to minimise the least-squares between y and $\hat{y}$, for all instances belonging to dataset D. This objective is called the loss function (Eq. 2).

$$\sum_{(X,y)\in D} (y - sign\{W \cdot X\}) \tag{2}$$

The loss function is defined over the whole dataset X, the weights W are updated with the learning rate α, and the algorithm iterates over the entire dataset until it converges. This procedure is refered to as stochastic gradien-descent, also expressed by Eq. 3 [3].

$$W \Leftarrow W + \alpha E(X)X \tag{3}$$

A multilayer neural network features multiple computational layers, called the hidden layers. The name refers to the black box nature of those layers, as the computations are obscured from the users point of view. The data is fed from the input layer to the successive layer with adequate computations along the way, and then fed to another layer and so on until it reaches the output layer. This mechanism is referred to as the feed-forward neural network [3]. The particular count of neuron nodes in the foremost hidden layer usually deviates from the number of inputs. The number of nodes and the number of layers depends on the complexity of the required model and on the availability of data [16]. While there are some instances where using a fully-connected layer of neurons is the norm, utilising hidden layers with the number of nodes below the number of inputs allows for a loss in representation, which actually betters the network's performance. This might come as a result of eliminating the noise in data [3].

Constructing a network with too many neurons tends to lead to overfitting. Overfitting, or overtraining, means that the model adjusted itself to very specific patterns of the training dataset, and therefore, not being general enough, will perform poorly on new, unforseen data [3].

7 Cross-Validation

The models created by various machine learning algorithms, including the ones utilised in this paper, come with a number of risks influencing their performance, including overfitting, or selection bias. The models experiencing those problems can perform outstandingly on the test set, but severely underperform when deployed on unforseen data. To mitigate this risk the models undergo a cross-validation procedure. K-fold cross-validation, or rotation estimation, is more effective in evaluating the effectiveness of machine learning models than simple dataset split methods, like random sub-sampling. The final result is an average of all the measures across all the folds, therefore it is a more fitting assessment of the used methods effectiveness. The k parameter in the very name of the procedure stands for the number of folds the method will use. A fold can be defined as a randomly sampled partition of the data, both equal in size and mutually exclusive with all the other partitions. The partitions are rotated as the test set for validating the model, with the remaining subsets constitute the training set. After repeating the procedure k-times the evaluation scores are averaged. The most common cross-validation procedure involves k = 10 folds, hence the name, 10-fold cross-validation. It is the method utilised in works [9,10].

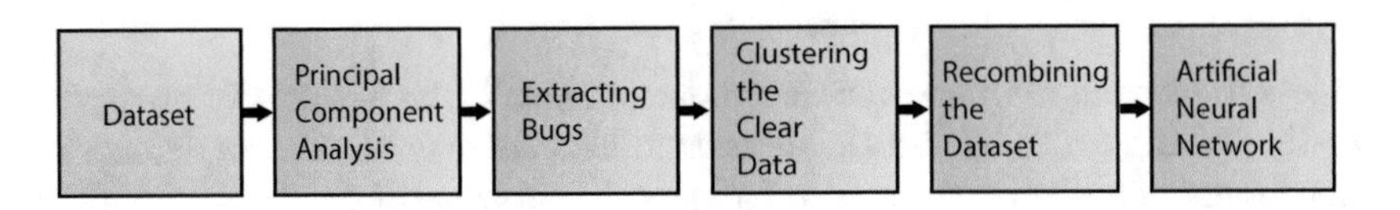

Fig. 1. The process pipeline

8 Experiments

The experiment was conducted using the 10-fold cross-validation methodology. The whole process performed in the research was presented in Fig. 1. The [8] dataset consists of 5 projects and an array of metrics dispersed throughout separate files. The files use comma-separated-values (CSV) format. That is a standard. For this particular approach, the bug-metrics, change-metrics, complexity-code-change and single-version-ck-oo datasets were used. Firstly, the datasets are pooled together. Since the metrics are gathered at the class-level, the datasets are merged with regard to the classname column. This results in a single, comprehensive dataframe containing vectors of 42 features (columns), the dependent variable and the classname, for each of the 5 projects in the dataset. Then those dataframes are concatenated to form an aggregare of all five projects. Thus, the dataframe containing the final dataset contains 44 columns and 5371 rows, out of which 853 are classes with after-release reported bugs. At this stage of research the algorithm will perform binary classification, deciding either if the class will be faulty, or not. For that reason the dependent variable is changed to a Boolean, with $Y > 0$ resulting in a $Y = True$. Secondly, after the dataframe has been formed, the data is scaled and a Principal Component Analysis is performed. The variance test reveals that in this dataset 30 features account for 99.2% of the variability. numberOfVersionsUntil - 0.00172. Thus, the feature vector can be safely reduced to 5, instead of 42 features. After performing PCA the bugged classes and the clear classes are manually divided, and then the clear classes are clustered using a k-means algorithm with the number of centroids matching the number of bugged classes. This balances the dataset, preventing selection bias.

Data prepared in this way is concatenated again, and the rows are randomly mixed. The dataset is then fed to an Artificial Neural Network, with 30 nodes on the input layer, two hidden layers - one fully connected, one with 15 nodes, and a single output node.

9 Performance Measures

A range of performance measures were employed to get an overview of the algorithm's performance.

- Accuracy - the ratio of correct predictions to all predictions
- Precision - Precision is the ratio of correctly predicted positive observations to the total predicted positive observations.
- Recall (Sensitivity) - Recall is the ratio of correctly predicted positive observations to all the observations in the actual class.
- f1-score is expressed with Eq. 4.

$$f1 - score = \frac{2 \cdot Recall \cdot Precision}{Recall + Precision} \tag{4}$$

Accuracy is the starting point for evaluating the robustness of machine learning algorithms. However, it is not a good metric because of the accuracy paradox - a situation where a model has high accuracy but poor classification ability. To combat this shortcoming additional measures of precision and recall are employed. Precision indicates how many of the classifications were relevant, ergo the quality of the classification. Recall depicts the number of relevant classifications in the sum total of the set, thus signifying the completeness of the model. Those two measures are interconnected, and should be taken into consideration together. Thus, a final measure born from combining precision and recall is introduced –the f1-score.

Macro average computes the metrics independently for the classes and then takes an average of the results, the micro average sums up all the contributions of all classes to come up with the average.

10 Results

The [8] datasets for all the projects were utilised. The dataset provides 42 features, 30 of witch were chosen through performing PCA without impacting the prediction accuracy. In the future data collected from tools such as GitLab or SonarQubeThe will be utilised, as presented in [11]. The data was severely unbalanced with collective 853 examples of classes with bugs in a total of 5371 records, constituting 15,89% of the dataset. The Artificial Neural Network (ANN) classified strongly in favour of the majority class, therefore two data balancing strategies were evaluated. A subsampling approach by matching the number of bug records with the same exact number of randomly picked samples of the clear dataset yielded and average accuracy of 0.553 in a 10-fold cross-validation test. The clustering approach achieved an accuracy of 0.643, proving an almost 10%

point increase in accuracy. Furthermore, we have noticed that the classifier relied heavily on the bug metrics features. Eliminating all the columns which referred to bugs found before release resulted in the ANN of improved precision and recall with similar accuracy of 0.656. In Tables 1 and 2 some more obtained results were presented.

Table 1. Model including the bug metrics

	Precision	Recall	f1-score	Support
False	0.62	0.71	0.66	87
True	0.65	0.55	0.59	84
Micro avg	0.63	0.63	0.63	171
Macro avg	0.63	0.63	0.63	171
Weighted avg	0.63	0.63	0.63	171

Table 2. Model not including the bug metrics

	Precision	Recall	f1-score	Support
False	0.72	0.63	0.67	84
True	0.68	0.76	0.72	87
Micro avg	0.70	0.70	0.70	171
Macro avg	0.70	0.69	0.69	171
Weighted avg	0.70	0.70	0.69	171

11 Conclusions

This paper evaluated the use of data-balancing methods in conjunction with artificial neural networks for the recognition of fault-prone software classes based on the bug prediction [8] dataset. There is room for future research in this area, in all the touched-upon aspects. Firstly, in order to fulfil the requirement expressed by software house's senior staff to improve the understanding of the developed projects and improve their ability to plan and assign tasks and sprints in rapid software development processes data from platforms like GITlab and SonarQube should be utilised, as seen in [6]. There is room for improvement both in the clustering procedure, and the ANN architecture. However, the current results are promising, and we are already working on improving the presented algorithm.

Acknowledgements. This work has received funding from the European Union's Horizon 2020 research and innovation programme under Grant agreement no. 732253. We would like to thank all the members of the Q-Rapids H2020 project consortium.

References

1. Vetro' A, Torchiano M, Morisio M (2010) Assessing the precision of findbugs by mining Java projects developed at a university. In: Second international conference on intelligent computation technology and automation. 7th IEEE working conference on mining software repositories (MSR 2010), Changsha, Hunan, pp 110–113. https://doi.org/10.1109/MSR.2010.5463283
2. Adak MF (2018) Software defect detection by using data mining based fuzzy logic. In: Sixth international conference on digital information, networking, and wireless communications (DINWC), Beirut, pp 65–69. https://doi.org/10.1109/DINWC.2018.8356997
3. Aggarwal CC (2018) Neural networks and deep learning: a textbook. https://doi.org/10.1007/978-3-319-94463-0
4. Li X et al (2017) A mining approach to obtain the software vulnerability characteristics. In: Fifth international conference on advanced cloud and big data (CBD), pp 296–301. https://doi.org/10.1109/CBD.2017.58
5. Chawla NV, Bowyer KW, Hall LO, Kegelmeyer WP (2002) Smote: synthetic minority over-sampling technique. J Artif Int Res 16(1):321–357 http://dl.acm.org/citation.cfm?id=1622407.1622416
6. Choraś M, Kozik R, Puchalski D, Renk R (2019) Increasing product owners' cognition and decision-making capabilities by data analysis approach. Cogn Technol Work. https://doi.org/10.1007/s10111-018-0494-y
7. Chun-mei Z, Zhi-ling L (2010) Applying data mining techniques in software development. In: 2nd IEEE international conference on information management and engineering, Chengdu, pp 535–538. https://doi.org/10.1109/ICIME.2010.5477841
8. D'Ambros M, Lanza M, Robbes R (2010) An extensive comparison of bug prediction approaches. In: Proceedings of MSR 2010 7th IEEE working conference on mining software repositories. IEEE CS Press, pp 31–41
9. James G, Witten D, Hastie T, Tibshirani R (2013) An introduction to statistical learning. In: Cluster computing 2018. https://doi.org/10.1007/978-1-4614-7138-7
10. Kohavi R (1995) A study of cross-validation and bootstrap for accuracy estimation and model selection. In: Ijcai, pp 1137–1145
11. Kozik R, Choraś M, Puchalski D, Renk R (2019) Q-rapids framework for advanced data analysis to improve rapid software development. J Ambient Intell Humaniz Comput. https://doi.org/10.1007/s12652-018-0784-5
12. Lo J (2009) The implementation of artificial neural networks applying to software reliability modeling. In: 2009 Chinese control and decision conference, pp 4349–4354. https://doi.org/10.1109/CCDC.2009.5192431
13. Maimon O, Rokach L (2010) Data mining and knowledge discovery handbook, 2nd edn. Springer, Heidelberg
14. Moser R, Pedrycz W, Succi G (2008) A comparative analysis of the efficiency of change metrics and static code attributes for defect prediction. In: Proceedings of the 30th international conference on software engineering, ICSE 2008. ACM, New York, pp 181–190. https://doi.org/10.1145/1368088.1368114
15. Bassis S, Esposito A, Morabito FC, Pasero E (2016) Advances in neural networks. https://doi.org/10.1007/978-3-319-33747-0

16. da Silva IN, Spatti DH, Flauzino RA, Liboni LHB, dos Reis Alves SF (2017) Artificial neural networks a practical course. https://doi.org/10.1007/978-3-319-43162-8
17. Viji C, Rajkumar N, Duraisamy S (2018) Prediction of software fault-prone classes using an unsupervised hybrid SOM algorithm. In: Cluster computing 2018. https://doi.org/10.1007/s10586-018-1923-7

Segmentation of Scanned Documents Using Deep-Learning Approach

Paweł Forczmański(✉), Anton Smoliński, Adam Nowosielski, and Krzysztof Małecki

Faculty of Computer Science and Information Technology,
West Pomeranian University of Technology, Szczecin,
Żołnierska Str. 52, 71-210 Szczecin, Poland
{pforczmanski,ansmolinski,anowosielski,kmalecki}@wi.zut.edu.pl

Abstract. In the paper we present an approach to the automatic segmentation of interesting elements from paper documents i.e. stamps, logos, printed text blocks, signatures, and tables. Presented approach involves object detection by means of Convolutional Neural Network. Resulting regions are then subjected to integration based on confidence level and shape. Experiments performed on representative set of digitizsed paper documents proved usefulness and efficiency of the developed approach. The results were compared with the standard cascade-based detection and showed the superiority of the CNN-based approach.

Keywords: Document segmentation · Object detection · Convolutional Neural Network

1 Introduction

Paper documents have been one of the elementary means of human communication for a long time. Despite many differences, they often contain typical elements such as blocks of text, tables, stamps, signatures, tables, and logos. In the paper we present an algorithm making possible to segment characteristic visual objects from paper document which would make the document understanding easier. Such an approach may be used to transform the document into a hierarchical representation in terms of structure and content, which would allow for an easier exchange, editing, browsing, indexing, filling and retrieval [22].

Developed algorithm, being a part of a larger document managing system, may help in determining parts that should be processed further or to be a subject of enhancement or denoising (e.g. pictures, graphics, charts, etc. [10]). It could also serve as a pre-processing stage of a sophisticated content-based image retrieval system, or just a filter that could select specific documents containing particular elements [13] or sort them in terms of importance (colored vs. monochromatic documents [3]). It is important, that the presented approach is

R. Burduk et al. (Eds.): CORES 2019, AISC 977, pp. 141–152, 2020.
https://doi.org/10.1007/978-3-030-19738-4_15

document type independent, thus it can be applied to any class of documents, e.g. invoices, diplomas, certificates, newspapers, post cards, envelopes, bank checks etc.

The paper is organized as follows: first we shortly review related works, then we present the algorithm and finally we address selected experimental results.

2 Previous Works

Literature survey shows that the problem addressed in this paper has been a subject of study for more than three decades. It is often referred as page segmentation and zone classification [15]. Page segmentation methods understood as global approaches can be divided into three categories: bottom-up, top-down and heuristic. Top-down methods are preferred in case of documents with initially known structure. In such strategy, the input document is decomposed into smaller elements such as blocks and lines of text, single words, characters, etc. On the other hand, bottom-up strategy begins with a pixel-level analysis, and then the pixels with common properties are grouped into bigger structures. Bottom-up techniques are preferred in case of documents with complex structure, but are often slower than top-down ones. Heuristic procedures often combine advantages of the above methods, i.e. robustness of top-down approaches and accuracy of bottom-up methods.

The most popular approaches in the group of bottom-up methods are connected component analysis [15] and sliding-window approach [19] combined with feature extraction based on textural features [12]. Such algorithms are quite robust to skew, but depending on selected measures, the processing cost may vary. Reported accuracy of text detection reaches 94%–99%

The top-down strategies rely often on run-length analysis performed on binarized, skew-corrected documents. Other solutions include usage of Gaussian pyramid in combination with low-level features or Gabor-filtered pixel clustering.

Heuristic methods combine bottom-up and top-down strategies. Good examples are XY-cuts joined with run-length matrix statistics, quad-tree adaptive split-and-merge operations connected with fractal signature [15].

A zone classification is associated with a multi-class discrimination problem. A classification of eight typical objects in the scanned documents was presented in [10]. This is a difficult task, as demonstrated by experimental results. Reported error rate is equal to 2.1%, but 72.7% of logos and 31.4% of tables were misclassified. Wang et al. [22] reported quite mean accuracy of 98.45%, yet 84.64% of logos and 72.73% of 'other' elements were misclassified.

In order to simplify the problem and increase the accuracy, one can use an individual approach, that focuses on single class detection and recognition. It is based on a detection of typical features, followed by "one versus all" classification [3]. For example, the detection of text blocks can be realized by means of statistical analysis [20], edge extraction [16], texture analysis [8,16]. Other authors used stroke filters [9], cosine transform [23] and LBP algorithm [14]. It should be also stressed, that the intraclass variance of many objects (especially tables and signatures) can be a huge problem [1,7,24].

The analysis of the literature shows, that many algorithms involve image pre-processing techniques (e.g. document rectification), deal with a limited forms of documents (e.g. bank cheques) and employ handcrafted features together with multi-tier approaches [6]. The other observation is that there is a lack of methods aimed at the detection of all possible classes of interesting objects in scanned documents. It is caused by the characteristics of objects that have a lot of variability in how they appear.

To overcome above limitations, the Convolutional Neural Networks were introduced, which are capable of dealing with all these issues within a single model. Such alternative direction in research is aimed at creating effective classifiers not taking care of carefully crafted low-level features [25]. In such scenario we solve the task of mapping input data such as image features, e.g. brightness of pixels, to certain outputs, e.g. abstract class of object. The deep learning model, being a hierarchical representation of the perceptual problem, consists of sets of input, transitional and output functions. Such model encodes low-level, elementary features into high-level abstract concepts. The main advantage of such an approach is that it is highly learnable. It should be stressed, that the idea of learning is not completely new, since it comes from a classical neural networks. The input of a deep network is fed iteratively, and the training algorithm lets it compute layer-by-layer in order to generate output, which is later compared with the correct answer. The output error moves backward through the net by back-propagation in order to correct the weights of each node and reduce the error. Such strategy continuously improves the model during computations. The key computational problem in a CNN is the convolution of a feature detector with an input signal. Thus, the features flow from elementary pixels to certain primitives like horizontal and vertical lines, circles, and color areas. In opposition to the classical image filters working on single-channel images, CNN filters process all the input channels simultaneously. Since convolutional filters are translation-invariant, they create a strong response in areas where a specific feature is discovered.

Hence, in the proposed approach, we do not apply any pre-processing but employ a very efficient CNN detector that deals with probably all possible object types that can be found in documents, which has got no significant representation in the literature.

The second element of the addressed detection algorithm is block integration, performed in order to join neighbouring regions of the same class, since CNN-based detectors often mark objects with a regular, often rectangular, box.

In our research we compared traditional feature-based detectors supported by AdaBoost algorithm with CNN-based detector in order to confirm the high robustness and accuracy of the later.

3 Algorithm Description

We propose to use a Convolutional Neural Network to solve a problem of multi-class classification, originating from NeoCognitron [5], advanced by LeCun et al. as LeNet [11]. In our case we employed a DarkNet/DarkFlow implementation [18].

The developed algorithm consists of two subsequent stages. The first one is a rough detection of candidates (classification of detected regions of interest), while the second one is a verification/integration of found objects.

The first stage of processing is based on an effective YOLOv2 algorithm [17], which results in significantly high number of true positives. Then it is supported with a region integration stage. At the training stage we learned the net with five individual classes of objects, namely: stamps, logos/ornaments, texts, tables, and signatures. Exemplary objects are presented in Table 1.

Table 1. Exemplary training samples

Class	Exemplary objects
Stamp	
Logo	
Text	
Signature	
Table	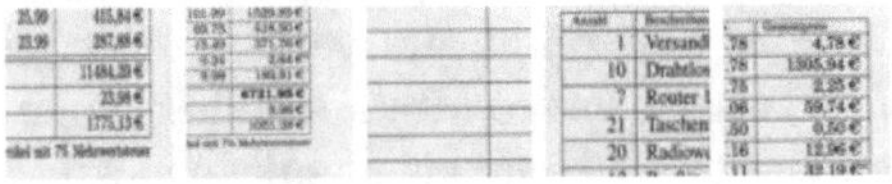

The YOLOv2 algorithm inputs a full image into a single neural network. The image is rescaled to a pre-defined constant size. The net divides the image into regions and predicts bounding boxes and probabilities for each region. These bounding boxes are weighted by the predicted probabilities. The scheme of processing is presented in Fig. 1. After the detection, we perform an integration of regions, for each class, individually, taking into consideration the confidence values returned by YOLOv2 and perform the normalization.

In our approach we employed a DarkFlow (Python-based implementation of DarkNet, using GPU-enabled TensorFlow) for implementing Convolutional Neural Networks. The CNN used in our algorithm consists of 28 layers. The input layer (which actually contains whole input image) is 416×416 elements. The net structure is presented in the Table 2, where *Layer type* represents the type of a layer (Convolutional computes the output of neurons that are connected to local regions, Max Pooling and Avg Pooling perform downsampling operation along the spatial dimensions; Soft Max performs SoftMax regression).

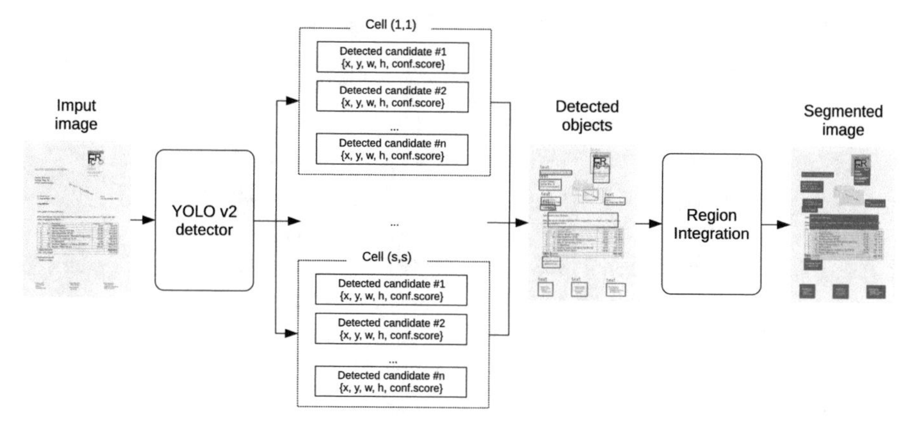

Fig. 1. Scheme of processing.

The *No. Filters* column describes the number of convolution filters and the *Size/Stride* is responsible for the window size. The *Output dim.* represents the spatial resolution of the map at each layer.

After a successful detection, the YOLOv2 returns boxes with confidence values for all classes. We take the boxes and, in case of overlapping regions, sum the confidences and perform *logical or* operation on regions. The confidence values are normalized to one. The comparison of the unaltered result and the integrated/normalized ones, are presented in Fig. 2. In this way we get the regions that are pointed out by the detector with higher (cumulative) confidence. Sometimes it leads to the falsely increased confidence, in the areas that do not contain respective objects, yet in most cases it increases the overall clarity of the results.

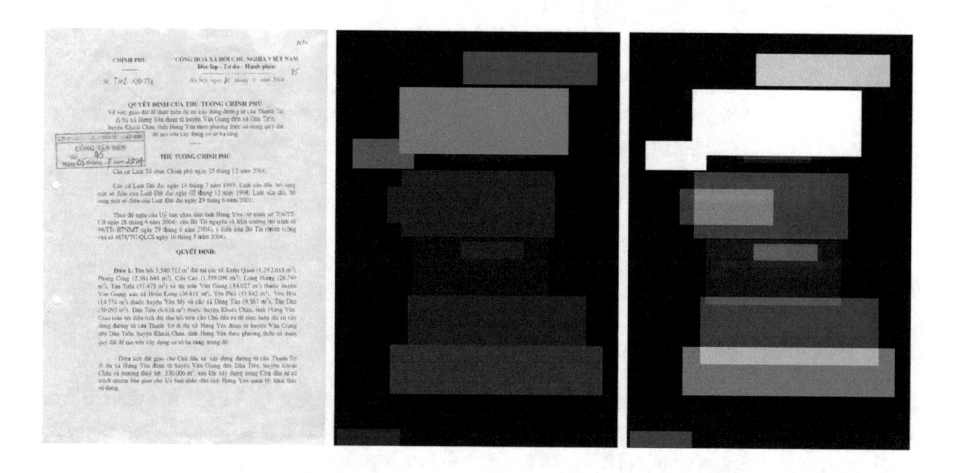

Fig. 2. Region integration with normalization.

Table 2. A structure of the CNN used in the experiments (based on YOLO v2)

Layer type	No. filters	Size/stride	Output dim.
Convolutional	32	$3 \times 3/1$	416×416
Max Pooling	—	$2 \times 2/2$	208×208
Convolutional	64	$3 \times 3/1$	208×208
Max Pooling	—	$2 \times 2/2$	104×104
Convolutional	128	$3 \times 3/1$	104×104
Convolutional	64	$1 \times 1/1$	104×104
Convolutional	128	$3 \times 3/1$	104×104
Max Pooling	—	$2 \times 2/2$	52×52
Convolutional	256	$3 \times 3/1$	52×52
Convolutional	128	$3 \times 3/1$	52×52
Convolutional	256	$3 \times 3/1$	52×52
Max Pooling	—	$2 \times 2/2$	26×26
Convolutional	512	$3 \times 3/1$	26×26
Convolutional	256	$1 \times 1/1$	26×26
Convolutional	512	$3 \times 3/1$	26×26
Convolutional	256	$1 \times 1/1$	26×26
Convolutional	512	$3 \times 3/1$	26×26
Max Pooling	—	$2 \times 2/2$	13×13
Convolutional	1024	$3 \times 3/1$	13×13
Convolutional	512	$1 \times 1/1$	13×13
Convolutional	1024	$3 \times 3/1$	13×13
Convolutional	512	$1 \times 1/1$	13×13
Convolutional	1024	$3 \times 3/1$	13×13
Convolutional	1024	$3 \times 3/1$	13×13
Convolutional	1024	$3 \times 3/1$	13×13
Convolutional	55	$1 \times 1/1$	13×13
Avg Pooling	55	Global	6
SoftMax	—	—	—

4 Experiments

4.1 Benchmark Datasets

The experiments were performed on a database consisting of 701 digitized documents of various origin gathered, among other, from the Internet. It was already used in our previous works [3,4]. It contains both color and grayscale certificates, diplomas, invoices, post-cards, envelopes, letters, and other official and unofficial documents written in different languages, with diverse background, and quality. The spatial resolution varies from 188×269 to 1366×620 pixels, however it

Fig. 3. Exemplary images used in the experiments.

should be noted that the YOLOv2 detector resamples the images to a constant dimensions of 416×416 elements with an information about image spatial proportions. Exemplary documents are shown in Fig. 3 and the number of samples in both training and testing sets are presented in Table 3.

4.2 Experimental Setup

We randomly selected 351 objects (whole documents) belonging to a training set, which is completely different from the testing set (350 documents) in order to show that the algorithm is unbiased for the training dataset.

The detector's accuracy is estimated with the help of Intersection over Union - *IoU* - considered reliable detection quality estimate in many challenges such as popular PASCAL VOC challenge [2]. An *IoU* score is in range 0–1. During

Table 3. Number of samples in the benchmark dataset

Object/class	Documents	Stamp	Logo	Text	Signature	Table
Training no.	351	458	216	2528	153	214
Testing no.	350	460	211	2808	165	232

experiments we set a *IoU threshold*, which is responsible for the false positive and false negative rates. The lower the threshold is, the larger number of falsely detected regions can appear. In parallel, the larger the threshold is, the false negative rate increases. The default threshold value proposed by YOLOv2 developers is equal to 0.25 [17]. In our case, we investigated it's value in the range from 0.02 to 0.93, in order to increase the detection accuracy and calculate the average *IoU* values for all classes.

4.3 Results

Several experiments have been carried out in order to verify the quality of page segmentation in scanned documents. The detection involved testing the detector with different thresholds values. The results, compared to our previous method (involving Haar-like features and AdaBoost learning i.e. Viola-Jones detector [21] implemented in OpenCV library), are presented in Table 4. Presented average and maximal values are for various thresholds of the detector.

As it can be seen, the CNN-based detector is superior to more classical Haar+Adaboost detector, except the text class (for which is almost equally accurate). The significantly higher detection accuracy can be observed in case of highly variable stamp and signature classes. It should be stressed out, that both algorithms were trained on the same datasets. Several examples of page segmentation are presented in the following figures (Figs. 4, 5 and 6).

Table 4. Detection accuracy [%]

Class	Stamp	Logo	Text	Signature	Table
Our method (max values)	97.00	99.41	91.33	99.31	97.79
Our method (avg values)	96.47	99.26	89.33	99.13	96.67
Haar+Adaboost [4]	60.04	84.01	91.63	29.23	94.75

If we take a closer look at the *IoU* values (see Fig. 7) we discover, that the YOLOv2 threshold parameter has only a slight impact on the detection accuracy. It seems, that the threshold around 0.5 gives the most optimal results, since no further increase in the mean *IoU* can be observed for its lower values.

Fig. 4. Results of the segmentation for document no. 147: original image with bounding boxes (ground truth) and the maps for signature class, stamp class and text class, respectively.

Fig. 5. Results of the segmentation for document no. 168: original image with bounding boxes (ground truth) and the maps for signature class, stamp class and text class, respectively.

Fig. 6. Results of the segmentation for document no. 201: original image with bounding boxes (ground truth) and the maps for stamp class, table class and text class, respectively.

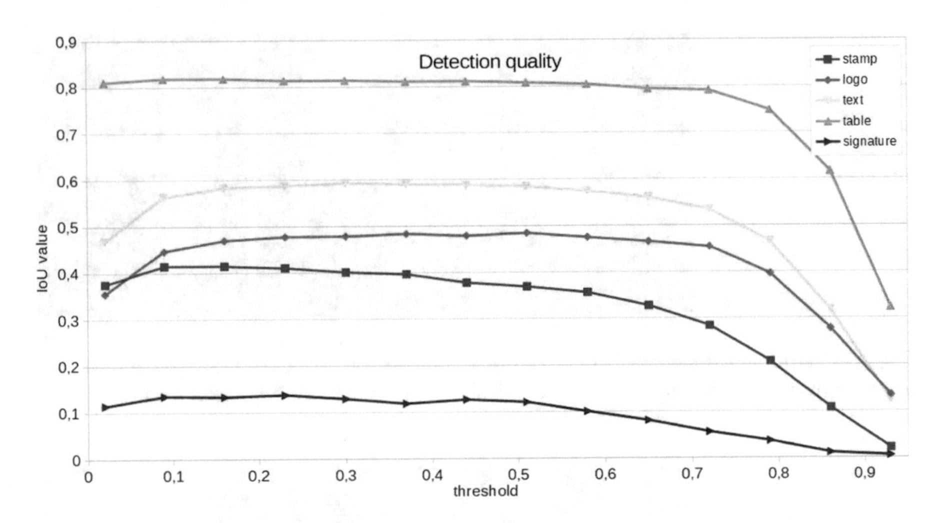

Fig. 7. IoU values for different YOLOv2 thresholds

4.4 Comparison with State-of-the-Art Methods

Direct comparison of obtained results with other, state-of-the-art methods, is not easy, since the benchmark can differ a lot. What is more, the direct comparison with individual methods is not justified because such individual methods often employ class-specific approaches, which are tuned for particular object types. Hence, in this section, a not entirely meaningful comparison with certain, selected global approaches is provided. The average detection accuracy in case of our algorithm is equal to 96.18%. The weighted value (taking into consideration different number of test samples), is equal to 91.57%. In [10], the authors obtained an average detection accuracy equal to 81.84%, however when we consider only classes, that are similar to our case (however, without stamp class), the accuracy drops to 72.95%. The main problem with that approach is a high number of false rejections in case of tables. On the other hand, our algorithm detected tables with a very high accuracy. In [22] the mean accuracy for 9 classes is equal to 84.38%. When we reduce the set in order to be similar to the one in our case (also without stamp class), it is equal to 89.11%. The highest detection rate was obtained for printed text class, while the most difficult class is logotype.

As it was shown, our approach is superior to the state-of-the art methods, while featuring very intuitive computational flow and high processing speed (thanks to the utilization of GPU-enabled CNN computations). The average processing time for a single document is lower than one second (we used Nvidia GeForce GTX 1070M). It also takes into consideration classes, that are not analysed by many existing algorithms, i.e. stamps and signatures. It should be also stressed out, that if we extend the learning datasets and perform extra training iterations, the accuracy could be even higher.

5 Summary

We have presented a novel approach to the extraction of visual objects from digitized paper documents. Its main contribution is a two-stage detection/verification idea based of deep learning. As opposite to other, known methods, the whole framework is common for various classes of objects. It also detects classes, that are not considered in other global approaches, namely signatures and stamps. Performed extensive experiments showed, that the whole idea is valid and may be applied in practice.

References

1. Cüceloğlu İ, Oğul H (2014) Detecting handwritten signatures in scanned documents. In: Proceedings of the 19th computer vision winter workshop, pp 89–94
2. Everingham M, Eslami SMA, Van Gool L, Williams CKI, Winn J, Zisserman A (2015) The PASCAL visual object classes challenge: a retrospective. Int J Comput Vis 111(1):98–136
3. Forczmański P, Markiewicz A (2015) Stamps detection and classification using simple features ensemble. Math Prob Eng. Article ID 367879
4. Forczmański P, Markiewicz A (2016) Two-stage approach to extracting visual objects from paper documents. Mach Vis Appl 27:1243
5. Fukushima K (1980) Neocognitron: a self-organizing neural network model for a mechanism of pattern recognition unaffected by shift in position. Biol Cybern 36(4):193–202
6. Gerdes R, Otterbach R, Kammüller R (1995) Fast and robust recognition and localization of 2-D objects. Mach Vis Appl 8(6):365–374
7. Hu J, Kashi R, Lopresti D, Wilfong G (2002) Evaluating the performance of table processing algorithms. Int J Doc Anal Recogn 4(3):140–153
8. Jain AK, Zhong Y (1996) Page segmentation using texture analysis. Pattern Recogn 29(5):743–770
9. Jung C, Liu Q, Kim J (2009) A stroke filter and its application to text localization. Pattern Recogn Lett 30(2):114–122
10. Keysers D, Shafait F, Breuel MT (2007) Document image zone classification - a simple high-performance approach. In: 2nd international conference on computer vision theory and applications, pp 44–51
11. Lecun Y et al (1998) Gradient-based learning applied to document recognition. Proc IEEE 86(11):2278–2324
12. Lin M-W, Tapamo J-R, Ndovie B (2006) A texture-based method for document segmentation and classification. S Afr Comput J 36:49–56
13. Marchewka A, Pasela A (2014) Extraction of data from Limnigraf chart images. In: Advances in intelligent systems and computing, vol 233, pp 263–269
14. Ojala T, Pietikäinen M, Mäenpää T (2000) Gray scale and rotation invariant texture classification with local binary patterns. In: Proceedings of the 6th European conference on computer vision, pp 404–420
15. Okun O, Doermann D, Pietikäinen M (1999) Page segmentation and zone classification: the state of the art. Technical Report: LAMP-TR-036/CAR-TR-927/CS-TR-4079, University of Maryland, College Park

16. Pietikäinen M, Okun O (2001) Edge-based method for text detection from complex document images. In: Proceedings of the sixth international conference on document analysis and recognition, pp 286–291 (2001)
17. Redmon J, Farhadi A (2017) YOLO9000: better, faster, stronger. In: 2017 IEEE conference on computer vision and pattern recognition (CVPR)
18. Redmon J (2013–2016) Darknet: open source neural networks in C. http://pjreddie.com/darknet/. Accessed 16 Feb 2018
19. Sauvola J, Pietikäinen M (1995) Page segmentation and classification using fast feature extraction and connectivity analysis. In: 1995 Proceedings of 3rd International Conference on Document Analysis and Recognition, ICDAR 1995, pp 1127–1131
20. Su C, Haralick MR, Ihsin TP (1996) Extraction of text lines and text blocks on document images based on statistical modeling. Int J Imaging Syst Technol 7(4):343–356
21. Viola P, Jones MJ (2004) Robust real-time face detection. Int J Comput Vis 57(2):137–154
22. Wang Y, Phillips TI, Haralick MR (2006) Document zone content classification and its performance evaluation. Pattern Recogn 39(1):57–73
23. Zhong Y, Zhang H, Jain AK (2000) Automatic caption localization in compressed video. IEEE TPAMI 22(4):385–392
24. Zhu G, Zheng Y, Doermann D, Jaeger S (2009) Signature detection and matching for document image retrieval. IEEE TPAMI 31(11):2015–2031
25. Krizhevsky A, Sutskever I, Hinton G (2012) Imagenet classification with deep convolutional neural networks. In: Advances in neural information processing systems, vol 25, pp 1106–1114

cGAAM – An Algorithm for Simultaneous Feature Selection and Clustering

Izabela Rejer(✉)

Faculty of Computer Science and Information Technology,
West Pomeranian University of Technology Szczecin, Szczecin, Poland
irejer@wi.zut.edu.pl

Abstract. In this paper the modified version of cGAAM (a genetic algorithm for feature selection for clustering) is introduced. As it can be shown, the algorithm is able to find significant subsets of features in data sets that differ in size and number of classes. The common feature of the sets that were used to test the cGAAM is that the examples are provided with class labels. Due to this, although the clustering process was performed without the class labels, the chosen feature sets could be compared with feature subsets returned by Lasso method in terms of classification accuracy. The most important observation from the results presented in the paper is that the classification accuracy obtained with feature subsets returned by cGAAM was not only comparable with accuracy obtained with feature subsets returned by Lasso but almost always was higher than 80% (*ionsphere* dataset) and 90% (*humanactivity* dataset).

Keywords: Feature selection · Clustering · c-means · cGAAM · Genetic algorithm

1 Introduction

One of the characteristic features of the nowadays datasets is a high dimension of their feature space. Since we can easily gather data describing more and more detailed characteristics of the problem at hand, we do it simply because they might be useful in further analysis. Next, when we generate features from all these different types of data, we end up with a data set that incorporates too much features to be reasonable analyzed with any statistical or machine learning methods. This issue is especially vivid in all life experiments involving human beings or animals where we are able to gather a huge number of neurophysiological factors describing the user mental or physical state but we cannot acquire enough trials to make use of even a small fraction of all these factors. The problem is that with the increasing number of trials the quality of data drops due to the habituation phenomenon, dropping level of concentration and attention, subject boredom etc. As a result, before we start the analysis, we need

R. Burduk et al. (Eds.): CORES 2019, AISC 977, pp. 153–163, 2020.
https://doi.org/10.1007/978-3-030-19738-4_16

to discard most of the previously gathered factors, keeping only some of them. Obviously, the process of feature space reduction is not a random one, we try to keep those features that are the most significant for the problem at hand and discard those of lower importance.

The problem of feature selection has been extensively researched, especially in the supervised classification field. A lot of methods that can be used to deal with this task have been developed so far. Three different groups of methods can be distinguished here: filters (e.g. Information Gain, ReliefF, Correlation-based Feature Selection (CFS), and Consistency-based Feature Selection [4]), wrappers (e.g. stepwise selection, genetic algorithms and random selection [2]) and embedded methods (e.g. Recursive Feature Elimination, elastic net, and LASSO [10]).

While a significant number of solutions for the feature selection problem under the supervised classification scheme have been provided so far, the range of methods that can be applied under the unsupervised classification scheme is much smaller. Of course, all filter methods, it is the methods that rely only on features characteristics and do not use the information about the class labels can be used here. One such example is provided in [5] where the method based on measuring similarity between features by a maximum information compression index is introduced. Also search strategies used in most wrapper approaches can be applied to deal with this task after defining the appropriate search criteria. For example, in [3] sequential forward selection with two different performance criteria for evaluating candidate feature subsets: scatter separability and maximum likelihood was used. An important feature of the proposed algorithm was that it not only searched for the best feature subset but also for the best number of clusters. Another example of feature selection method for clustering process is CLIQUE algorithm introduced in 1998 by Agrawal et al. [1].

One of the feature search strategies that works very well under the classification scheme is a heuristic strategy called GAAM (Genetic Algorithm with Aggressive Mutation) [6]. In [7] and [8] it was shown that this strategy outperforms three other popular feature selection methods (Lasso, forward selection, and ReliefF) and also four heuristic search strategies based on genetic algorithms. Since only the feature subset evaluation measure changes in the unsupervised classification and not the search strategy, GAAM should also provide good results when used for feature selection under the clustering scheme. This thesis was partially verified in [9] where the cGAAM (clustering GAAM) algorithm was used for finding features of EEG (electroencephalographic) signal allowing to define clusters specific for different levels of gamers' involvement. As it was shown in the paper the features found by the cGAAM were highly consistent across subjects and provided more compact clusters than clusters built over the feature subsets returned by the forward selection search strategy. However, since the task of the mentioned paper was not to verify cGAAM method but to demonstrate how to make the experts life easier, no specific validation of the algorithm results, or evaluation on more data sets were provided there.

The aim of this paper is to close this gap and to show that cGAAM algorithm is able to find significant subsets of features in data sets of different characteristics. Moreover, also two modifications that was introduced lately to the original cGAAM in order to make its performance more stable and reliable are presented. Three data sets from Matlab repository were used in the analysis: *fisheriris*, *ionosphere*, and *humanactivity*. The common feature of all of these sets is that they are accompanied by the classification labels. Due to this, the feature subsets returned by cGAAM during the clustering process could be compared in terms of classification accuracy with feature subsets returned by one of the well-known feature selection methods (LASSO).

2 cGAAM

The cGAAM algorithm is a GAAM (Genetic Algorithm with Aggressive Mutation) adapted to the clustering scheme requirements. The original GAAM algorithm was introduced to enable the feature selection process in the case when only a few features of the highest discriminative capabilities have to be chosen from the whole feature set. The GAAM algorithm was described in details in [7]. In short it can be summarized as follows. At first a random population of individuals is created. Each individual is composed of fixed number of genes, where each gen codes an index of one of the features from the feature set. Both, the number of feature-coding-genes (M) and the population size (N) are the algorithm parameters that are provided by the user. Individuals from the initial population take part in two genetic operations, classic one-point crossover and aggressive mutation. The crossover operation is performed according to the well-known scheme – two randomly chosen parent individuals are cut at the same crossing point and their after-crossing parts are exchanged to create two off-springs. Hence, the two parent individuals produce always two off-springs. When it comes to the mutation operation, the situation is much different. Here one parent individual produces a set of M off-springs, each created by mutating another gene of that individual. The mutation is done by replacing the feature index assigned to the mutated gene with another index randomly chosen from the feature set. After the reproduction step, the mother population is created, which is composed of: N individuals from the initial population, N individuals created in the crossover process, and NM individuals created in the mutation process. Next, all individuals from the mother population are evaluated with the given fitness function (F), verifying the individuals' classification capabilities. Finally, N individuals (feature subsets) of the highest value of the fitness function are chosen for the initial population of the next generation and the reproduction process starts again.

In order to apply GAAM in the clustering process, its fitness function has to be changed. Since in the clustering task, the class labels are usually unknown, the fitness function has to be independent from them. This problem was approached by evaluating the individuals' fitness using the same criterion as that used for building a set of clusters in the feature space, the clusters' compactness. According to that criterion those individuals were evaluated higher which coded the feature subsets allowing for building more compact clusters in the feature space. Hence, the sum of the distances between the samples from the data set and the clusters' centres (found with the classic k-means algorithm with Manhattan distance as a distance measure) was used as the fitness function:

$$F = \frac{1}{L} \sum_{j=1}^{L} \sum_{i=1}^{m_j} |x_{ij} - c_j|, \tag{1}$$

where: c_j – centre of cluster j ($j = 1 \ldots L$), m_j – number of samples in cluster j, x_{ij} – sample i contained in cluster j ($i = 1, ..., m_j$).

During the experiments performed with cGAAM it occurred that although the fitness function given in (1) works quite well two modifications might be done to make the whole algorithm more robust and better fitted to the clustering scheme. Both of them were introduced in order to tackle the outliers' problem.

It is well-known that outliers can severely violate the clustering process. Usually they are eliminated from the feature set during the data preprocessing stage, it is before the real work with the data will even start. Such a solution cannot be applied in the feature selection task because 1. eliminating outliers individually from each feature could easily remove all the data from the data set (especially when the number of features is greater than number of samples); 2. eliminating outliers from the whole feature set would be ineffective because the feature subsets change at each algorithm step. Bearing these two facts in mind the process of outliers' elimination was incorporated into the fitness evaluation. After this modification the process of individuals' fitness evaluation consisted of three parts:

1. eliminating outliers (individually for each feature) from the feature subset codded in the individual,
2. normalizing the feature subset according to min-max normalization,
3. calculating the value of the fitness function, given in (2).

The outliers are eliminated in cGAAM according to the slightly modified IQR (interquartile range) method. The modification is very simple, instead of using the 25th and 75th percentiles and 1.5 multiplier to define the valid data range, just the lower and upper percentiles (parameters LP and UP, respectively) chosen by the user are used. The 5th and 95th percentiles are set as the default values of these two cGAAM parameters but of course their values should be adapted to the required number of features. The more features are allowed, the lower LP and higher UP should be used.

The second modification of the preliminarily cGAAM scheme was the change of the clustering function from the hard k-means to a more flexible c-means minimizing the objective function given as:

$$arg \max_{c} \sum_{i=1}^{N} \sum_{j=1}^{L} w_{ij}^{q} \|x_j - c_j\|^2, \tag{2}$$

where: N – number of samples in a dataset, q – level of cluster fuzziness ($q \geq 1$; larger q results in smaller membership values w_{ij} and fuzzier clusters; $q = 1$ implies a crisp partitioning), w_{ij} - degree of membership of sample i to cluster j, given as:

$$w_{ij} = \frac{1}{\sum_{k=1}^{L} \left(\frac{\|x_i - c_j\|}{\|x_i - c_k\|} \right)^{\frac{2}{q-1}}} \tag{3}$$

The main difference between k-means and c-means algorithms is that while the first algorithm requires that each sample belongs to only one cluster, the second one allows a sample to belong to all clusters with a degree given by w_{ij}. Due to applying the soft c-means clustering the impact of outliers on the algorithm results is further reduced. Even if some outliers survive the elimination process, their w_{ij}'s are much smaller than those of the samples lying in the clusters' centres and hence they are not able to drag the clusters to the borders of the data distribution.

3 Methods

To validate the cGAAM algorithm three datasets from Matlab repository were chosen: *fisheriris*, *ionosphere*, and *humanactivity*. The characteristic feature of all these datasets is that they are used to verify the classification algorithms and hence they are provided with class labels. Thanks to this feature, the quality of the final feature sets returned by cGAAM could be assessed by calculating the classification accuracy. The classification accuracy calculated for feature sets returned by GAAM was compared with the classification accuracy obtained for feature sets returned by the well-known feature selection algorithm used in the classification scheme – the Least Absolute Shrinkage and Selection Operator (Lasso), proposed by Tibshirani [10].

The first set (*fisheriris*) is a very small set, containing only 150 examples described by 4 features and one of the three classes providing the flower names (setosa, versicolor, virginica). This set was just a toy example for cGAAM and was used only to give some visual insight into the clustering process and to show that the idea of using the cluster compactness as a criterion for evaluating feature subsets is worth considering. In fact, the whole cGAAM scheme was even not used with this example – only its fitness function. The real task of this part of the study was to evaluate all the possible 1-, 2-, and 3-feature subsets and choose the best one from the cGAAM point of view. To deal with this task, the cGAAM parameters were set as follows: N - 4, M - 1 (for 1 feature set), N - 6, M - 2 (for 2 features set), and N - 4, M - 3 (for 3 features set); number of generations - 1. Also, none genetic operations or outliers' correction were performed here, only the individual evaluation and selection.

The second set (*ionosphere*) is much bigger. It contains 351 examples described by 34 features and one of the two classes providing the information about the free electrons in the ionosphere ('g' – there is an evidence of some type of structure in the ionosphere, 'b' there is no such evidence, signals pass through the ionosphere). Out of the total 34 features, the two first features (both categorical with two labels only) were removed from the set. Hence, the feature set used in the study was composed of 32 features (3–34 features from the original data set).

The last set (*humanactivity*) contains 24075 examples described by 60 features and one of the five classes providing information about the human activity (sitting, standing, walking, running, dancing). Also here, before starting the feature selection process three categorical features of the smallest number of unique labels (features no.: 8, 11, and 14) were removed from the feature set. Hence, the final feature set was composed of 57 features (features no.: 1–7, 9, 10, 12, 13, 15–60) from the original set). The cGAAM parameters were set at the same levels for *ionosphere* and *humanactivity* datasets: N - 20, M – number of features used in the given algorithm run; LP - 5, UP - 95; number of generations - 200.

After the preprocessing described above the analysis proceeded with the same scheme for each dataset. First, the cGAAM algorithm was run 20 times (12 for the *fisheriris* dataset) in order to find 1, 2, 3, 4, or 5 significant features using 2, 3, 4, or 5 clusters. Since there were only 4 features in the *fisheriris* dataset, the cGAAM algorithm was run only for 1, 2, and 3 features here. The feature set returned at each algorithm run was then used to train a binary classification tree (CART). The matlab *fitctree* function with a default set of parameters was used to deal with this task. The tree was trained ten times according to the 10-fold cross validation scheme. At the end of each validation loop the accuracy on the test set was recorded. The mean accuracy from all 10 test sets was calculated and reported as the final accuracy of the given feature set. Next, the Lasso algorithm was run to find five sets containing 1, 2, 3, 4, or 5 features most significant from the classification point of view. Also these feature sets were introduced to the tree classifier trained and tested exactly with the same scheme as in the case of feature sets returned by cGAAM.

4 Results and Discussion

Table 1 presents the results of the feature selection in the *fisheriris* dataset. The succeeding columns of this table (and also columns of Tables 2 and 3) define: 1 - the number of features in the feature subset; 2 - the number of clusters used in the cGAAM fitness function; 3 and 5 – the feature vector returned by cGAAM and Lasso, respectively; 4 and 6 the mean classification accuracy (calculated over the test sets) of the 10 tree classifiers trained for the feature set from the column 3 and 5, respectively. The most important fact that can be noticed in the table is that all the feature subsets chosen as the best ones according to the cluster compactness criterion for different number of features and different number of clusters have very similar classification accuracy as those calculated for feature subsets returned by the Lasso algorithm. That allows to assume that the cluster compactness criterion is a correct one. Of course, the issue that should be addressed here is that maybe the classification accuracy is similar because all the possible feature sets provided the same high accuracy. Figure 1 resolves such doubts. It compares the data distribution in the two-feature space composed of features returned by the clustering algorithm as the most important ones (features no. 3 and 4) with the three other two-feature spaces that were rejected by the algorithm. As it can be noticed in the figure the clusters chosen by the clustering algorithm have the most compact distribution.

Table 1. Feature subsets chosen according to the cluster compactness criterion from the *fisheriris* dataset for different number of features and different number of clusters.

Number of features	Number of clusters	cGAAM		Lasso	
		Feature vector	Classification accuracy [%]	Feature vector	Classification accuracy [%]
1	2, 3, 4	[3]	94.67	[3]	94.67
1	5	[4]	95.33		
2	2, 3, 4, 5	[3 4]	96.00	[3 4]	96.00
3	2	[2 3 4]	96.67	[2 3 4]	96.67
3	3, 4, 5	[1 3 4]	96.00		

The *fisheriris* dataset was used only to present that the distance criterion used by fuzzy c-means algorithm can be useful when deciding on the feature significance. Since there were only a few possible combinations of features for each size of the feature set, the cGAAM algorithm was not really needed here. The two other examples were completely different. Of course, the number of possible feature subsets to evaluate was still small for one (32 and 57) and two (496 and 1596) feature subsets but it grew exponentially with each additional feature (3-feature subsets: 4960 and 29.260; 4-feature subsets: 35.960 and 395.010; 5-feature subsets: 201.376 and 4.187.106). Therefore, the process of feature selection was performed with cGAAM for both datasets. The results are presented in Tables 2 and 3 for the dataset *ionosphere* and *humanactivity*, respectively.

The first observation that can be made on the basis of the results gathered in Tables 1 and 2 is that the classification accuracy obtained with all feature subsets returned by cGAAM (except for some one-feature subsets) was very high, always

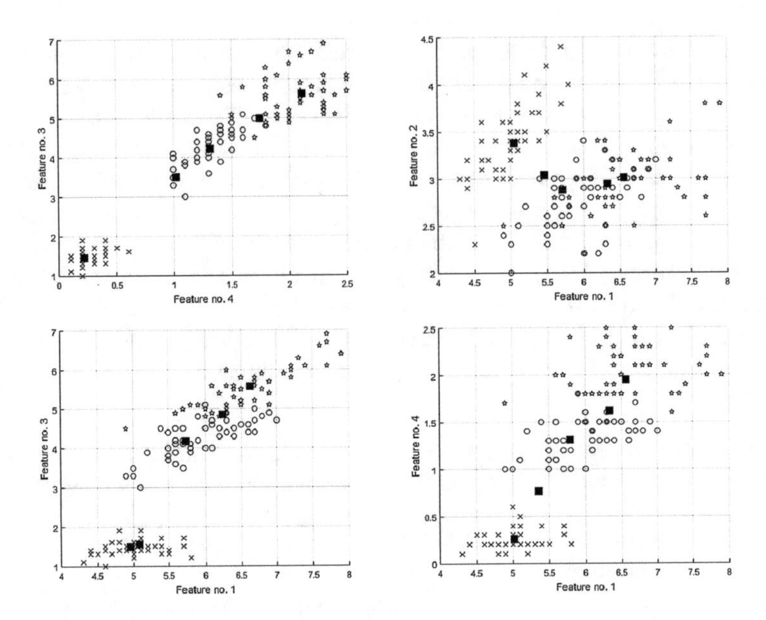

Fig. 1. Clusters found in four two-feature subspaces of the fisheriris dataset; first subfigure presents the feature subspace (features no. 3 and 4) chosen according to the cluster compactness criterion; different shapes (crosses, circles and pentagrams) denote examples from different classes.

higher than 80% regardless of the number of clusters used. Moreover, although different feature subsets were chosen by cGAAM for different number of features and different number of clusters, some of the features were consistently selected in most settings (e.g. features no. 6, 11, 13, 21 (Table 2) or 9, 32, 33 (Table 3)). Hence, the cGAAM results were stable both in terms of classification accuracy and the chosen features across the different number of clusters. Only for one-feature subsets, there were high differences in the classification accuracy obtained with features selected for different number of clusters.

Comparing the feature subsets found by cGAAM with those returned by the Lasso algorithm it can be noticed that completely different feature subsets were chosen by both methods. However, despite of this, the classification accuracy was comparable (at least for one number of clusters). It was even slightly higher in most cases (for cGAAM) for the *humanactivity* dataset.

Table 2. Feature subsets chosen by cGAAM from the *ionosphere* dataset for different number of features and different number of clusters.

Number of features	Number of clusters	cGAAM		Lasso	
		Feature vector	Classification accuracy [%]	Feature vector	Classification accuracy [%]
1	2, 3	[6]	83.20	[3]	81.49
1	4	[13]	75.21		
1	5	[3]	81.49		
2	2, 3, 4	[6 8]	81.79		
2	5	[11 13]	80.00	[1 3]	84.33
3	2	[2 6 8]	81.77	[1 3 5]	86.30
3	3	[4 6 8]	83.48		
3	4, 5	[11 13 21]	82.89		
4	2	[2 6 8 10]	84.07	[1 3 5 6]	92.31
4	3, 4, 5	[11 13 19 21]	85.19		
5	2	[2 4 6 8 10]	83.46	[1 3 5 6 20]	91.18
5	3	[2 11 13 19 21]	87.21		
5	4	[11 13 17 21 29]	85.49		
5	5	[2 11 13 15 21]	88.90		

One more observation that should be commented here is connected with the relation between the number of clusters and selected features. Both, Tables 2 and 3, show that for most numbers of features different feature subsets were chosen for different number of clusters. However, there were no clear dependency between the number of clusters and the classification accuracy. Usually the classification accuracy was the lowest when the two clusters were applied and slightly higher for other number of clusters but it was not a strict rule. This observation points that there is a need to further adapt the cGAAM algorithm by incorporating into its fitness function a criterion allowing for choosing the number of clusters providing the best subset of features when the true number of clusters is unknown in advanced.

Table 3. Feature subsets chosen by cGAAM from the *humanactivity* dataset for different number of features and different number of clusters.

Number of features	Number of clusters	cGAAM		Lasso	
		Feature vector	Classification accuracy [%]	Feature vector	Classification accuracy [%]
1	2	[32]	83.02	[5]	89.49
1	3	[33]	84.49		
1	4, 5	[2]	76.11		
2	2	[10 36]	89.00	[5 41]	90.32
2	3	[32 33]	89.70		
2	4, 5	[9 53]	91.28	[5 17 41]	90.60
3	2	[32 33 54]	90.77		
3	3, 5	[9 32 33]	92.72		
3	4	[32 33 53]	92.60		
4	2, 4	[32 33 34 53]	93.26	[3 5 17 41]	92.00
4	3	[9 32 33 36]	93.32		
4	5	[9 32 33 53]	92.91		
5	2	[10 32 33 34 36]	93.42	[3 5 16 17 41]	92.45
5	3, 4. 5	[9 32 33 34 53]	93.62		

5 Conclusion

There is a broad range of algorithms that can be used to select features that are important from the classification point of view. However, when the class labels are not provided in the dataset, this range shrinks significantly. Of course, regardless of the type of the problem at hand always filter methods can be applied. Although they provide satisfactory results, usually wrappers are more effective, at least under the classification scheme. The problem with the wrappers that makes them difficult to use for feature selection in the clustering problems is that they need an external criterion deciding on feature importance. In the classification scheme this criterion is based on the classifier results but since there are no class labels in the clustering problem this criterion has to be changed.

The cGAAM algorithm presented in this paper incorporates the cluster compactness criterion to the wrapper feature selection scheme. As it was shown via the results obtained for the three repository datasets (*fisheriris*, *ionsphere*, and *humanactivity*) both, this general concept and the cGAAM scheme, successfully deal with the feature selection task. The classification accuracy obtained with feature sets returned by cGAAM was not only comparable with accuracy obtained with feature sets returned by Lasso but almost always was higher than 80% (for the *ionsphere* dataset) and higher than 90% (for the *humanactivity* dataset).

References

1. Agrawal R, Gehrke J, Gunopulos D, Raghavan P (1998) Automatic subspace clustering of high dimensional data for data mining applications, vol 27, no 2. ACM, pp 94–105
2. Burduk R (2012) Recognition task with feature selection and weighted majority voting based on interval-valued fuzzy sets. In: Computational collective intelligence, technologies and applications. Springer, Berlin, pp 204–209
3. Dy JG, Brodley CE (2000) Feature subset selection and order identification for unsupervised learning. In: ICML
4. Koprinska I (2010) Feature selection for brain–computer interfaces. In: New frontiers in applied data mining, pp 106–117
5. Mitra P, Murthy CA, Pal SK (2002) Unsupervised feature selection using feature similarity. IEEE Trans Pattern Anal Mach Intell 24(3):301–312
6. Rejer I (2013) Genetic algorithms in EEG feature selection for the classification of movements of the left and right hand. In: Proceedings of the 8th international conference on computer recognition systems CORES 2013. Springer, Heidelberg, pp 579–589
7. Rejer I (2015) Genetic algorithm with aggressive mutation for feature selection in BCI feature space. Pattern Anal Appl 18(3):485–492
8. Rejer I (2015) Genetic algorithms for feature selection for brain-computer interface. Int J Pattern Recogn Artif Intell 29(5):1559008 (World Scientific Publishing Company)
9. Rejer I, Twardochleb M (2018) Gamers' involvement detection from fig EEG data with cGAAM-A method for feature selection for clustering. Expert Syst Appl 101:196–204
10. Tibshirani R (1996) Regression shrinkage and selection via the lasso. J Roy Stat Soc Ser B 58(1):267–288

Online Adaptation of Language Models for Speech Recognition

Dang Hoang Vu[1], Van Huy Nguyen[1,2], and Phuong Le-Hong[1,3(✉)]

[1] FPT Technology Research Institute, Hanoi, Vietnam
VuDH5@fpt.com.vn

[2] Thai Nguyen University of Technology, Thai Nguyen, Vietnam
huynguyen@tnut.edu.vn

[3] College of Science, Vietnam National University, Hanoi, Vietnam
phuonglh@vnu.edu.vn

Abstract. Hybrid models of speech recognition combine a neural acoustic model with a language model, which rescores the output of the acoustic model to find the most linguistically likely transcript. Consequently the language model is of key importance in both open and domain specific speech recognition and much work has been done in adapting the language model to the speech input. We present an efficient pipeline for hybrid speech recognition where a domain-specific language model is selected for each utterance based on the result of domain classification. Experiments on public speech recognition datasets in the Vietnamese language show improvements in accuracy over the baseline speech recognition model for little increase in running time.

Keywords: Language model adaptation · Vietnamese speech to text · Domain classification · Lattice rescoring · Topic identification

1 Introduction

Generally in the construction of a speech recognition (SR) system, an acoustic model and a language model are trained on all available speech and text corpora. The accuracy of the system is limited by the topic and style of these corpora, which should ideally match the topic and style of the input speech at inference time. Assuming that topic-specific data is available, the acoustic and language models can be re-trained to fit the expected input speech. However that is so computationally expensive as to be impractical for most applications of open-domain speech recognition, since the topic is constantly changing. The most successful approach to deal with this variety is n-gram language model (LM) adaptation, where a new topic-adapted LM is created by adapting a background LM to the topic-specific or domain-specific application, then used to rescore the decoding lattice of the input utterance. Many proposed techniques demonstrated improvements using this approach such as [2,4,7,9]. In order to produce a topic-adapted LM, a generic background LM and topic-specific LMs are first created.

R. Burduk et al. (Eds.): CORES 2019, AISC 977, pp. 164–173, 2020.
https://doi.org/10.1007/978-3-030-19738-4_17

The number and nature of these specific topics can be determined by topic modeling or manual assignment. Depending on how the topic-adapted LM is created, the aforementioned techniques can be broadly categorized into two groups. In the first group, trained topic-specific LMs or interpolated LMs are used as topic-adapted LMs. In the second group, the LM is dynamically adapted to the topic of the input utterance, based on an initial transcript produced by decoding with the background LM. The common feature of these existing techniques is that all input utterances (or at least a sufficiently large set of sentences) need to be transcribed at first in order to determine the topic in question, before rescoring can be carried out. This is not suitable for online applications where the system has to produce the final transcription for each sentence in real time or near real time. In addition a conversation may involve many different topics, in which case using only one topic-specific LM the whole a set of sentences is not optimal.

In this work we propose a new approach of supervised language model adaptation for every single utterance, based on the assumption that a sentence contains enough information to detect its topic. The first-pass decoding transcript first goes through a topic identification model. If the score is above a certain threshold, the corresponding topic-adapted LM will be used for rescoring to get the final transcript, otherwise the output is set to the first transcript. This adaptation technique is thus suitable for online applications since there is no need to decode a whole set of sentences before rescoring.

This paper is structured as follows: In Sect. 2 we give a very brief description of lattice rescoring in speech recognition. In Sect. 3 we outline our proposal for an efficient pipeline for speech recognition with language model adaptation based on the aforementioned approach. Section 4 details our experiment with the proposed approach on two Vietnamese speech recognition datasets.

2 Background on Language Model Adaptation for Speech Recognition

The most common approach for LM adaptation is lattice rescoring. Firstly a background LM and an acoustic model (AM) are used to generate a first-stage lattice, then the LM scores in this lattice are removed. The rescoring step recalculate the sentence scores of the n-best list with a new LM which is relevant to the target topic. The new LM is usually selected based on the initial recognition result. Finally the n-best list is reordered according to the new scores to get the final output [8].

The n-best list of $arcs = a_1, ..., a_n$ such a transducer T from the initial state to a final state for a given utterance can be formulated as follow:

$$\max_a P(x[0,\tau], a) = \max_{a, t_1, \ldots, t_{n-1}} \prod_{i=1}^{n} P(x[t_{i-1}, t_i] | I(a_i)) P(a_i) \tag{1}$$

where $P(x[t_{i-1}, t_i]|I(a_i))$ represents the likelihood assigned by the AM to the phone $I(a_i)$ given the observed feature vectors $x[t_{i-1}, t_i]$ from time t_{i-1} to time t_i. $P(a_i)$ represents the LM probability for the arc in T.

In the rescoring step, the LM probability $P(a_i)$ in is removed and replaced by the probability $P_{LM^*}(a_i)$ given by the new LM. After that, the best path a of arcs $a_1, ..., a_n$ is determined by searching the best scoring path from all possible paths $B(s) = \{a_1...a_k\}$ which is defined as (2):

$$\max_a P(x[0,\tau], a) = \max_{a \in B(s), t_1, ..., t_{k-1}} \prod_{i=1}^{k} P(x|t_{i-1}, t_i|a_i) P_{LM^*}(a_i) \quad (2)$$

The advantage of this approach is that the recognition for the given utterance is only performed in the first stage, and the adaptation is only applied to LM in the second stage for rescoring. It reduces computations for the decoder and make it suitable for real time applications.

3 Our Proposed Method

3.1 General Pipeline

Compared to the baseline speech recognition system, our proposed pipeline involves extra steps in both training and decoding.

In the *training phase* we first fix a large text corpus and a set of topics. We then construct a topic classifier which accepts individual sentences as input. This classifier serves two purposes: to select the training corpora for our topic-specific LMs and to select the correct topic-adapted LM to rescore each sentence. A fixed set of topic-specific LMs are subsequently trained from subsets of the original text corpus, where each sentence is selected with the aforementioned topic classifier. In addition we train a generic background LM on the whole original corpus, to be used in the baseline system for first-pass decoding. For each topic, we generate a topic-adapted LM by a weighted combination of the topic-specific LM and the background LM.

In the *decoding phase*, an utterance is first decoded using the background LM in conjunction with the acoustic model. The decoded output is passed through a topic classifier:

- If the confidence score for the most likely topic meets a fixed threshold, the sentence is rescored using the topic-adapted LM for the corresponding topic.
- Otherwise the original decoded output is returned.

A chart of the decoding process can be found in Fig. 1.

3.2 Constructing Topic-Adapted Language Models

In general our proposed pipeline does not depend on any specific language modeling method. For this work we use 3-gram language models trained with the SRILM toolkit [13]. The models are trained with mono-syllables as the basic linguistic unit instead of full words since the acoustic model outputs predictions for mono-syllables. This also reduces the vocabulary size since there are only about

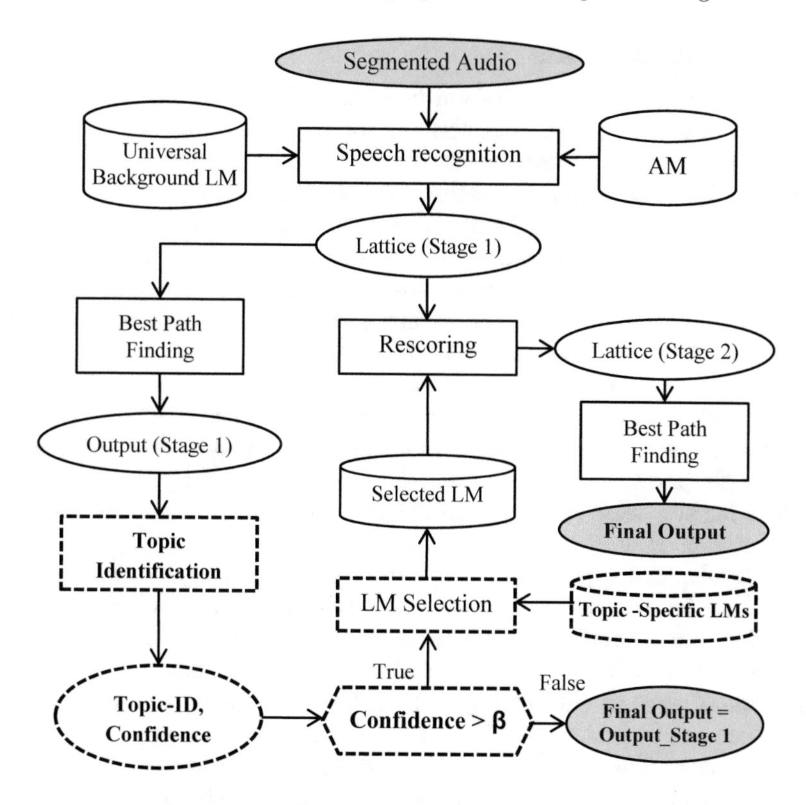

Fig. 1. Decoding architecture using the pipeline LM adaptation method

14000 mono-syllables in total. We will describe the way to get this vocabulary in Sect. 4.4.

First a background LM is trained on the entire available corpus, to be used for both first-pass decoding and interpolation with topic-specific LMs. We further select a number of subsets of our corpus to train topic-specific LM as follows: Each sentence in the original corpus is passed through the topic classifier. If the predicted probability for a topic passes a fixed threshold, we assign that sentence to the training set for the corresponding topic-specific model.

Topic-adapted LMs are created by interpolating the background LM with the topic-specific LMs by the following linear interpolation:

$$P(w) = \lambda P_{LM1}(w) + (1 - \lambda) P_{LM2}(w) \tag{3}$$

where P_{LM1} and P_{LM2} are probabilities assigned by the background LM and a topic-specific LM respectively to a given word sequence w and λ is the interpolation weight.

3.3 Topic Identification

We approach the topic identification problem as a sentence classification task. Namely we select a number of predefined topics and a corpus of documents on the selected topics. The documents are used to train a sentence classifier to predict the topic of individual sentences. Since the classifier is to be used on speech recognition output, all non-phonetic information such as capitalization, punctuation and tokenization is removed from the training data.

In this work, we use the common convolutional neural network (CNN) model to detect the most likely topic of each input sentence. We build our CNN model upon that of [5] which is originally proposed for sentence classification. Our CNN consists of six main layers: (1) a look-up layer to encode words in sentences by their embeddings, (2) a convolutional layer to recognize w-grams, (3) a non-linear layer with the rectifier activation function, (4) a max pooling layer to determine the most relevant features, (5) a fully connected layer with drop-out and (6) a logistic regression layer (a linear layer with a softmax at the end) to perform classification.

Specifically, let $\mathbf{s} = [w_1, w_2, \ldots, w_n]$ be a sentence of length n, where w_i is the i-th word of the sentence. Each word w_i is represented by its word embedding $\mathbf{x}_i$ which is a row vector of d dimensions. The sentence $\mathbf{s}$ can now be viewed as a tensor $\mathbf{X} = [\mathbf{x}_1, \mathbf{x}_2, \ldots, \mathbf{x}_n]^\top$ of size $n \times d$. This matrix is fed into the convolutional layer to extract higher level features. Given a window size w, a filter is seen as a weight tensor $\mathbf{F}$ of size $o \times d \times w$, where o is the output frame size of the filter. The core of this layer is obtained from the application of the convolutional operator on the two tensors $\mathbf{X}$ and $\mathbf{F}$. The output layer of the convolutional layer is precisely computed as

$$\mathbf{Y}_{ti} = \sum_{j=1}^{d} \sum_{k=1}^{w} \mathbf{F}_{ijk} * \mathbf{X}_{t-1+k,j} + b_i,$$

for all $t = 1, 2, \ldots, n - w + 1, \forall i = 1, 2, \ldots, o$, where $\mathbf{b} = [b_1, b_2, \ldots, b_o]$ is the bias tensor of size o. Then a rectifier linear unit layer is applied element-wise on the output layer to produce score tensor.

The pooling is then applied to further aggregate the features generated from the previous layer. The popular aggregating function is max as it bears responsibility for identifying the most important features. More precisely, the max pooling layer produces $\mathbf{z} = [z_1, z_2, \ldots, z_o]$, where $z_i = \max_{1 \le t \le n-w+1} \mathbf{Y}_{ti}$. This feature vector is then fed into a fully connected layer of standard FNN. Following the previous work [5], we execute a dropout for regularization by randomly setting to zero a proportion p of the output elements. Finally, this feature vector is fed into a logistic regression layer to perform classification.

4 Experiments

4.1 Dataset for Topic Identification

We select the text for our topic-specific language models from a corpus articles from VNExpress so-called VNE-Corpus, a leading news portal in Vietnam.

The articles are organized into broad categories, five of which are selected for our experiments: Business (64,000 articles, so-called Bus-Corpus), Entertainment (65,000 articles, so-called Ent-Corpus), Education (13,000 articles, so-called Edu-Corpus), Health (27,000 articles, so-called Hea-Corpus), and Digital (43,000 articles, so-called Dig-Corpus). The data for our topic classifier consists of the articles from all five categories, where we remove sentences shorter than 10 words. The test set is selected at random from the training data and set aside during training. The training corpora for our topic-specific LMs are selected by re-running the topic classifier on the whole VNE-Corpus with a fixed probability threshold of 0.8 as described in Sect. 3.2. To train the generic background LM, we augment the VNE-Corpus with speech transcripts and various collected texts from daily news and online forums. Detailed information on these corpora can be found in Table 1.

4.2 Training Dataset for Speech Recognition

The dataset used to trained AM and LMs is developed by FPT Technology Research Institute (FTRI). Speech corpus is a reading speech with sentences chosen from sentences of daily news and forum websites. There are 1113 speakers including male and female from Northern, Central, and Southern dialects of Vietnam. All audio files are stored and converted to wave format with sample rate of 16 kHz and analog/digital con-version precision of 16 bits for training AM. Text corpora described in the previous section are used for training LMs. Where, the background LM is trained by using a Big-corpus including VNE-Corpus, and topic-specific LMs are trained based on the categorized corpora.

Table 1. Dataset for training and testing

Corpus	Type	Sentences	Duration	Speakers	Domain
Speech corpus	Reading speech	3,2M	1400h	1113	News
VLSP2018	Reading speech	796	2 h	Unkown	News
FPT-Test	Spontaneous speech	18596	20 h	53	Open
Big-Corpus	Text	89M			News, Story, Daily questions and answers,...
Bus-Corpus		1M			Business
Ent-Corpus		1,3M			Entertainment
Edu-Corpus		0,2M			Education
Hea-Corpus		0,5M			Health
Dig-Corpus		0,7M			Digital

4.3 Test Dataset for Speech Recognition

To evaluate the system, we used two test sets described in Table 1. The first set is VLSP2018 developed by research teams involved in Vietnamese language and speech processing [14]. It composed of continuous .wav files of news speech for a total duration of two hours. The speech was recorded in a non-noisy environment. There is no information about speakers. The proportion of dialects is 50%, 40% and 10% for Northern, Southern and Central respectively. The second test set is FPT-test developed by FTRI. The test set includes spontaneous speech sentences recorded in streets, office rooms,... without any limitation of noise.

4.4 System Training

In order to create the lexicon, a Vietnamese vocabulary was produced by selecting the most frequency words appearing in the Big-Coprus and the training transcript. There were about 200k unique words in the corpora, but we only selected words which appeared with frequency more than 400. That resulted vocabulary of 14k words. The lexicon was further created by applying a grapheme-to-phoneme algorithm as described in [10]. The toneme set containing 154 phones was used to create the lexicon.

The acoustic model with 3136 tri-phone tied states was built using the available Wall Street Journal (WSJ) training recipe of Kaldi Toolkit [12]. The model was a hybrid model of Hidden Markov Model (HMM) and Time delay Deep Neural Network (TDNN) as proposed in [11] which is the one of stage of the art models using Kaldi. We firstly evaluated many type of models, and this model gave the best WERs on the test sets. Basically, the acoustic model was trained in two phases. At the first phase, HMM-GMM models representing mono-phones were firstly initialised and trained on the training data. Once mono-phone models trained, tri-phone tied states were created and clustered into 3136 classes, and used to train tied states HMM-GMM models. The Gaussian mixtures for each stage was set to 30. The feature used for HMM-GMM models was 13 Mel-frequency cepstral coefficients (MFCCs) and their the first and second derivations. In the second phase, the alignment of the training data was created on the top of trained HMM-GMM models, and used as the target for training TDNN model. The input feature for TDNN was 11 stacked vectors of 40 coefficients which were transformed by applying a Linear discriminant analysis (LDA) [3] on a combination of 40 Mel-frequency cepstral coefficients (MFCCs) and 100-dimensional iVector [1]. The final TDNN model was obtained after 3 training epochs with the initial and final learning rates were 0,005 and 0,0005 respectively.

Our CNN network model for training the topic classifier has the following parameter settings. The look-up layer uses 50-dimensional pre-trained word embeddings as described in [6]. The temporal convolution layer has 256 hidden states. The temporal max-pooling layer has the window size of 3. The fully-connected layers all has 128 hidden units and a drop-out rate of 0.1. The optimization method adopted is Adagrad with the learning rate of 0.01 and the learning rate decay of 0.001. The model is trained for a fixed number of 100 epochs.

4.5 Experimental Results and Discussions

Evaluation results are presented in Word Error Rate (WER). Baseline WERs for the test sets were first obtained using the trained AM and background LM. The results of the proposed adaptation method are presented in Tables 2 and 3, where we experimented with different values of the parameters β and λ respectively. The proposed adaptation method improves WER relatively by 14,7% on VLSP2018 and 8,4% on FPT-Test, at $\beta = 0,7$ and $\lambda = 0,8$ for VLSP2018 and $\lambda = 0,6$ for FPT-Test. This result demonstrates that our single-sentence topic identification model works well enough; and our method is applicable to online speech recognition systems since the adaptation can be performed immediately after each spoken sentence. In contrast, with existing adaptation methods the adaptation process can only begin after the whole speech input is transcribed in the first pass, which is only suitable for offline systems.

Table 2. WER adaptation results on different interpolation factors β

	VLSP2018	FPT-Test
Baseline	9,85	17,24
$\beta = 0,4$	9,24	17,59
$\beta = 0,5$	9,16	17,44
$\beta = 0,6$	9,07	17,34
$\beta = 0,7$	**8,98**	**17,04**
$\beta = 0,8$	8,99	17,16
$\beta = 0,9$	9,08	17,97

Table 3. WER adaptation results on different confidence thresholds λ

	VLSP2018	FPT-Test
Baseline	9,85	17,24
$\lambda = 0,4$	8,98	17,04
$\lambda = 0,5$	8,93	16,83
$\lambda = 0,6$	8,91	**15,80**
$\lambda = 0,7$	8,33	15,81
$\lambda = 0,8$	**8,40**	16,11
$\lambda = 0,9$	9,10	17,20

5 Conclusion

In this paper we presented a new process for language model adaptation in online speech recognition, with the advantage that the adaptation step is performed immediately after first-pass decoding of each individual spoken sentence. We propose a full pipeline from corpus selection and language model training to utterance decoding, topic identification and rescoring. Our method is suitable for online speech recognition systems and improved WERs by 8,4% to 14,7% relatively on Vietnamese test datasets over a baseline system composed of the same acoustic model and a language model trained on the entire available text corpus. We believe it should be possible to combine topic identification and language modeling in one step using conditional generative neural networks; and this is the direction we are currently exploring to build efficient, domain-flexible speech recognition systems.

References

1. Dehak N, Kenny PJ, Dehak R, Dumouchel P, Ouellet P (2011) Front-end factor analysis for speaker verification. IEEE Trans Audio Speech Lang Process 19:788–798
2. Echeverry-Correa J, Ferreiros-López J, Coucheiro-Limeres A, Córdoba R, Montero J (2015) Topic identification techniques applied to dynamic language model adaptation for automatic speech recognition. Expert Syst Appl 42(1):101–112
3. Haeb-Umbach R, Ney H (1992) Linear discriminant analysis for improved large vocabulary continuous speech recognition. In: Proceedings of ICASSP-92: 1992 IEEE international conference on acoustics, speech, and signal processing, vol 1, pp 13–16. https://doi.org/10.1109/ICASSP.1992.225984
4. Herms R, Richter D, Eibl M, Ritter M (2015) Unsupervised language model adaptation using utterance-based web search for clinical speech recognition. In: CLEF
5. Kim Y (2014) Convolutional neural networks for sentence classification. In: Proceedings of EMNLP. ACL, Doha, pp 1746–1751
6. Le-Hong P, Nguyen TMH, Nguyen TL, Ha ML (2015) Fast dependency parsing using distributed word representations. In: Trends and applications in knowledge discovery and data mining, Lecture Notes in Artificial Intelligence, vol 9441. Springer, Heidelberg
7. Liu Y, Liu F (2008) Unsupervised language model adaptation via topic modeling based on named entity hypotheses. In: International conference on acoustics, speech and signal processing. IEEE, Las Vegas
8. Ljolje A, Pereira F, Riley M (1999) Efficient general lattice generation and rescoring. In: EUROSPEECH
9. Nanjo H, Kawahara T (2003) Unsupervised language model adaptation for lecture speech recognition
10. Nguyen VH, Luong CM, Vu TT (2015) Tonal phoneme based model for Vietnamese LVCSR. In: 2015 International conference oriental COCOSDA held jointly with 2015 conference on Asian spoken language research and evaluation (O-COCOSDA/CASLRE), pp 118–122
11. Peddinti V, Povey D, Khudanpur S (2015) A time delay neural network architecture for efficient modeling of long temporal contexts. In: INTERSPEECH

12. Povey D, Ghoshal A, Boulianne G, Burget L, Glembek O, Goel N, Hannemann M, Motlicek P, Qian Y, Schwarz P, Silovsky J, Stemmer G, Vesely K: The Kaldi speech recognition toolkit. In: IEEE 2011 workshop on automatic speech recognition and understanding. IEEE Signal Processing Society, IEEE Catalog No.: CFP11SRW-USB
13. Stolcke A (2002) SRILM – an extensible language modeling toolkit. In: Proceedings of the 7th international conference on spoken language processing, pp 901–904
14. VLSP: VLSP 2018 - automatic speech recognition. http://www.vlsp.org.vn/vlsp2018/eval/asr (2018). Accessed 19 Aug 2018

Pedestrian Detection in Severe Lighting Conditions: Comparative Study of Human Performance vs Thermal-Imaging-Based Automatic System

Adam Nowosielski(✉), Krzysztof Małecki, Paweł Forczmański, and Anton Smoliński

West Pomeranian University of Technology, Szczecin, Żołnierska Str. 52, 71-210 Szczecin, Poland
{anowosielski,kmalecki,pforczmanski,ansmolinski}@wi.zut.edu.pl

Abstract. The paper discusses the problem of human body detection in severe lighting condition from the driver perspective. Results of a study of threat situation recognition, defined as the sudden appearance of a pedestrian in the field of view, are presented. A human reaction efficiency and delay time are contrasted with the automatic detection based on thermal imagery.

Keywords: Thermal imagery · Thermovision · Pedestrian detection · Brake reaction time

1 Introduction

In the severe lighting conditions human eye and computer vision systems which operate in the visible spectrum are susceptible to malfunction. When both fail to function satisfactorily it is necessary to find an alternative. In contrast to human eye, a computer vision system can process ranges of electromagnetic radiation outside the visible spectrum, offering significant extension of the human visual senses. All objects with a temperature above absolute zero emit infrared radiation. It can be registered with the thermal cameras producing thermograms. Thermal sensors can be used to perform human body detection without the necessity to properly illuminate the subject. They offer stable images in context of diversified lighting and severe lighting conditions. Figure 1 presents examples of a pedestrian in a distance captured using standard and thermal camera. In the case of the visible spectrum the pedestrian is barely visible. In both cases of thermal images the person is easily distinguishable.

Many studies indicate man as the main factor responsible for road accidents (e.g. 92% [26]). Such a large indicators are caused mainly by driver fatigue [5] and limited visibility of the road and its surroundings [13]. As presented in Fig. 1 imaging techniques beyond the visible spectrum can increase perceptual abilities of human and facilitate the process of driving.

R. Burduk et al. (Eds.): CORES 2019, AISC 977, pp. 174–183, 2020.
https://doi.org/10.1007/978-3-030-19738-4_18

Fig. 1. Visibility of a pedestrian: visible spectrum (left) and thermal radiation (middle and right; two variants presenting various appearance of the scene depending on the ambient temperature).

In the paper we focus our attention on the problem of human body detection in severe lighting condition in the road environment. We presents results of a study of threat situation recognition both by human sensually, and by an automatic system using image recognition techniques.

The article is organized as follows. Next section discusses the related works with the emphasis on the concept of driver's reaction time. In Sect. 3 a proposed approach is presented and in Sect. 4 the experimental results are reported. The article ends with a discussion and conclusions.

2 Related Works

2.1 Driver's Reaction Time

Reaction time is understood as the period from the stimulus of the sensory organ to the appearance of a reflex reaction [24]. The reaction time can be divided into five components: time of arousal in the receptor, time to transfer stimulation to the central nervous system, time of excitation by nerve centers and formation of an executive signal, time of the signal from the central nervous system to the muscle and time of muscle stimulation. The reaction time in case of a human is dependent on many factors. Research conducted among patients of attention-deficit hyperactivity disorder (ADHD) and spectrum of autistic disorders (autism) showed very large fluctuations in the reaction time results [1]. Their overall results were assessed as slow. Reaction time is also worse in healthy but older people [2]. It is related to the loss of muscle mass and the decline in muscle function and what is related to it, physical deterioration [22]. Similar variations and weaker results of reaction time can be observed in people with brain damage [15]. Another factor that can affect negatively the response time is the number of tasks performed simultaneously. Research carried out on both sexes show that in the case of men, if they had more than one task to perform, their response time worsened considerably. In the case of women, a difference was observed, though it was not so significant [14].

The driver's reaction time is understood as the time interval from the moment the driver notices a potential danger to performing a preventive actions (pressing the brake pedal) [12]. Testing the driver's reaction time is an important aspect for the safety of road users. Consiglio et al. [4] investigated the impact of telephone conversations and other potential disturbances on reaction time in the braking response. The authors used a laboratory station for this purpose, which simulated the activity of the foot while driving. The test was conducted for 22 participants who were asked to release the accelerator pedal and depress the brake pedal as soon as possible after the activation of red brake lamp. The results indicated that the conversation, regardless of whether it was conducted in person or by a mobile phone, slowed the pace of reaction time, whereas listening to music over the radio was not important [4]. The research results of [18] indicate that tactile warnings are more effective than auditory warnings during both simple and complex conversations. Research is also conducted on delays of the reaction time after taking a medicine or alcohol [3]. The influence of high temperatures on the mood and response capability of bus drivers was also examined and differences were shown depending on the age of drivers and time of driving [27].

2.2 Technical Solutions

Today, many manufacturers equip their vehicles with the on-board ADAS systems (Advanced Driver Assistance Systems) to provide some assistance to the driver [8,17]. These systems are supported by many different sensors and matrix reflectors. One of the first examples of vehicles equipped with some sort of night vision systems were Cadillac DeVille, Toyota Land Cruiser, Lexus LX470 and Jaguar DCX [11,23]. Nowadays, most car manufacturers offer solutions that support drivers during the night driving. There are passive systems such as 'Night Vision Assistant' in Audi, 'BMW Night Vision' in BMW and active system like 'Night Vision Assist Plus' offered by Mercedes-Benz.

In the scientific literature, one of the first attempts to solve the problem of poor visibility of pedestrians was carried out in 2002 by Nanda and Davis [19]. Probabilistic templates have been used in the sliding window mode. After the acceleration the algorithm processed a 320×240 pixels video file in 11 fps. In 2005 Dalal and Triggs [6] have introduced a Histograms of Oriented Gradients (HOG) features to detect human silhouettes. HOG features have been applied with success to pedestrians detection in far-infrared (e.g. [20]). Many known feature extraction techniques like Haar-like features, LBP, wavelet transform, density of edge (DOE) and different classifiers have been used with success to pedestrian detection [21].

3 Methods and Material

Figure 2 presents the scheme of conducted research. First, the video data has been acquired using visible and thermal cameras. Then, a classifier has been trained to recognize human silhouette in thermograms and the embedded automatic night vision human detection system developed. The system has been subjected to thorough examinations. All video material captured in the visible spectrum has been processed and prepared for exhaustive studies with participants. The ability to detect the appearance of a pedestrian and the reaction time have been measured. Finally, the results have been compared, indicating the weak points of each approach.

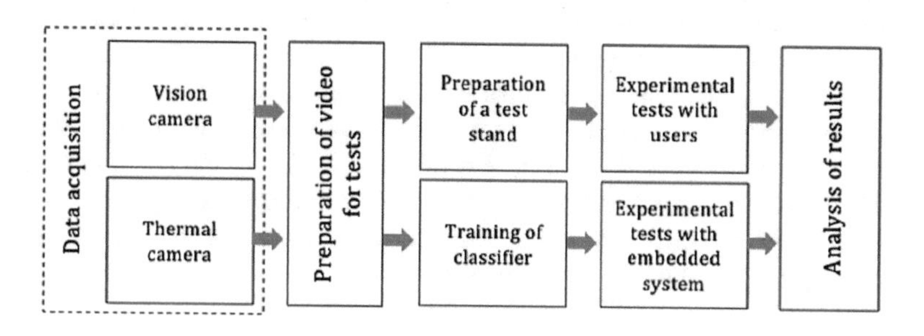

Fig. 2. The research scheme.

The experimental stand for evaluating human performance is presented in Fig. 3. The stand has been developed with the aim of multispectral data acquisition for the assessment of driver's fatigue [16]. It consist of five monitors, of which four imitate the vehicle windows showing real driving situations. There is a steering wheel and three cameras of different kind to capture driver's behaviours in selected spectral bands (visual image, depth map, thermal image). For the current task the stand has been enhanced with foot-operated do-it-yourself brake pedal.

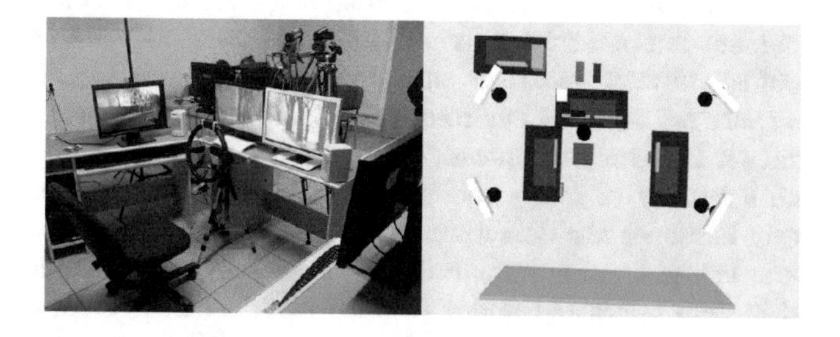

Fig. 3. The experimental stand for evaluating human performance.

3.1 Characteristics of the Obtained Test Material

The evaluation materials have been recorded in parallel with the use of two types of cameras. The scenes have been registered using the SONY HDR CX-550 video camera and IR camera FLIR A325sc. The first one is a standard Full HD camera operating in the visible spectrum light. The other is a thermal imaging camera, equipped with 16-bit sensor of 320×240 pixels. It offers $25 \times 18.8°$ FOV, works at 60 Hz and is interfaced by Ethernet [7].

Seven video recordings have been prepared for testing. They differed either in weather conditions, the number of people present and the surroundings. There are people walking alone and in groups. Their direction of movement is parallel or perpendicular to the direction of the car movement. The distance of the pedestrian from the car was measured with an accuracy of 10 m.

We acquired 17443 thermal frames for the evaluation of the automatic night vision human detection system. The corresponding visible spectrum video material has been processed and montaged producing 8 min 9 s continuous sequence. It has been used to examine human responses for comparison purposes. There are 23 events and an individual event denotes the situation when person or group of people appears in the camera field of view.

3.2 Study of Human Reaction

The study of the reaction time of people was carried out using the stand depicted in Fig. 3. The participant of the study was presented with the prepared video material in the visible light. The material simulated driving a car. The participant's task was to press the brake pedal as soon as he or she observes a pedestrian on the road or in its surroundings. The response time was measured. The total number of participants in the experiment was 31.

3.3 Automatic Recognition of Human Silhouette in Thermograms

Thermal imagery, beside its advantages, has its own specificity. Objects presented in thermograms can be visualized with different appearance depending on the environment temperature. The calibration process takes into account the temperature of the objects and their surroundings (the whole scene) producing various appearance for different environment settings (presented already Fig. 1). This phenomenon impede the process of automatic detection and during the learning stage a variety of images should be provided. The appearance variations negatively influence the detection of objects.

The automatic night vision human detection system have been developed as an embedded system operating under Ubuntu MATE control. We used classic cascading classifier based on Haar-like features [25]. The approach have been applied with success to many object detection tasks (e.g. face [25], fish [9], vehicles [10]).

The detector has been implemented using the Open Computer Vision library (OpenCV). The positive training samples for the cascading classifier consisted of images acquired with the FLIR A325sc camera in different environment settings and various temperature conditions (163 samples). Examples are presented in Fig. 4. The negative samples (599 in total) have been extracted from images captured in the thermal spectrum, collected from our camera and found in the Internet.

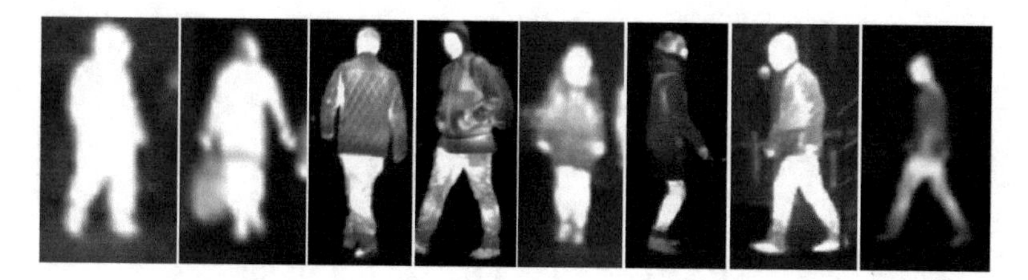

Fig. 4. Example positive training images.

4 Experimental Results

4.1 Human Performance

The human reaction time is presented graphically in Fig. 5. The timeline is shown on the ordinate. The red solid lines denote the time of individual events occurrence. Blue dots denote the time of pressing the brake pedal by each participant. As can be observed some of the events were easy to detect while some other constitute quite a challenge. Those problematic occurrences, as a result, have not been detected by some participants. Interestingly, some participants indicated the appearance of a pedestrian without the actual presence. In addition, for the correct detections of dangerous situations, the reaction time differed between the participants. The upper limit of 3 s has been set arbitrary meaning that when the reaction exceeded the threshold it was considered as not occurring. It must be observed that time interval of 3 s is quite a long and a car can travel 50 or 75 m moving with 60 or 90 km per hour, respectively.

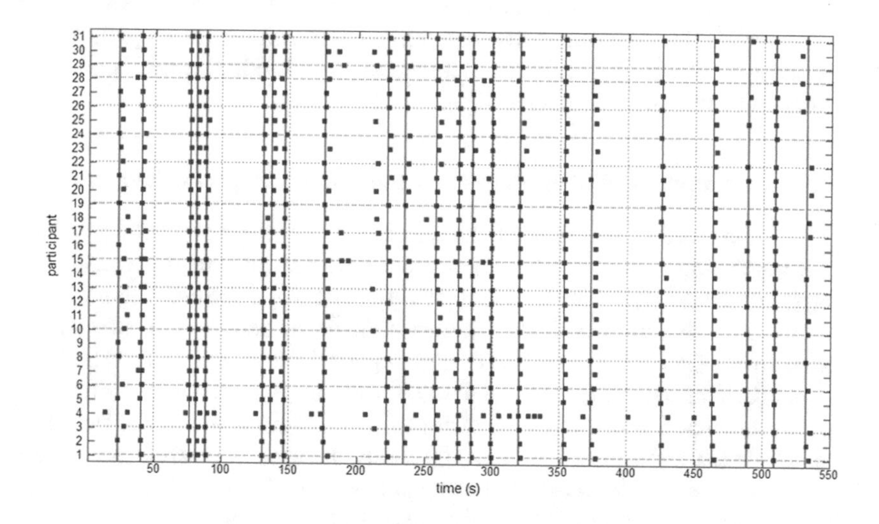

Fig. 5. Times of human reaction for individual events.

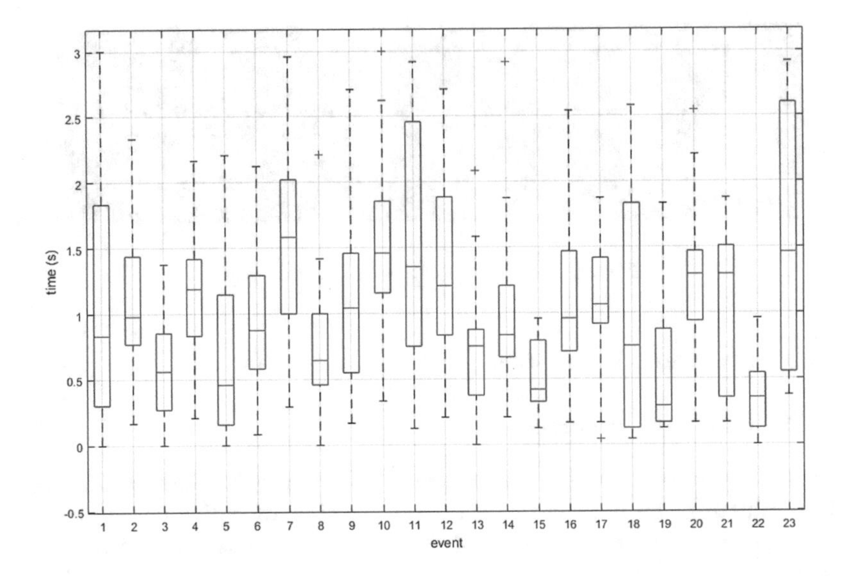

Fig. 6. Delay in human reaction time.

The variety of reaction times has been presented as a box-plot in Fig. 6. The significantly delayed reactions are excluded from the plot. The median is indicated with the central red mark in the box. The bottom and top edges of the box indicate the 25th and 75th percentiles, respectively. The whiskers extend to the most extreme reaction times not considered as outliers. The latter are marked using the '+' symbol. Considering the fixed reaction time limits, the recognition rate of potentially dangerous situation has been calculated and presented in two graphs in Fig. 7. The measured average human recognition rate equals 71.39% with a large discrepancy both in the detection of individual events and in the performance of individual persons.

4.2 Performance of Automatic Recognition Using Thermograms

Figure 8 presents a few examples of the automatic detection pictorially. When a person appears alone on the scene, the system does not have much problem with the detection achieving 98.52%. When group of people is present the individual persons overlap and form irregular shapes difficult to detect. The effectiveness decreases to 50.43%. The overall accuracy equals 76.95% with 17.29% of false detection. A relatively high false detection rate in the case of an automatic system can be significantly reduced with the accumulation process of individual frames. While testing thermal data all the frames have been treated individually. With the frame rate of 25 most of the erroneous detections in individual frames can be successfully filtered reducing the false detection rate.

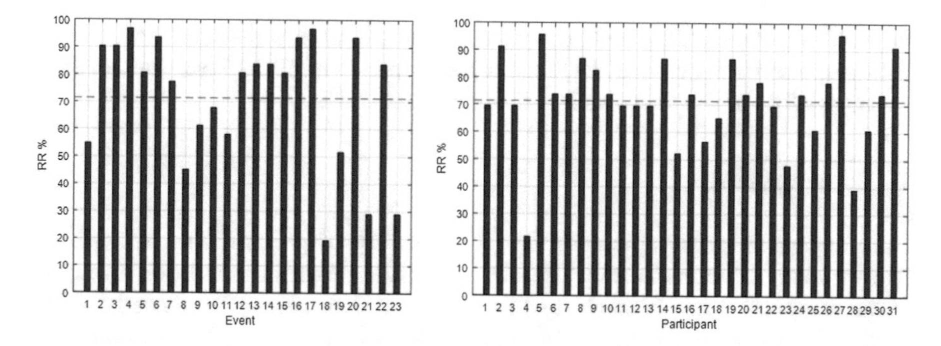

Fig. 7. The effectiveness of dangerous situations recognition by humans, having regard to events (left) and participants (right).

Fig. 8. Examples of automatic pedestrian detection on thermal images.

5 Discussion

Human performance has been assessed using almost 10 min video material of real situations containing pedestrians crossing or moving along the road, from the driver's perspective. The recording contained scenes lit with dipped beam and traffic lights, and pedestrians were passing at different distances from the vehicle. Such a large variety of existing conditions proved to be problematic for some participants. Many numerous delays of the reaction time to the occurring events have been observed. The worst participant was able to recognize the appearance of a pedestrian within 3 s only in 21.74% of cases. Two best participants achieved 95.65%. In many cases, participants did not notice the pedestrian at all. A false press of a brake pedal, for non-existent events, was also noticed. Taking into account the individual events, the event no. 18 proved to be the most challenging with only 19.35% of appropriate driver reactions.

The results of an automatic night vision human detection system proved to be slightly better (overall accuracy of 76.95%) compared with the measured human performance (71.39%). Despite the better visibility of pedestrians using the thermal spectrum the resolution of images (320×240 pixels) prove to be too small to detect human silhouette with success from a distance. While for a distances less than or equal to 50 m, a performance of 89.18% was noted, for a distances of 60 m the effectiveness dropped to 23.33%.

6 Conclusion

The paper addresses topics related to the safety of pedestrians on the road. The problem of human body detection in severe lighting condition from the driver perspective has been studied in the paper. Human performance has been compared with an automatic night vision system based on the thermal imagery. The automatic solution has demonstrated accuracy in difficult lighting conditions and systems of this type can offer the extension of the human visual senses.

References

1. Adamo N, Hodsoll J, Asherson P, Buitelaar JK, Kuntsi J (2019) Ex-Gaussian, frequency and reward analyses reveal specificity of reaction time fluctuations to ADHD and not autism traits. J Abnorm Child Psychol 47(3):557–567. https://doi.org/10.1007/s10802-018-0457-z
2. Arnold P, Njemini R, Vantieghem S, Gorus E, Pool-Goudzwaard A, Buyl R, Bautmans I (2018) Reaction time in healthy elderly is associated with chronic low-grade inflammation and advanced glycation end product. Exp Gerontol 108:118–124
3. Boult TE, Lewis R (2011) System and method for driver reaction impairment vehicle exclusion via systematic measurement for assurance of reaction time. US Patent App. 12/980,899
4. Consiglio W, Driscoll P, Witte M, Berg WP (2003) Effect of cellular telephone conversations and other potential interference on reaction time in a braking response. Accid Anal Prevent 35(4):495–500
5. Cyganek B, Gruszczyński S (2014) Hybrid computer vision system for drivers eye recognition and fatigue monitoring. Neurocomputing 126:78–94
6. Dalal N, Triggs B (2005) Histograms of oriented gradients for human detection. In: 2005 IEEE computer society conference on computer vision and pattern recognition (CVPR 2005). IEEE
7. FLIRInstruments (2010) Thermovision sdk user's manual, 2.6 sp2 edition
8. Forczmański P, Małecki K (2013) Selected aspects of traffic signs recognition: visual versus RFID approach. In: International conference on transport systems telematics. Springer, pp 268–274
9. Forczmański P, Nowosielski A, Marczeski P (2015) Video stream analysis for fish detection and classification. In: Wiliński A, Fray IE, Pejaś J (eds) Soft computing in computer and information science. Springer, Cham, pp 157–169. https://doi.org/10.1007/978-3-319-15147-2_14
10. Frejlichowski D, Gościewska K, Nowosielski A, Forczmański P, Hofman R (2015) Application of cascades of classifiers in the vehicle detection scenario for the 'SM4Public' system. In: Jackowski K, Burduk R, Walkowiak K, Wozniak M, Yin H (eds) Intelligent data engineering and automated learning - IDEAL 2015. Springer, Cham, pp 207–215
11. Holz M, Weidel E (1998) Night vision enhancement system using diode laser headlights. Technical report, SAE Technical Paper

12. Johansson G, Rumar K (1971) Drivers' brake reaction times. Hum Factors 13(1):23–27
13. Källhammer JE (2006) Night vision: requirements and possible roadmap for FIR and NIR systems. In: Photonics in the automobile II, vol 6198. International Society for Optics and Photonics, p 61980F
14. Kaur M, Nagpal S, Singh H, Suhalka M (2014) Effect of dual task activity on reaction time in males and females. Indian J Physiol Pharmacol 58(4):389–394
15. Lange B, Hunfalvay M, Murray N, Roberts CM, Bolte T (2018) Reliability of computerized eye-tracking reaction time tests in non-athletes, athletes, and individuals with traumatic brain injury. Optometry & Visual Performance
16. Małecki K, Nowosielski A, Forczmański P (2017) Multispectral data acquisition in the assessment of driver s fatigue. Commun Comput Inf Sci 715:320–332. https://doi.org/10.1007/978-3-319-66251-0_26
17. Małecki K, Watróbski J (2017) Mobile system of decision-making on road threats. Proc Comput Sci 112:1737–1746
18. Mohebbi R, Gray R, Tan HZ (2009) Driver reaction time to tactile and auditory rear-end collision warnings while talking on a cell phone. Hum Factors 51(1):102–110
19. Nanda H, Davis L (2002) Probabilistic template based pedestrian detection in infrared videos. In: Intelligent vehicle symposium, vol 1. IEEE, pp 15–20
20. O'Malley R, Jones E, Glavin M (2010) Detection of pedestrians in far-infrared automotive night vision using region-growing and clothing distortion compensation. Infrared Phys Technol 53:439–449
21. Piniarski K, Pawłowski P (2016) Multi-branch classifiers for pedestrian detection from infrared night and day images. In: Signal processing: algorithms, architectures, arrangements, and applications (SPA). IEEE, pp 248–253
22. Pires IM, Marques D, Pombo N, Garcia NM, Marques MC, Flórez-Revuelta F (2018) Measurement of the reaction time in the 30-s chair stand test using the accelerometer sensor available in off-the-shelf mobile devices. In: ICT4AWE, pp 293–298
23. Schreiner K (1999) Night vision: infrared takes to the road. IEEE Comput Graph Appl 19(5):6–10
24. Sternberg S (1969) Memory-scanning: mental processes revealed by reaction-time experiments. Am Sci 57(4):421–457
25. Viola P, Jones MJ (2004) Robust real-time face detection. Int J Comput Vis 57(2):137–154. https://doi.org/10.1023/B:VISI.0000013087.49260.fb
26. Weller G, Schlag B (2007) Road user behavior model, deliverable D8 project RIPCORD-ISERET, 6th framework program of the European union
27. Xianglong S, Hu Z, Shumin F, Zhenning L (2018) Bus drivers mood states and reaction abilities at high temperatures. Transp Res Part F: Traffic Psychol Behav 59:436–444

Feature Extraction and Classification of Sensor Signals in Cars Based on a Modified Codebook Approach

Hawzhin Hozhabr Pour[1(✉)], Lukas Wegmeth[1], Alexander Kordes[1], Marcin Grzegorzek[2], and Roland Wismüller[1]

[1] Operating Systems and Distributed Systems, University of Siegen, Siegen, Germany
hawzhin.hozhabrpour@uni-siegen.de
[2] Medical Informatics, University of Lübeck, Lübeck, Germany
http://www.bs.informatik.uni-siegen.de, http://www.imi.uni-luebeck.de

Abstract. In this paper we indicatively address the problem of classifying sensor signals transmitted on the Controller Area Network (CAN bus) inside cars. The aim of this work is to solve the problem of dealing with large amounts of CAN bus data, finding out the semantics of the most important signals inside CAN buses in cars, independently of the cars' model and manufacture. For the purpose, we are aiming at classifying vehicle CAN bus data into basic sensor signals like speed, brake, steering-angle and throttle that assume a prior role in finding driving behaviors. In particular, our approach is starting with two feature extraction methods, a handcrafted feature selection (HFS) and an automated feature selection (AFS). Our automated method is based on a codebook approach, which extracts histogram-based features characterizing the shape of signals. Since some of the signal data are behaving very similar but with different semantics, in order to gain a better result, not only a one-vs-one classifier but also a multi-class classification algorithm has been applied. The result of our approach is a comparison between the applied methods, which validates the use of innovative AFS.

1 Introduction

Nowadays, due to the large demand of mobility, high increase in the number of vehicle crashes resulting in fatalities and injured drivers, driving style analysis, recognizing abnormal driving patterns and preventing aggressive inappropriate driving behaviors have been a major scientific concern. On the other side the contribution of these analysis plays an important role in the road safety, solving major traffic problems, ecological issues and these results are also beneficial to insurance and rental car companies. These firms have started incorporating the results of such analysis to provide their customers with "pay as you drive" insurance and rental plans, which decide the initial premium based on the driver's past history of crashes and violations [17]. We are working within the research

R. Burduk et al. (Eds.): CORES 2019, AISC 977, pp. 184–194, 2020.
https://doi.org/10.1007/978-3-030-19738-4_19

project LEICAR. For this project our goal is to accomplish an affordable, reliable, ecological and environmentally compatible mobility through car-sharing concepts, which is the main goal of LEICAR.

In order to analyze driving patterns, access to the car signals and their semantic should be provided. There are few approaches to acquire data for driving study analysis purposes. Most existing works on driving behavior monitoring and analyzing are based on smart phone sensor data. Because they are ubiquitous, small, and cheap [4,11]. Though smartphones have a myriad of sensors, only few signal modalities are useful for this purpose. For instance, when it comes to processing and interpreting certain driving events, a phone's gyroscope might be utilized, though motion direction is hardly accessible and reliable. Besides for recognizing aggressive driving behaviors, smart phone sensor signals do not acquire any information on the pressure of misusing vehicle's pedals like brake or throttle, which are the most common sign of aggressive driving acts. In addition, the data captured by a smartphone is strongly dependent on its placing inside the vehicle and its battery state.

In some works, like [1] computer vision based techniques are used to analyze driving style. Analyzing data through visual modalities is effective for drowsiness detection since the facial appearance, head position and eye activity of a driver are directly correlated to sleepiness and fatigue. However, driving style recognition based on visual data, not only requires much work in visual analysis but also not all vehicles provide camera in their modalities. Nonetheless, data privacy is another con of this category of data acquisition.

On the other, hand regarding driving analysis, the most robust, efficient and reliable data source is the vehicle bus. Nowadays CAN-base (Controller Area Network) data is the most available data source inside cars to study driving behaviors. The CAN bus is a standard message-based protocol in vehicles which allows automotive components and microcontrollers to communicate on a single or dual-wire networked data bus up to 1 Mbps. Like all signal modalities for capturing driving data it has own pros and cons. First of all, the amount of redundant data available in the CAN bus inside a car is enormous. For that reason, it needs much work to filter out the required signal data. Secondly every car company uses their own protocol of writing CAN bus data. Owing to car industry competition the access to the semantic of these protocols are mostly impossible.

In this context, LEICAR project provided access to a large variety of car sensor values from the CAN buses in cars. In this paper, we are aiming at solving the problem of dealing with large amount of CAN bus data, finding out the semantic of most important signals inside CAN buses in cars, independently of the cars' model and manufacturer. By classifying the most appropriate signals into speed, brake, steering-angle and throttle classes for the purpose of driving behavior analysis we will ease the procedure of data acquisition for the driving studies purposes (Fig. 1).

In order to achieve an accurate signal classification, we are proposing two methods of feature extraction which are going to be explained in Sect. 4. The

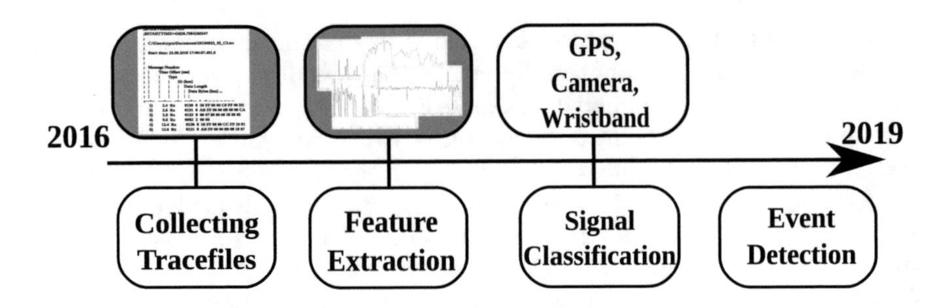

Fig. 1. Proposed methodology.

first one is a so-called handcrafted feature selection (HFS) based on our prior knowledge and statistical attributes of distinctive features of these signals; the second one is an innovative automated feature selection (AFS) based on the codebook approach. Eventually, we are classifying signals based on both feature selection methods in Sect. 5 and comparing the results in Sect. 6. Our future work plan and ideas will conclude this paper.

2 Related Work

Nowadays, the standardization of the CAN bus and the increase of the electronic component units in modern cars offer a large range of sensor data that make possible a more reliable and direct characterization of driving styles. Although, data collection and integration using multiple sensors itself is a major effort in many such studies involving accident analysis [6,16]. Ever since the first vehicle was produced featuring a CAN-based multiplex wiring system in 1988, different car companies have been using their own writing systems based on individual sensors applied in their cars. Considering a general sensor data acquisition extracted and decoded from CAN buses inside the cars, not much has been done in this area. Nevertheless in [12] the structure of a CAN bus architecture, through hardware and software, to the interface and to the application are described in detail.

Once the data are collected, the next step is to reduce these data and generate the epoch files [3], which will be analyzed by different study groups for further behavioral studies [7]. Therefore, this necessitates the data preprocessing step, which not only reduces the amount of redundant data but also generates information at the mid-level of abstraction.

Based on CAN bus data, the following papers have investigated their work on driving analysis. In [9], they propose the concept of driving DNA, looking at individual driving behavior as a global resultant of single easy-to-measure characteristics (genes) that describe some specific aspects of driving attitudes. More in general, the purpose of their work is to show that (1) it is possible, thanks to the CAN technology, to compute and visualize individual driving attitudes, in form of driving DNA; (2) different people have different driving behaviors (DNAs) and it is possible to visualize and compare them; (3) ultimately, driving DNA is useful for educative purposes in order to achieve a better driving style.

Based on CAN bus data, [20] proposes SafeDrive approach for detecting driving anomalies. SafeDrive detects abnormal driving behaviors by employing a state graph (SG) - a model based on both contextual relations between statuses of the same type of data, such as speed, at different timings and correctional relations between statuses of different types of data, such as the vehicle revolutions per minute (RPM) and gear position and then an online detection based on the comparison of the real-time driving data stream with the SG.

[14] investigated the need for automated drive analysis to reduce the time and effort needed for data reduction on driving analysis. The proposed driving analysis techniques are shown to synergistically fuse the data from a forward-looking camera, an in-vehicle CAN bus, an IMU, and a GPS to extract lane positions, the vehicle localization in the lane, the types of lane markers, lane-change events, the speed information, and the vehicle trajectory curvature information. These mid-level semantics were further fused in different ways to automatically determine a set of 23 semantics about the driving trace file.

With our work, we are aiming at easing the procedure of data acquisition for driving studies purposes, dealing with large amount of CAN bus data, finding out the semantic of the most important signals inside CAN buses in cars, independently of the car model or the year of manufacture. We were able to decode CAN bus protocols and filter out the most prior signals for driving behavior analysis. Therefore, for the future work, we have this opportunity to analyze driving traces independently of the car model and the manufacturer.

3 Data

In order to access CAN bus data inside cars, we used an OBD adapter directly connected to the car. Working within collaboration with INVERS company under LEICAR project we could collect enough ground-truth data to label the training data set. In this way we could access 13 car models' CAN-based data. In total, we had 30 driving traces consisting of four car signal types: speed, brake, steering-angle and throttle. The duration of driving traces is between 30 to 90 min and each signal, depending on the vehicle's CAN bus protocol, has different timestamps in milliseconds.

In order to have a synchronized data set to work on, we re-sampled the time-series into 100 ms sample-rate. Then we used two types of normalization. For handcrafted feature selection we normalized all signal time-series with min-max normalization and for the automated feature selection we used zero-mean, unit variance normalization technique. For the classification we are classifying our signal data into four classes of speed, brake, steering-angle and throttle.

4 Feature Extraction

Before starting to extract features based on the vehicle type and individual sensors, we should note that our data set varies to a big extent in range, sampling rate and type. To give a better view on our data set, in Fig. 2 four types of signals are shown. To tackle this problem in order to extract most efficient features, representing the characteristic of our signals, we needed to apply down-sampling and normalization on our data set. Therefore, we used a sampling rate of 100 ms and zero-mean unit-variance normalization. For the concept of feature extraction of CAN bus signals in cars, we are introducing and comparing two feature learning techniques: one relying on handcrafted features, and the other on automated feature extraction using a codebook approach. Both are explained in the following subsections.

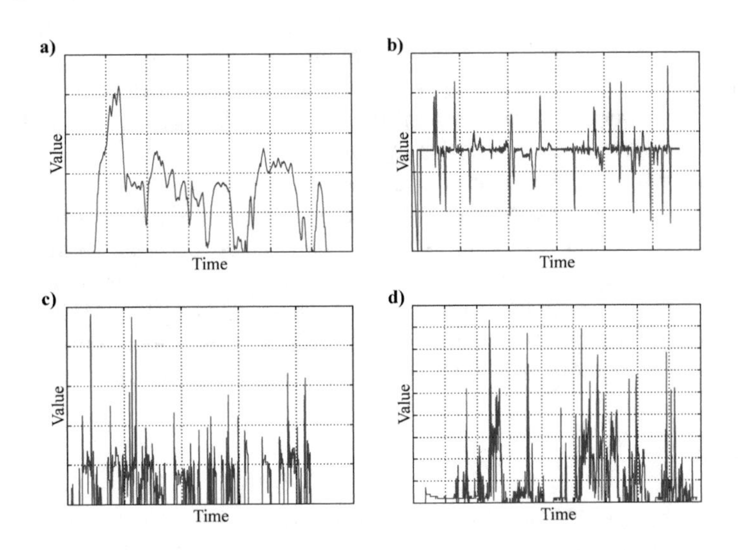

Fig. 2. (a) Speed, (b) Steering-angle, (c) Brake, (d) Throttle.

4.1 Handcrafted Feature Extraction

For the classification with manually selected features we computed the following 14 feature values:

- The signal's first value (i.e., its value at time 0) and the absolute difference of its first and last value, both relative to the signal's maximum and minimum value. These two features provide some coarse information about the overall shape of the signal.
- A histogram of the values that the signal assumes, using five equally sized bins between the signal's minimum and maximum value. This allows to distinguish between signals that show clear peaks (e.g., brake) and signals that assume all values with comparable probability (e.g., speed).
- The median of the signals' values, relative to its minimum and maximum value. This should allow to distinguish between, e.g., brake (just positive

peaks, i.e., median close to the minimum) and steering angle (peaks in both directions, i.e., median in the middle between minimum and maximum).
- The sum of the absolute differences of each signal value and the median value. This feature provides information about the variability of the signal.
- The signal's amplitude in five selected frequency bands (<0.2 Hz, 0.2–0.5 Hz, 0.5–1 Hz, 1–5 Hz, >5 Hz). The coarse spectrum helps to distinguish short peaks (e.g. brake) from smoother signals (e.g. throttle and speed).

4.2 AFS Based on the Codebook Approach

Unlike the common representations of feature extraction, which are mostly based on prior knowledge of the data, the codebook approach provides features obtained in an unsupervised way, without the need to consider previous knowledge of the data.

The codebook approach used in our study is based on [15] with some significant changes implemented. In our case, using the same approach does not lead to desired results in finding distinguishable features, since the similarity of sensor data collected from the CAN bus is very high. On the other hand, compared to the handcrafted features, this approach has a long computation time.

Figure 3 illustrates an overview of our modified codebook approach. In this overview the steps of the codebook-based feature extraction after having our data set down-sampled and normalized are as follows:

- Divide the signals (referred to as sequences) into overlapping sub-sequences (referred to as sub-windows).
- By using K-Means clustering based on [18], group all sub-windows into a number of clusters and set the cluster-centers in a so-called codewords set.
- By assigning each sub-window to the most similar cluster-center (using a soft assignment method) and representing them in a histogram with the frequency of codewords appeared in a signal, features of any specific signals is extracted. This histogram has the same dimension as the number of clusters and is considered as a point in a multi-dimensional space.

Some remarks should be considered in implementing the codebook approach. The window-size w for dividing each signal into sub-windows, the overlapping size l and the number of clusters should be chosen carefully. Small values of w and l can cause the approach to learn high frequency noise for codewords.

By a tiny w and l, the codewords are basically high frequency noises. In the last step of our approach, to assign each sub-window to the most similar codeword, we used a soft-assignment approach based on Gaussian kernel density estimation, from [19].

The frequency of each cluster center c_n in a given signal is represented as $F(c_n)$ formulated as follows:

$$F(c_n) = \frac{1}{S} \sum_{s=1}^{S} \frac{K_\sigma(D(x_s, c_n))}{\sum_{n'=1}^{N} K_\sigma(D(x_s, c_{n'}))} \tag{1}$$

where

$$K_\sigma(D(x_s, c_n)) = \frac{1}{\sqrt{2\pi}\sigma} \exp(\frac{-D(x_s, c_n)^2}{2\sigma^2}) \tag{2}$$

Here D(x_s,c_n) is the Euclidean distance. K_σ is the Gaussian kernel density estimation with smoothness parameter sigma.

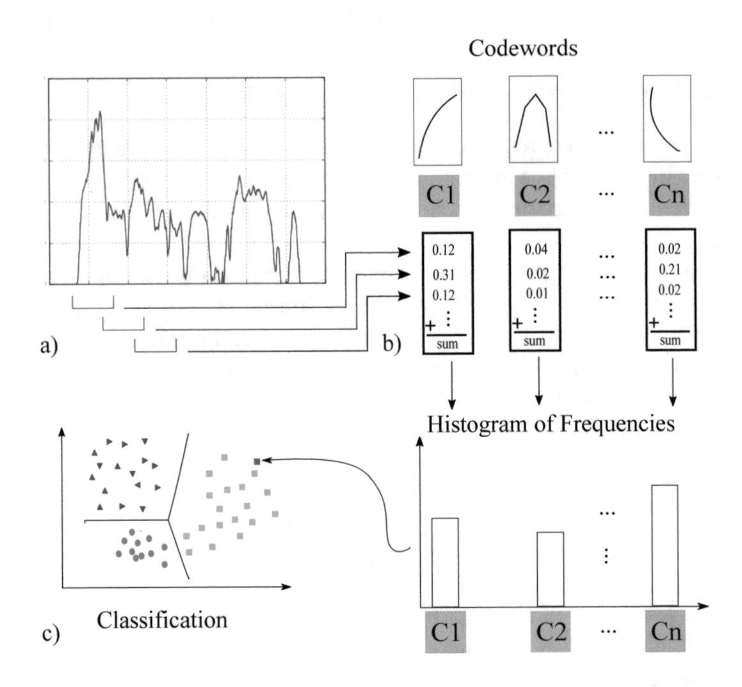

Fig. 3. An illustration of modified codebook approach, (a) Sub-window extraction, (b) Similarity computation between codewords and sub-windows, (c) Classification.

In our modified codebook approach the K-means algorithm based on paper [18] is extended by adding well-known radial basis function (rbf) kernel [8].

5 Classifier

Either using handcrafted or codebook-based feature extraction, the aim is to classify sensor data received from CAN bus system inside cars. Thus, like other machine learning methods, we need to divide our data into training and testing sets. Afterwards, we have applied SVM classifier based on both linear and RBF kernels from papers [8] and [2].

Since the sequences captured from throttle and brake signals are very alike, we have decided to use not only one-vs-one classifiers, but also multi-class learning approach based on [5] which makes a significant improvement in our results.

For all of the classification tasks, Scikit-learn library implemented in [13] is used in our procedure and finally, for tuning parameters in SVM classification, [10] has been applied to our application as well.

6 Results

Our data set is based on CAN bus traces from 30 rides with 13 different car models. We have an overall of 230 sequences of brake, speed, steering-angle and throttle signals. The main parameters to be considered in our approach are window-size, overlapping factor and number of clusters. In order to choose an appropriate window-size to capture not tiny noises, but a significant raise and/or fall in signal values, we have chosen a range from 5 to 15 samples per sub-window with a sampling-rate of 100 ms. The overlapping factor is chosen to be half of the window-size. For the number of clusters in K-Means clustering we set the range from 5 to 10 clusters.

Concerning the importance of choosing the above-mentioned parameters, in Fig. 4 two sets of codewords based on different window-size, overlapping factors and number of clusters are illustrated.

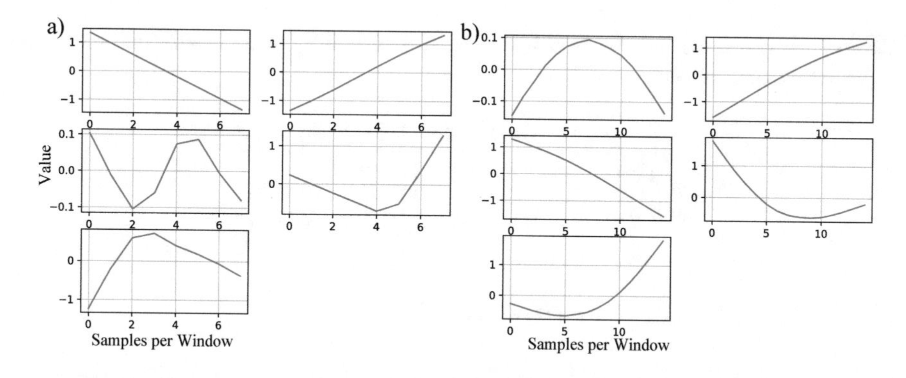

Fig. 4. (a) A set of codewords with window-size 8 and 5 number of clusters, (b) A set of codewords with window-size 15 and 5 number of clusters.

In Fig. 5, the classification score of both one-vs-one and one-vs-all classifiers, with RBF and linear kernels with different window-size is shown. Here we conclude that the bigger window-size with a multi-class classification achieves a significantly better result in our approach.

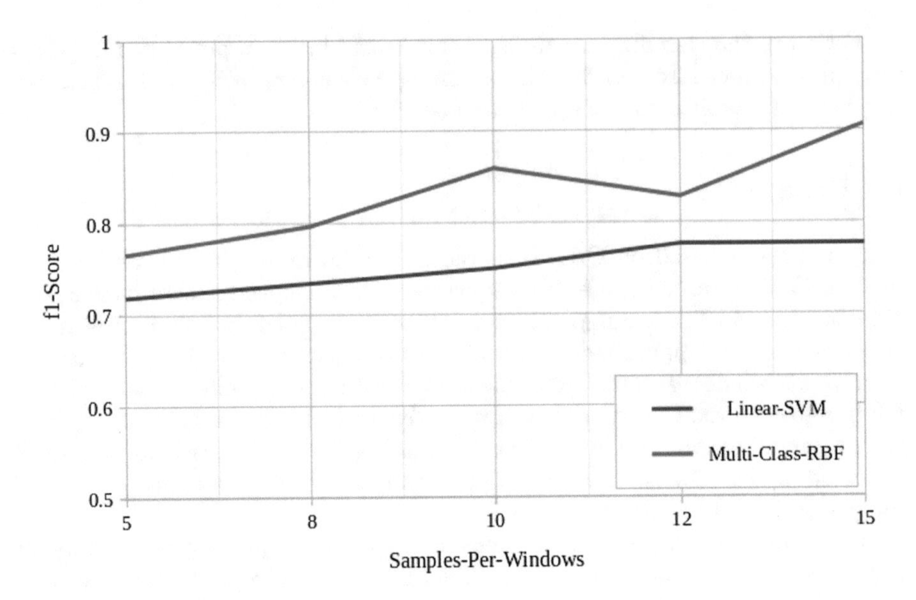

Fig. 5. Accuracies of different classifiers based on AFS with different number of samples per window.

Table 1. Performance comparison between AFS and HFS methods with different classifiers based on accuracy.

	Linear SVM score	RBF SVM score
AFS	0.776	0.906
HFS	0.928	0.934

7 Conclusion and Future Work

In Table 1 and Fig. 4 the result of different car signals' classification based on HFS and AFS are demonstrated. It is clear that in our case using handcrafted features gives a slightly better classification score by 93%. In spite of achieving higher classifier score based on HFS compared with the AFS, an efficient and accurate feature selection based on prior knowledge is labor and time intensive.

On the other hand, the classification result by choosing appropriate parameters based on AFS with the help of the codebook approach has gained a significant 90% score. Like nearly all automated approaches, the computation time of our modified codebook approach is noticeably high compared to the HFS.

After having most prior CAN bus signals classified independently of car models, our next future plan is to detect specific events related to unsafe driving situations in the signals. These events are time intervals containing abnormal changes in the signal values, which mostly occur during unsafe driving situations. We are investing our research not only on the car signals but also on the drivers' physical signals like heart rate and stress level in order to point out any irregular events during driving.

For this reason, a strong cross-correlated signal behavior in a specific time sub-sequences among drivers' physical signals and car signals, is predicted to be used in the learning phase of our machine learning process.

Apart from event detection within driving data records, learning the general driving style e.g., aggressive, ecological, normal and etc., is another goal to be addressed in our future work.

References

1. Al-Sultan S, Al-Bayatti AH, Zedan H (2013) Context-aware driver behavior detection system in intelligent transportation systems. IEEE Trans Veh Technol 62(9):4264–4275
2. Amari S, Wu S (1999) Improving support vector machine classifiers by modifying kernel functions. Neural Netw 12(6):783–789
3. Brach A (2012) SHRP 2 naturalistic driving study what we have learned so far. In: Third international symposium on naturalistic driving research, pp 1–12
4. Chen Z, Yu J, Zhu Y, Chen Y, Li M (2015) D^3: abnormal driving behaviors detection and identification using smartphone sensors. In: 2015 12th annual IEEE international conference on sensing, communication, and networking (SECON). IEEE, pp 524–532
5. Dietterich TG, Bakiri G (1995) Solving multiclass learning problems via error-correcting output codes. J Artif Intell Res 2:263–286
6. Dingus TA, Klauer SG, Neale VL, Petersen A, Lee SE, Sudweeks J, Perez MA, Hankey J, Ramsey D, Gupta S, et al (2006) The 100-car naturalistic driving study. Phase 2: results of the 100-car field experiment. Technical report, United States. Department of Transportation. National Highway Traffic Safety Administration
7. Ebe K (2012) Development of distracted driving database. In: Third international symposium on naturalistic driving research, pp 1–17
8. Filippone M, Camastra F, Masulli F, Rovetta S (2008) A survey of kernel and spectral methods for clustering. Pattern Recogn 41(1):176–190
9. Fugiglando U, Santi P, Milardo S, Abida K, Ratti C (2017) Characterizing the "driver DNA" through can bus data analysis. In: Proceedings of the 2Nd ACM international workshop on smart, autonomous, and connected vehicular systems and services, CarSys 2017. ACM, New York, pp 37–41. https://doi.org/10.1145/3131944.3133939
10. Golub GH, Heath M, Wahba G (1979) Generalized cross-validation as a method for choosing a good ridge parameter. Technometrics 21(2):215–223
11. Júnior JF, Carvalho E, Ferreira BV, de Souza C, Suhara Y, Pentland A, Pessin G (2017) Driver behavior profiling: an investigation with different smartphone sensors and machine learning. PLoS One 12(4):e0174959
12. Lawrenz W (2011) CAN Controller area network: grundlagen, design, anwendungen, testtechnik. VDE Verlag
13. Pedregosa F, Varoquaux G, Gramfort A, Michel V, Thirion B, Grisel O, Blondel M, Prettenhofer P, Weiss R, Dubourg V et al (2011) Scikit-learn: machine learning in Python. J Mach Learn Res 12(Oct):2825–2830
14. Satzoda RK, Trivedi MM (2015) Drive analysis using vehicle dynamics and vision-based lane semantics. IEEE Trans Intell Transp Syst 16(1):9–18

15. Shirahama K, Köping L, Grzegorzek M (2016) Codebook approach for sensor-based human activity recognition. In: Proceedings of the 2016 ACM international joint conference on pervasive and ubiquitous computing: adjunct. ACM
16. Takeda K, Hansen JH, Boyraz P, Malta L, Miyajima C, Abut H (2011) International large-scale vehicle corpora for research on driver behavior on the road. IEEE Trans Intell Transp Syst 12(4):1609–1623
17. Tefft BC, Horrey WJ, Yang C (2018) Reducing crash risk and improving traffic safety: research on driver behavior and performance. ITE J 88(8):30–34
18. Teknomo K (2006) K-means clustering tutorial. Medicine 100(4):3
19. Terrell GR, Scott DW (1992) Variable kernel density estimation. Ann Stat 20:1236–1265
20. Zhang M, Chen C, Wo T, Xie T, Bhuiyan MZA, Lin X (2017) Safedrive: online driving anomaly detection from large-scale vehicle data. IEEE Trans Ind Inf 13(4):2087–2096

A Deep Learning Approach to Recognition of the Atmospheric Circulation Regimes

Victor Luferov and Ekaterina Fedotova(✉)

Global Energy Problems Laboratory, Moscow Power Engineering Institute, Krasnokazarmennaya 14, 111250 Moscow, Russia
lyferov@yandex.ru,
e.v.kasilova@gmail.com
http://github.com/Luferov
http://github.com/ekatef

Abstract. A supervised deep learning approach has been developed to automate recognition of the large-scale atmospheric circulation patterns. The approach is based on an application of the convolution neural network. The reanalysis meteorological fields were used as an input dataset. The dataset was labeled according to the circulation calendar constructed using the subjective Dzerdzeewski classification. One of the key issues for the success of the modeling was found to be a proper data preprocessing. The developed approach has demonstrated an accuracy compared with the more detailed regional classification methods that currently are being widely used for automated synoptic analysis.

Keywords: Convolutional Neural Networks · Climate data · Classification · Image analysis

1 Introduction

The global climate change remains among the most persistent global problems during the last forty years. The impact of the climate change on the everyday life and economics is still evident around the world and will be only more pronounced during the whole twenty-first century at least. A urgent need for mitigation and adaptation measures determines nowadays motivation of the climate scientists towards the better understanding of the climate processes. The investigation of the interconnections between the local and the global climate processes is on the cutting-edge of the modern climate science. That is exactly the link which determines the impact of the climate change and variability on the climate of each specific region.

A plenty of the concepts and tools has been developed to encounter regional manifestation of the global climate processes. There are recent findings introducing the theoretical approaches [11,16], computational techniques [22] and

R. Burduk et al. (Eds.): CORES 2019, AISC 977, pp. 195–204, 2020.
https://doi.org/10.1007/978-3-030-19738-4_20

numerous empirical methods [3,4,10] addressing this problem. However, manual expert analysis for analysis of the climate processes still cannot be replaced by any formal method due to an incredibly complicated nature of the climate system.

An invaluable advantage of the expert analysis is a human ability to see the system as the whole and recognize poorly-formalized associations between its elements. That is why meteorologists around the world often over-perform automated weather forecasts. The aim of our work was to get advantage of the expert knowledge for the development of an automated method intended to recognize the large-scale atmospheric patterns. Such atmospheric circulation regimes are the very regional manifestations of the global climate processes and are of key importance to comprehend the impacts of the climate change on the everyday life.

2 Survey of the Field

2.1 Machine Learning in Meteorology

The most common application of the deep learning methods in the climate science is processing of the remote sensing data. Deep learning models give very good results for a wide range of applications, among which are e.g. analysis of the surface cover and land use [14], investigation of the ocean dynamics [1], study of clouds [28] and hazards detection [12].

Applications of the deep learning methods to other classes of the climate problems are rather few. A certain deficit of the meteorological data has surely contributed to this gap. The case is that one of the fundamental requirements of the meteorology is the homogeneity of the measured time-series. Thus, operation of the meteorological stations should adhere as close as possible to the principles developed for meteorological observations more than a century ago before any sight of the data-driven concept. The meteorological stations of the official observation systems nowadays are taking measurements just a few times daily. At the same time, the deep learning methods demonstrate good forecast skills using meteorological data monitored on much finer time resolution (minutes to hours) [20,26], which makes rather difficult a direct application of the deep learning methods to process the meteorological time series.

A much more successful area of the machine learning utilization to the meteorological parameters is the pattern recognition. The principal component analysis (PCA) has thirty years ago revolutionized understanding of the climate system allowing to identify the global-scale persistent atmospheric patterns. An idea of the atmospheric teleconnections has emerged [2] and became one of the most powerful tools of the modern climate science to quantify the natural variability of the global climate. The weak point of the teleconnections concept (as for each statistical concept) is a need to reveal the physical mechanisms which are responsible for transfer of this variability to the regional level [4,9,10,22].

The frameworks of the atmospheric circulation classification methods seem to have a great explanatory potential which may be used to resolve the mentioned issue. Application of diverse classification methods was demonstrated by a number of the recent works both for the current [3,4,9,10] and the projected climate change [22].

2.2 Atmospheric Circulation Classifications

The main elements of the atmospheric circulation structure are cyclones and anticyclones. Both are quite compact structures with a characteristic fingerprint in the fields of the meteorological parameters. This fact makes different clustering methods a natural choice for construction of an atmospheric circulation classification scheme. There are examples of successful applications of the hierarchical [17], the fuzzy-rule [19,27] and the regression-based [6] clustering methods to classify the atmospheric circulation regimes.

In general, some dozens of the automated classification approaches are available nowadays [15]. They all are focused on the local to regional scale synoptic processes and have contributed significantly to understanding of the regional climate dynamics. However, the amount of the accounted by them local details necessarily mean a sacrifice with the links to the global processes. That may easily lead to an inconsistency between the different regional classification methods. The studies of one of the largest modern circulation archive have resulted in a conclusion that different regional classification schemes often disagree in trends for a certain circulation type even if this type has a very similar flow configuration in all the schemes [5,15].

The climate system seems to be too complicated and too non-linear to be well described with quite straightforward clustering methods. A solution of this issue may be obtained when using a global-scale classification approaches. The thing is that the large-scale structures of the atmospheric turbulence are much more robust as compared with the local and regional ones [11].

The global-scale classifications of the atmospheric circulation have been first introduced almost a century ago. However, thus far these classifications are being used as purely subjective approaches which are based solely on an expert opinion. This subjectivity ultimately decreases an explanatory potential of the global-scale circulation approaches. We were determined to develop a deep-learning model to catch the main features of the global atmospheric classification which could replace an expert. To authors' best knowledge this is the first attempt to apply the modern automated recognition methods to automate the global-scale classifications.

3 Modeling

The purpose of our work was to implement a supervised learning approach for recognition of the global-scale atmospheric circulation patterns. The convolution neural network (CNN) has been used as a computational method. The

CNN seems to be is a very reasonable approach for pattern recognition of the atmosphere circulation field due to its excellent ability of to extract features from the noisy images which has been proved by a wide number of the recent works [18,28].

3.1 Data

We have used a calendar of the hemispheric circulation regimes labeled according to the Dzerdzeewski classification [13,25]. The calender comprises the time span between 01.01.1899 and 31.12.2017 and was developed using the results of a manual analysis of the complex synoptic maps[1].

The Dzerdzeewski classification is based on an extraction of the stable elements of the atmospheric turbulence structure in the Northern Hemisphere [13]. This scheme has been initially developed in 1930th to increase an accuracy of the weather forecasts provided for the Arctic expeditions. Extensive improvements has been done during 1960-70th to account for the main synoptic processes on the hemispheric scale. The latest version has been shown to demonstrate an outstanding performance for the study of the regional manifestations of the global climate processes [13,25].

The classification is based on an analysis of the atmospheric conditions over the Northern Hemisphere (NH). Five groups of the circulation processes are considered (Table 1). The examples of the global atmospheric pressure fields corresponding to the each circulation type are shown in the Fig. 1.

Table 1. Parameters of the labeled dataset of the atmospheric circulation regimes

Circulation type	Number of days	Atmospheric processes in the NH
Zonal	3015	Developed westerlies, anticyclone in the polar area, no blocking events
Zonal disturbed	10435	Anticyclone in the polar area, one northern blocking process
Meridional North	23482	Anticyclone in the polar area, 2–4 northern blocking processes
Meridional South	5927	Cyclone in the polar area
Transitional	604	Not included in any other group

The latest 2c version of the NOAA's 20th Century Reanalysis [8] was used as a model input. At the moment, that is the largest available data set on the atmospheric circulation. It contains the global fields of the main meteorological parameters reconstructed by the observation data using techniques of the climate computational fluid dynamics. The fields of the sea level pressure were selected to be used as a base for model training, as the atmospheric pressure is known to be the most closely associated with the circulation regime.

[1] Available via atmospheric-circulation.ru.

3.2 Model Architecture

The model was developed using the convolution neural network (CNN) and implemented with the Keras machine learning library [7]. In fact, we process a meteorological field in the same way as images are processed in the computer vision problems, which allows us to use a wide range of experience and approaches established in this field.

The CNNs are similar to the conventional artificial neural networks with the direct signal propagation, but have a different architecture in terms of communication between the neurons in the adjacent layers [18]. The advantage of the CNNs in comparison with the classical architecture is the specific spatial-local correspondence between the network elements. The neurons are connected only with a small area of the previous layer, so called receptive field. This is achieved by the use of the filter in a convolutional layer (CL) over all the surfaces of the output volume of the previous layer. A feature map is generated as an output after each convolution operation.

The model structure was based on the VGG16 architecture [21] which was fitted for solving our problem. The model consists of a number of the CLs with the max-pooling layers (PL) following each of them. The shallow CL is responsible for detection of the image robust patterns while the deeper ones are aimed to look for finer details [24]. The output of each roll-up operation is a feature map. The final full-connections layers use the soft-max activation function to produces a classification output. The number of the convolution layers and the number of the filters in each of them were selected to ensure a satisfactory model accuracy (Table 2) [23]. The selected model configuration consists of two CL of rather modest resolution (Fig. 2).

Table 2. Optimization of the CNN structure

Optimized parameter	Considered range
Number of layers	1; 2; 3
Number of filters in layer	32; 64; 128; 264
Filter size	2

The model takes a matrix of the meteorological parameter as an input and returns a sequence of the label predictions as outputs, that represent the probability of each circulation type.

3.3 Model Training

The stochastic gradient descent algorithm was used by model training. One of the model performance issues was a selection of the batch size that is a size of a subsample being processed during a single algorithm step. The batch size increase leads to a growth of the memory consumption during the model training,

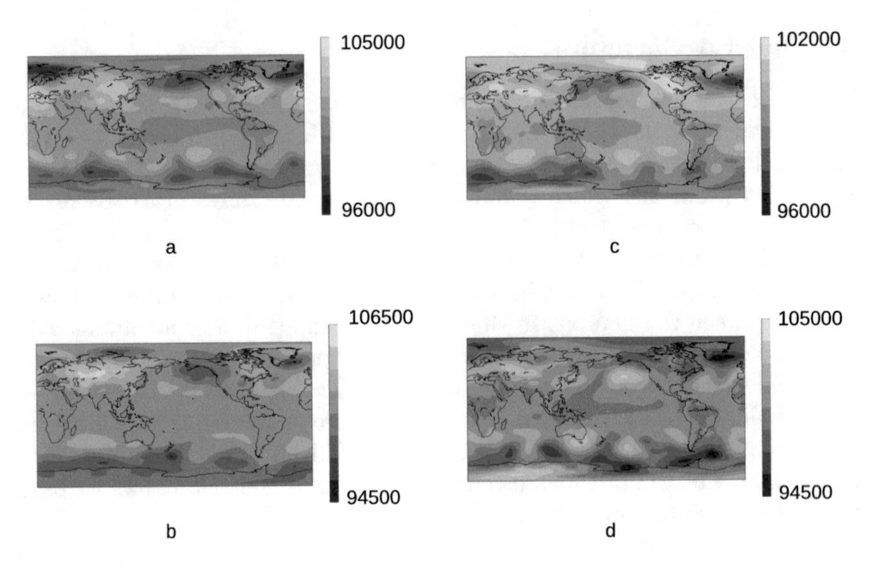

Fig. 1. Examples of the seasonal sea-level pressure fields [Pa] representing different circulation groups: a – zonal, b – zonal disturbed, c – meridional north, d – meridional south. Authors' calculations using [8] data.

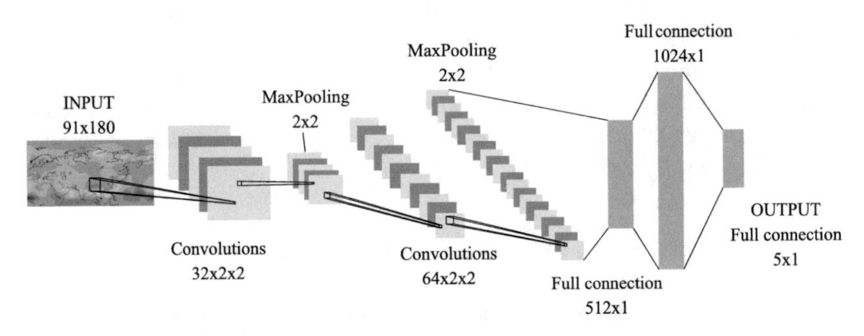

Fig. 2. The architecture of the developed CNN model

but decreases a number of the iterations needed. A batch size 32 was taken for modeling after some testing and according to our image processing experience.

The whole available dataset of the reanalysis pressure fields was divided into a training set and a test set. A training set was constructed by a random selection of an equal number samples for each circulation group.

The data preprocessing step was found to be of the key importance for a success of the modeling. The case is, the atmospheric pressure has some seasonal patterns (Fig. 3) which are overlaid with the daily-scale atmospheric dynamics that we are analyzing. A removal of these patterns was found to be critical for the success of the model. The used preprocessing procedure was similar to a usual calculation of the climate parameter anomalies:

$$p_{anom}(x_i, y_i, t_i) = p(x_i, y_i, t_i) - \frac{\sum_{i=1}^{i=N} p(x_i, y_i, t_i)}{N}. \quad (1)$$

where $p_{anom}(x_i, y_i, t_i)$ is a departure of the parameter value from its seasonal average in a location with the x_i and y_i coordinates for the time moment t_i, p(x_i, y_i, t_i) is the parameter value, N is the number of the days in a certain season.

The visualized accuracy metric was defined as a share of the circulation patterns which were classified correctly by the model (Fig. 4). The learning dynamics testifies the convergence of the model. Somewhat surprisingly, the developed approach demonstrates accuracy that is higher than the regional classification combined with the clustering methods [27] and is compatible with the CNN-model applied to a much more comprehensive remote sensing datasets [28].

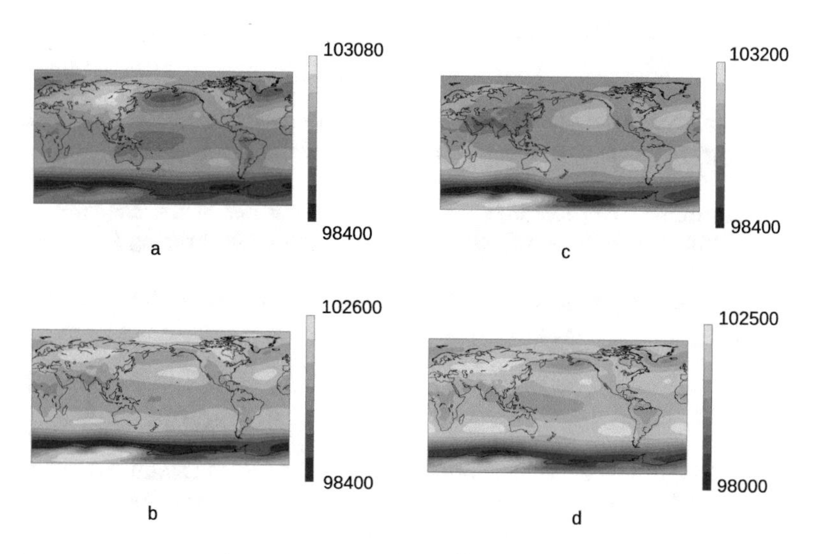

Fig. 3. The multi-annual seasonal pressure patterns [Pa] for the whole considered time span (01.01.1899 to 31.12.2014): a – winter, b – spring c – summer, d – autumn. Authors' calculations using [8] data

4 Results and Discussion

The proposed approach allows to take an advantage of the expert understanding of the synoptic processes when using an automated deep learning method. The developed CNN-based model justifies application of the subjective classification approach to analysis of the atmospheric circulation processes as an explanatory tool for the climate problems. The CNN seems to suit very well for recognition of the large-scale atmospheric turbulence structures in virtue of the CNN's ability to extract features from noisy images. At the same time, we have shown that the removal of the seasonal patterns is essential for the modeling success. That

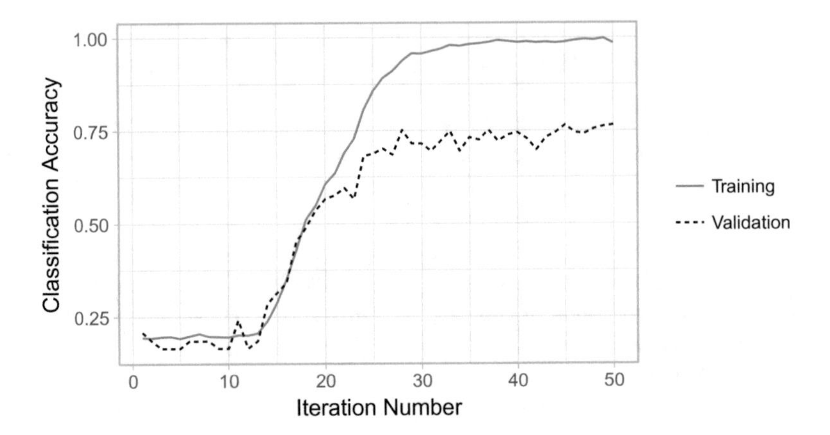

Fig. 4. Dynamics of the model training and validation

means that preprocessing should be considered very carefully when applying the deep learning methods to the atmospheric circulation problems.

Acknowledgments. Authors highly appreciate discussions of the meteorologic processes with Prof. N.K. Kononova that have been highly encouraging for a presented work.

The work was supported by the Russian Science Foundation (project no. 18-79-10255).

References

1. Ashkezari M, Hill C, Follett C, Forget G, Follows M (2016) Oceanic eddy detection and lifetime forecast using machine learning methods. Geophys Res Lett 43:12234–12241. https://doi.org/10.1002/2016GL071269
2. Barnston AG, Livezey RE (1987) Classifications, seasonality, and persistence of low-frequency atmospheric circulation patterns. Mon Weather Rev 115:1083–1126. https://doi.org/10.1175/1520-0493(1987)115⟨1083:CSAPOL⟩2.0.CO;2
3. Bartoszek K (2017) The main characteristics of atmospheric circulation over East-Central Europe from 1871 to 2010. Meteorol Atmos Phys 129:113–129. https://doi.org/10.1007/s00703-016-0455-z
4. Bednorz E, Czernecki B, Tomczyk A, Polrolniczak M (2018) If not NAO then what?—regional circulation patterns governing summer air temperatures in Poland. Theoret Appl Climatol 1–13 (in press). https://doi.org/10.1007/s00704-018-2562-x
5. Cahynova M, Huth R (2016) Atmospheric circulation influence on climatic trends in Europe: an analysis of circulation type classifications from the COST733 catalogue. Int J Climatol 36:2743–2760. https://doi.org/10.1002/joc.4003
6. Cannon A (2012) Regression-guided clustering: a semisupervised method for circulation-to-environment synoptic classification. J Appl Meteorol Climatol 51:185–190. https://doi.org/10.1175/JAMC-D-11-0155.1
7. Chollet F et al (2015) Keras. https://keras.io

8. Compo G, Whitaker J, Sardeshmukh P, Matsui N, Allan R, Yin X, Gleason B, Vose R, Rutledge G, Bessemoulin P, BroNnimann S, Brunet M, Crouthamel R, Grant A, Groisman P, Jones P, Kruk M, Kruger A, Marshall G, Maugeri M, Mok H, Nordli O, Ross T, Trigo R, Wang X, Woodruff S, Worley S (2011) The twentieth century reanalysis project. Q J Roy Meteorol Soc 137:1–28. https://doi.org/10.1002/qj.776
9. Fleig A, Tallaksen L, James P, Hisdal H, Stahl K (2015) Attribution of European precipitation and temperature trends to changes in synoptic circulation. Hydrol Earth Syst Sci 19:3093–3107. https://doi.org/10.5194/hess-19-3093-2015
10. Gerlitz L, Steirou E, Schneider C, Moron V, Vorogushyn S, Merz B (2018) Variability of the cold season climate in Central Asia. Part I: weather types and their tropical and extratropical drivers. J Clim 31:7185–7207. https://doi.org/10.1175/JCLI-D-17-0715.1
11. Hannachi A, Straus D, Franzke C, Corti S, Woollings T (2017) Low-frequency nonlinearity and regime behavior in the Northern Hemisphere extratropical atmosphere, 55:199–234. https://doi.org/10.1002/2015RG000509
12. Kadavi P, Lee C, Lee S (2018) Application of ensemble-based machine learning models to landslide susceptibility mapping. Remote Sens 10:1252–1269. https://doi.org/10.3390/rs10081252
13. Kononova N (2018) Type of global atmospheric circulation: results of monitoring and observations for 1899–2017 yy. Fundam Pract Climatol 3:108–123. https://doi.org/10.21513/2410-8758-2018-3-108-123 (in Russian)
14. Korycki L, Krawczyk B (2018) Combining active learning and self-labeling for data stream mining. In: Kurzynski M, Wozniak M, Burduk R (eds) Proceedings of the 10th International Conference on Computer Recognition Systems, CORES 2017. Advances in Intelligent Systems and Computing, vol 578. Springer, Cham. https://doi.org/10.1007/978-3-319-59162-9_50
15. Kucerova M, Beck C, Philipp A, Huth R (2017) Trends in frequency and persistence of atmospheric circulation types over Europe derived from a multitude of classifications. Int J Climatol 37:2502–2521. https://doi.org/10.1002/joc.4861
16. Mammadov A, Rajabov R, Hasanova N (2018) Causes of periodical rainfall distribution and long-term forecast of precipitation for Lankaran, Azerbaijan. Meteorol Hydrol Water Manag 6(2):1–5. https://doi.org/10.26491/mhwm/89763
17. Nojarov P (2017) Genetic climatic regionalization of the Balkan Peninsula using cluster analysis. J Geogr Sci 27(1):43–61. https://doi.org/10.1007/s11442-017-1363-y
18. Park K, Kim D (2018) Accelerating image classification using feature map similarity in convolutional neural networks. Appl Sci 9(1):108. https://doi.org/10.3390/app9010108
19. Pringle J, Stretch D, Bardossy A (2014) Automated classification of the atmospheric circulation patterns that drive regional wave climates. Nat Hazards Earth Syst Sci 14:2145–2155. https://doi.org/10.5194/nhess-14-2145-2014
20. Rodrigues E, Gomes A, Gaspar A, Henggeler Antunes C (2018) Estimation of renewable energy and built environment-related variables using neural networks - a review. Renew Sustain Energy Rev 94:959–988. https://doi.org/10.1016/j.rser.2018.05.060
21. Simonyan K, Zisserman A (2014) Very deep convolutional networks for large-scale image recognition, CoRR. arXiv:1409.1556
22. Stryhal J, Huth R (2018) Classifications of winter atmospheric circulation patterns: validation of CMIP5 GCMs over Europe and the North Atlantic. Clim Dyn 7:1–24. https://doi.org/10.1007/s00382-018-4344-7

23. Szegedy C, Liu W, Jia Y, Sermanet P, Reed S, Anguelov D, Erhan D, Vanhoucke V, Rabinovich A (2015) Going deeper with convolutions. In: Proceedings of the IEEE conference on computer vision and pattern recognition, Boston, MA, USA. https://doi.org/10.1109/CVPR.2015.7298594
24. Cohen TS, Welling M (2016) Group equivariant convolutional networks. arXiv: 1602.07576
25. Tarabukina L, Kononova N, Kozlov V, Innokentiev D, Shafer Y (2018) Analysis of atmospheric circulation condition during severe thunderstorms in Yakutia in 2009–2016. In: Miloch WJ, Vodinchar GM, Shevtsov BM (eds) 62 Proceedings of the 10th 9th International Conference Solar-Terrestrial Relations and Physics of Earthquake Precursors, STRPEP 2018. EDP Science. https://doi.org/10.1051/e3sconf/20186201001
26. Wan J, Ren G, Liu J, Hu Q, Yu D (2016) Ultra-short-term wind speed prediction based on multi-scale predictability analysis. Cluster Comput 19:741–755. https://doi.org/10.1007/s10586-016-0554-0
27. Woyciechowska J, Ustrnul Z (2011) Fuzzy logic circulation types based on the Osuchowska-Klein classification system created for Poland. Theoret Appl Climatol. https://doi.org/10.1007/s00704-010-0366-8
28. Zhang J, Liu P, Zhang F, Song Q (2018) CloudNet: ground-based cloud classification with deep convolutional neural network. Geophys Res Lett 45:1–8. https://doi.org/10.1029/2018GL077787

Deep Learning for Object Tracking in 360 Degree Videos

Ahmad Delforouzi[1(✉)], David Holighaus[1], and Marcin Grzegorzek[2]

[1] Pattern Recognition Group, University of Siegen, Siegen, Germany
ahmdel@gmail.com
[2] Institute of Medical Informatics, University of Lübeck, Lübeck, Germany
http://www.pr.informatik.uni-siegen.de,
https://www.imi.uni-luebeck.de/de

Abstract. Object tracking is used to locate the position of an object over a period of time using the association of an object of interest over consecutive frames. In the last years, several methods were proposed to track objects in rectangular videos. This paper presents is an object tracking method within 360-degree videos using a state-of-the-art tracking-by-detection paradigm. This method uses two trackers namely Kalman filter and Lucas-Kanade methods to handle challenges in the 360-degree videos. The proposed method uses a deep learning object detector for extraction of prior information of the object of interest. The information is then used, to track the object of interest using a combination of the two trackers of the Kalman filter and Lucas-Kanade. The experiments show that this combination improves the tracker stability.

Keywords: Object tracking · 360-degree videos · Lucas-Kanade · YOLO

1 Introduction

Many years have been devoted to research on object tracking. Object trackers often find a place in surveillance systems to ensure security and to protect individuals at crowded locations. The 360-degree camera is a type of camera which provides a high-resolution 360-degree view. By a wider field of view, the required number of installed cameras can be significantly decreased in comparison with the case of commonly used normal cameras. When the conventional security systems are replaced by 360-degree cameras, there will be a significant reduction of hardware costs, software license and maintenance costs. The wider field of view for conventional cameras increase their applications as well. For instance, the application of the new sensor of 360-degree cameras is growing in various areas such as robotics, car traffic control, and intelligent surveillance system [8]. The difference of the researched methods is the format or the size of the field of view of the video data set. In methods presented in [5,15], authors have focused on multi-object trackers in low- or medium-density crowd sequences.

R. Burduk et al. (Eds.): CORES 2019, AISC 977, pp. 205–213, 2020.
https://doi.org/10.1007/978-3-030-19738-4_21

These methods are able to track instead of hundreds of people, tens of people and some of them use data association. Since the scenes contain fewer objects, detectors yield better results than in the previous case involving hundreds of people in a scene. The data association problem is formulated as finding the maximum-weight independent set of a graph. To create the graph of tracklets, an object detector is first applied to all images of a video, in order to extract object information [5]. This information is then used, to create tracklets for each object by consideration of two consecutive frames. The paper proposes a method to associate the tracklets to the corresponding object by finding the maximum-weight independent set of the graph. The approach of Liu *et al.* [10] has used the tracking-by-detection paradigm to track multiple objects in a dataset, generated by a Spherical Panoramic Camera. To successfully apply an object detector and reach better detection results, the images must first be rendered in a rectangular format and distorted regions at the top and bottom are removed by cropping. The extracted object information is then provided to the Kalman Filter for tracking multiple objects and the Hungarian algorithm solves the data association. Real time face and object tracking as a component of a perceptual user interface was proposed by Bradski *et al.* [4] In [14], the authors propose a tracking approach to track moving objects by combining the Kalman Filter and the Camshift method. In this method, the Kalman Filter is used for predicting a linear system and Gaussian distributed system. The Camshift method is used for a general purpose related to moving object tracking. In the method presented in [6], authors use a PTZ (pan, tilt, zoom) camera for data collection and is able to detect and track multiple objects. Another tracking-by-detection approach proposed by Chen *et al.* [7] uses to track multiple people in dense and complex scenarios even when the amount of objects is unknown. In this paper, an object tracking method on 360-degree video using a tracking-by-detection technique is presented. The videos images are first unwrapped and they are fed to two object trackers. To track the desired object, a combination of two object tracking methods namely the Kalman filter and the Lucas-Kanade is proposed. This paper starts with the current Section of introduction which shows some of the related work. The proposed method is explained in Sect. 2. Section 3 shows the experiments of the proposed method and finally, Sect. 4 concludes the paper.

2 Proposed Object Trackers

The proposed tracker first rectifies 360-degree images, then apply YOLO detector on the images. In the next step, it applies Kalman filter on the results of the YOLO and finally, the proposed tracker applies the Lucas-Kanade tracker on the rectified images and combined it with the YOLO-Kalman tracker. The Kalman filter needs the previously extracted information to determine the object state. The Lucas-Kanade method, however, requires no prior information for object tracking. In the following, first the procedure of the two mentioned tracker will be explained and then the combination method will be presented. The aforementioned methods are explained in the following subsections.

2.1 Image Transformation

The outcome of 360-degree cameras are sequences of polar images with a 360° field of view as shown in Fig. 1. Most of the object detection systems fail to detect objects directly in polar images. This is related to the fact, that they were trained on rectangular image datasets. In 360° images, the pixels are arranged in a circle, where the top side of a scene is mapped to the center of the polar image. The circular orientation of the pixels causes objects to be distorted. To overcome the object detection difficulties related to the polar object representation, polar image pixels are transformed to a rectangular arrangement. The rectangular version of the polar image can be obtained by converting polar coordinates to Cartesian coordinates. For this aim, an unwrapping approach [9] is used. First, a polar coordinate system is created by shifting the image origin from the top left towards the center. In the next step, corresponding polar coordinates which describe the points are created. In this case, the reference point is the image center. Image pixels are described using an angle, θ and a distance p between origin and the pixels in the image. The distance and angle are given by the following equations:

$$p = sqrt(x'^2 + y'^2) \tag{1}$$

$$\theta = arctan(x'/y') \tag{2}$$

The Cartesian coordinates can be obtained according to the following equation:

$$x = \alpha * (R_{max} - p), y = \beta * (p * \theta) \tag{3}$$

Where R_{max} is the polar image radius, α and β are scaling factors which affects the outcoming resolution. Note that α and β are in the range of [0,1], where 1 means that the output image has the maximum resolution. After unwrapping step, the out-come image has a rectangular form, however, it is still deformed. To overcome the deformation Eq. 4 is used to disperse the pixels along the rectangle.

$$x = \alpha * (R_{max} - p), y = \beta * (R_{max} * \theta) \tag{4}$$

2.2 Object Detection

To overcome challenges like occlusion or disappearance of the desired object in the data-set, the proposed system uses the state-of-the-art tracking-by-detection paradigm. Obscuration and disappearances of objects often arise in more complex videos, due to the movement of multiple objects or a non-stationary camera. The recognition of objects in a video is done by a state-of-the-art object detector YOLO You Only Look Once [12] which was proposed by Redmon *et al.* YOLO refers to the evaluation for the prediction of bounding boxes and class probabilities over the entire image. YOLO uses a single neural network to predict objects bounding boxes and class probabilities in one evaluation whereas previous object detectors are applied to several locations on the image to evaluate the appearance of an object. Thus, YOLO is able to predict objects presence and

Fig. 1. If YOLO is applied directly to the polar image, it can not extract any object information, due to the objects orientation.

location significantly fast. Another benefit of the single Convolutional Neural Network is the avoidance of a complex pipeline. In addition, YOLO takes a look at the entire image once and rates it globally to decide whether the object is a part of the background. For detection, the network uses pre-trained features to simultaneously predict all bounding boxes in one image. YOLO is composed of 24 convolutional layers followed by 2 fully connected layers. A convolutional layer basically convolves an image in a kernel to form a filtered image. The kernel is a matrix which contains specific object features. By convolving the original image with the kernel, the output is a filtered image which contains statements about where a possible object can be and how confident the system is. Fully connected layers are then used for voting whether an object exists at a certain position. The pure tracker (i.e., Kalman filter) used in the proposed system can not generate any information about the object position and size by themselves. Therefore, the tracker needs the information extracted by YOLO in order to allow the later tracking process. The visual output of YOLO is shown in Fig. 2.

2.3 Kalman Filter

Kalman Filter is a linear state estimator, that receives measurements over a time period to describe the object state. It is an *optimal recursive data processing algorithm* [11]. The proposed system uses the Kalman Filter to predict the object future location using its history of location and speed. The Kalman filter uses

Fig. 2. An example of the YOLO visual output with correct labeling.

the location information extracted by YOLO. The proposed system uses a six-dimensional state vector, with the following values: x, y, w, h, v_x, v_y where x and y are the position coordinates, v_x, v_y are the object's velocity in the horizontal and vertical direction and w, h is the object size. There is also a four-dimensional measurement vector containing $\{i_x, i_y, i_w, i_h\}$ which is only used when the object of interest provides information about the true nature, during the update phase.

2.4 Lucas-Kanade

Lucas-Kanade is a well-known tracker which relies on the optical flow of a video sequence [13]. It assigns a two-dimensional movement vector to a point(s) of interest to track it from one frame to the next. In order to calculate this vector first, unique points are found. These points are edges or corners within the image. Lucas-Kande considers a small region around the desired point and assumes that the flow within this region is constant. This method works best when object moves slowly. We used this method as a baseline for our comparisons. This method can not be compared to the deep learning based object detector namely YOLO. It is object tracker not object detector. We will combine this tracker to a combination of YOLO-Kalman tracker and we will achieve better results as it will be shown in experiments section.

2.5 Combination of Kalman Filter and Lucas-Kanade

Both of the object tracking methods of Kalman filter and Lucas-Kanade have their advantages and disadvantages. The Kalman filter has very good results, but only if the YOLO detector regularly detects the object of interest and provides associated information. That is why the results strongly depend on YOLO. Thus, if the detector misses the object, the overall tracking performance falls down. The quality of the image influences the results of the Lucas-Kanade method. If the points of interest over several images can't be clearly identified, Lucas-Kanade might either track a wrong point or lose the object of interest. The different combinations of Lucas-Kanade and Kalman filter were tested. In the first scenario, Lucas-Kanade detects the object by continuously calculating the optical flow and the Kalman filter runs in the background and predicts the object's size and position. Since the position has already been determined by the Lucas-Kanade, only the size information of the filter is needed from the Kalman filter. In the second scenario, YOLO-Kalman tracks the object. When this tracker fails the object, the Lucas-kanade revive the lost object and tracks the object. If none of the trackers can follow the object, it is considered to be occluded or left the scene.

3 Results

The proposed system was tested on 14 different 360-degree videos [1]. Each video has a diverse number of frames containing different desired objects to track. The objects of interest are humans, cars, and aircraft. All videos have been captured with 30 frames per second. Except for the NYC video, they were filmed by a non-stationary camera. All polar images have a resolution of 752×752 pixels, except for rob which has a resolution of 720×720. To generate individual images from the videos, FFmpeg [2] was used. After rectification and cropping of the polar images, a resolution of 720×250 remained. The videos mainly differ in the number of displayed objects and the complexity of the movements. To evaluate the different object tracking methods, we used Matlab for manual selection of ground truth. Suppose *(x, y)* are coordinates and *width* and *height* are the size of the object of intersect. The acquired ground truth is then used to calculate the precision, recall, and F-measure of the individual methods. Precision indicates the proportion of correctly tracked objects, while recall is the fraction of the tracked objects over the total amount of relevant objects. F-measure is the combination of both, precision and recall. In order to calculate the precision, recall and F-measure, *TP (true positive)*, *FP (false positive)* and *FN (false negative)* are needed. Positive in this context means, that the tracker was able to track something. The keyword true or false gives information about whether the object really existed or not. In other words, a true positive exists if the tracker keeps track of the desired object (with the true object's location and size), whenever the object is seen.

In the case of a *false positive*, the tracker follows another object that does not exist or is not the object of interest. *False negative* occurs if the tracker does not track the object of interest. Suppose the position and size of the ground truth and the tracked bounding box are overlapped, the area of the overlapping region is calculated and compared with the total area. If the overlapped area is $>50\%$, the object is tracked correctly and it is a *TP* and *FP* occurs when the value is $<50\%$. If the tracker does not provide bounding box information, even though there is a ground truth, then there is a *FN*. Based on the previously defined values, precision $P = TP/(TP+FP)$, recall $R = TP/(TP+FN)$, and f-measure $F = 2*(P*R)/(P+R)$ are calculated. The trackers were implemented in C++ using the image processing library called OpenCV [3]. The result calculation and ground truth determination were made in Matlab. According to the information of Fig. 5, the Kalman filter outperforms the Lucas-Kanade method. The main reason is that the YOLO object detector provides good results in terms of the information extraction of the desired object. Due to the fact that the Kalman filter does not rely on object information continuously and estimates the future object state even without measured values, it provides good results in case of the detector failure. Even if an object is occluded over a short period, the filter predicts the future position based on the linear model and can update the measurements as soon as the object becomes visible again. A visual output of the filter is shown in Fig. 3. Figure 4 also shows a case which YOLO misses the object but, the Kalman filter is still tracking the object of interest.

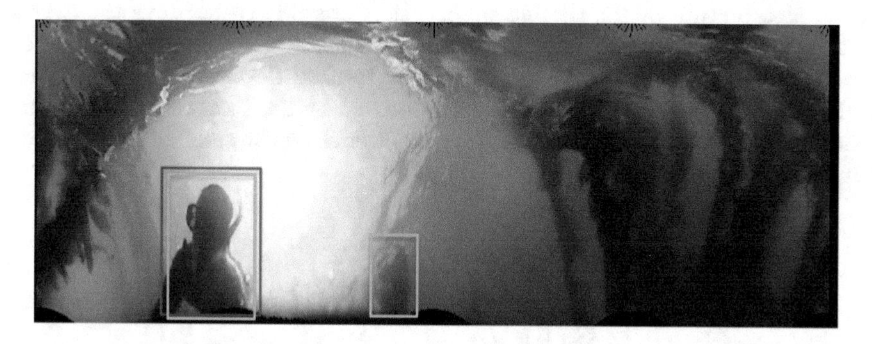

Fig. 3. The visual output of the Kalman filter. The green rectangles show the YOLO output and red rectangles are the Kalman filter estimation.

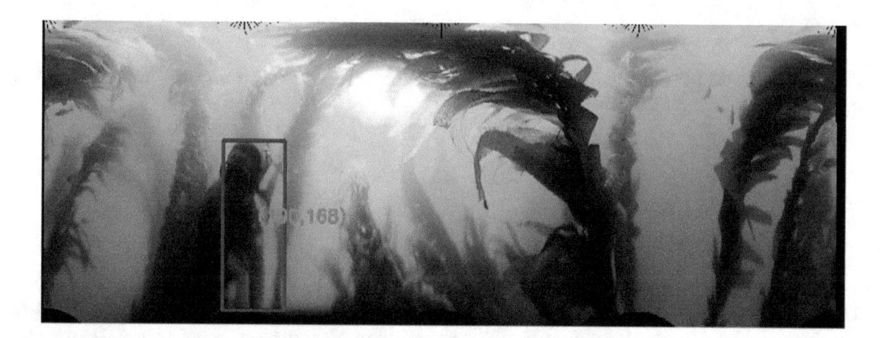

Fig. 4. YOLO failed to extract object position. Nevertheless, the position of the object estimated with a slight deviation using Kalman filter.

The other reason is the surface texture of the objects is uniform and it is not possible for the Lucas-Kanade method to find an interesting point within the object. In addition, the objects may have large leaps in position within a few frames due to fast movements and non-stationary cameras. These problems can be seen in the videos *Snowboarding, Shredding* and *Harley*. However, there are also videos in which the Lucas-Kanade method performs slightly better than the Kalman filter, For instance, in the case of the *Balloon* video, YOLO can not detect the object category *balloon* and therefore, does not give any object information to the Kalman filter. For the *Park* and *Rob* videos, YOLO does not give continues information about the object of interest, to get good results from. The Lucas-Kanade method, on the other hand, is able to track the object due to the constant movement of the desired objects. Both methods have their own advantages and disadvantages. The Kalman filter is bound to the results of YOLO, while the Lucas-Kanade method is attached to the movement speed and the surface texture of the desired object. It is possible to improve the average performance of both methods by combining them. The combination made it possible for the desired object to be better tracked despite poor results from YOLO, as sees in *Rob* video, which got improved by the Lucas-Kanade method. Videos with good results using the Kalman filter show the same or better results using the new combination.

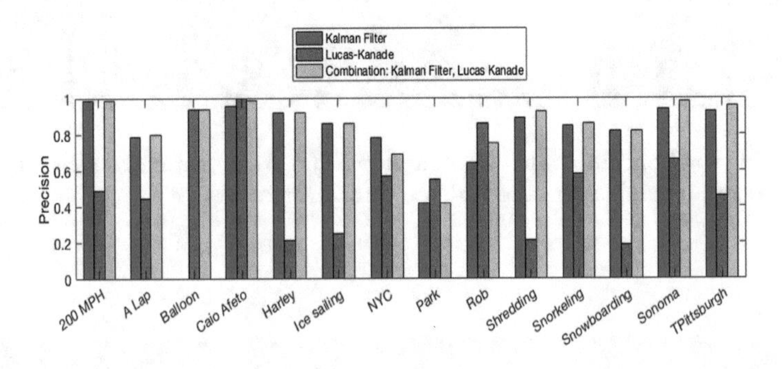

Fig. 5. Results of precision measures for Kalman filter, Lucas-Kanade and the combination of both.

4 Conclusion

In this paper, a combination of the Kalman filter and the Lucas-Kanade method was proposed. This method tracks a single object in a 360-degree video consisting of polar images. Using the state-of-the-art tracking-by-detection paradigm, prior object information was extracted from the images using a deep learning object detection method. By combining both methods namely the Kalman filter and the Lucas-Kanade method, the benefits were combined in favor of the tracking results.

The results show that the proposed system outperforms the individual Lucas-Kanade method and the Kalman filter and is able to overcome challenges like occlusion or disappearance of the desired object. The proposed tracker is still bound to the information from YOLO but can compensate quickly for a possible lack. For the future work, the combination of the proposed tracker with KCF (Kernelized Correlation Filters) tracker is planned.

References

1. https://360fly.com/videos . Accessed 30 Sept 2010
2. http://ffmpeg.org/. Accessed Aug 2018
3. Bradski G (2000) The OpenCV library. Dr. Dobb's J Softw Tools 25:120–125
4. Bradski GR (1998) Real time face and object tracking as a component of a perceptual user interface. In: Proceedings fourth IEEE workshop on applications of computer vision, WACV 1998 (Cat. No. 98EX201), pp 214–219. https://doi.org/10.1109/ACV.1998.732882
5. Brendel W, Amer M, Todorovic S (2011) Multiobject tracking as maximum weight independent set. In: 2011 IEEE conference on computer vision and pattern recognition (CVPR). IEEE, pp 1273–1280
6. Chavda HK, Dhamecha M (2017) Moving object tracking using PTZ camera in video surveillance system. In: 2017 international conference on energy, communication, data analytics and soft computing (ICECDS). IEEE, pp 263–266
7. Chen L, Wang W, Panin G, Knoll A (2015) Hierarchical grid-based multi-people tracking-by-detection with global optimization. IEEE Trans Image Process 24(11):4197–4212
8. Delforouzi A, Tabatabaei SAH, Shirahama K, Grzegorzek M (2018) A polar model for fast object tracking in 360-degree camera images. Multimed Tools Appl 2018:1–23
9. El Kadmiri O, Masmoudi L (2011) An omnidirectional image unwrapping approach. In: 2011 international conference on multimedia computing and systems (ICMCS). IEEE, pp 1–4
10. Liu KC, Shen YT, Chen LG (2018) Simple online and realtime tracking with spherical panoramic camera. In: 2018 IEEE international conference on consumer electronics (ICCE). IEEE, pp 1–6
11. Maybeck PS (1982) Stochastic models, estimation, and control, vol 3. Academic Press, Cambridge
12. Redmon J, Farhadi A (2018) Yolov3: an incremental improvement. CoRR abs/1804.02767. http://arxiv.org/abs/1804.02767
13. Rojas R (2010) Lucas-kanade in a nutshell. Freie Universit at Berlinn, Department of Computer Science, Technical Report
14. Swalaganata G, Affriyenni Y Moving object tracking using hybrid method
15. Wen L, Li W, Yan J, Lei Z, Yi D, Li SZ (2014) Multiple target tracking based on undirected hierarchical relation hypergraph. In: Proceedings of the IEEE conference on computer vision and pattern recognition, pp 1282–1289

Texture Features for the Detection of Playback Attacks: Towards a Robust Solution

Maciej Smiatacz(✉)

Faculty of Electronics, Telecommunications and Informatics,
Gdańsk University of Technology,
Narutowicza 11/12, 80-233 Gdańsk, Poland
slowhand@eti.pg.edu.pl

Abstract. This paper describes the new version of a method that is capable of protecting automatic speaker verification (ASV) systems from playback attacks. The presented approach uses computer vision techniques, such as the texture feature extraction based on Local Ternary Patterns (LTP), to identify spoofed recordings. Our goal is to make the algorithm independent from the contents of the training set as much as possible; we look for the descriptors that would allow the method to detect attacks performed in an environment entirely different from the training one and with the use of the equipment that differs considerably from the devices that captured the training samples. The final form of our method, based on the previously presented proof of concept, performs significantly better than the reference Textrogram algorithm.

Keywords: Playback detection · Antispoof algorithms · Biometrics

1 Introduction

After decades of development automatic speaker verification (ASV) [1] can now be considered a mature biometric technology. The accuracy, effectiveness and usability of modern, commercially available ASV systems reached the level that allows for their application in real-world, mass-market solutions such as e-banking. However, despite the very good performance of the state-of-the-art speaker verification systems [2], their vulnerability to spoofing attempts discourages many potential users. The very basic type of attack, based on the playback of previously recorded speech, does not require any extraordinary skills, specialized knowledge nor sophisticated equipment, yet still can be extremely dangerous. Although application of the challenge-response scenario might seem to be a natural solution to the problem, such approach not only eliminates the protection related to the knowledge of a speaker-specific passphrase but could also deteriorate the performance of the core biometric identification module [3]. To address these difficulties, one of the speech recognition companies created a two

R. Burduk et al. (Eds.): CORES 2019, AISC 977, pp. 214–223, 2020.
https://doi.org/10.1007/978-3-030-19738-4_22

stage algorithm [4]: (1) the user is prompted to utter the first challenge phrase selected from the enrollment vocabulary so that the features extracted from the voice signal could be compared with the voice print stored in the database; (2) in case of positive identity verification, the speaker is prompted to utter the second challenge phrase, selected from a vocabulary different than the enrollment to obtain the second voice signal. Speech recognition is necessary to confirm that the speaker indeed uttered the correct phrase. Finally, the algorithm must check if both challenge phrases were uttered by the same person. This solution, although complicated, is probably effective, as the impostor would have to obtain a huge collection of utterances and then combine them in real time to compose the second challenge phrase. Still, a method capable of detecting the playback attempts just by means of sound analysis could simplify the process significantly.

Additionally, the problem of playback detection itself is certainly an interesting challenge. As a consequence, there have been many attempts to solve it. The early countermeasures tried to analyze the noise patterns [5] and other effects related to the extended distance between the speaker and the recording device, or to detect the pop noise, a distortion that appears when human breath reaches a microphone [6]. Another group of methods uses a simple assumption that if the database already contains a sample that resembles the new recording too closely, we probably experience a playback attack. One of the first attempts to implement this idea in practice was reported in [7]. The authors were well aware of the challenges that stemmed from the fact that during subsequent transmissions noise and variations in channel characteristics would work in the intruder's favor. Therefore, they proposed a set of features created from the indices of the voiced frames of the speech waveform and the FFT bin numbers of the n highest peaks of the magnitude spectrum for each voiced frame. Such feature sets can be visualized as *peakmaps* and compared using the value of normalized cross-correlation function. This method was applied again to playback attack detection by the authors of [8], who treated the pattern representing the constellation of connected peaks as a description of the voice signal.

Recently, typical visual features (i.e. Local Binary Patterns, describing the image's texture) applied to spectrograms proved to be relatively effective in detecting playback attacks, even though they express only the micro-differences between images (which can be difficult to interpret). In our previous work [9] we have proposed significant improvements to the latest solution based on this approach, the so-called Textrogram method [10]. Yet, we have clearly stated that the problem remains open and decided to continue our research in this area. Below we present the results achieved by improved algorithms during the newly designed experiments.

Our goal is still the same: to find a set of texture features, which are detectable in the visual representation of every playback recording, but can rarely be found in the equivalent representation of genuine speech signals. The use of deep learning techniques, currently viewed as the optimal solution to similar problems in other areas, could be challenging in this case, since it is very difficult to create a representative and large enough collection of all possible playback attack examples. Smaller or less diverse datasets, such as those used during experiments reported in [11], can easily lead to overfitting. The authors of the recent study [12] state that CNN-based playback attack detectors still lack the ability to generalize across databases and are unable to detect unknown attacks well.

It is fairly simple to detect playback attacks, even by means of straightforward algorithms, when the training set contains examples of spoofing attempts similar to those included in the testing set (e.g. when the equipment used to record the samples from both sets was comparable). On the other hand, finding common features for playback attacks performed in different ways appears to be challenging. In our opinion, the feature engineering supported by the use of external knowledge, that we propose instead of deep learning, can be practically effective in the case of the playback attack detection, although we are aware that the attempts reported in this paper cannot be treated as the ultimate solution to the problem.

2 Databases, Protocols and Data-Related Problems

While our goal is to find the general features, which are common for as many playback attacks as possible, there is a high risk that the classifier will treat unimportant patterns, present only in a particular dataset, as those which define the difference between authentic and spoofed recordings. In order to prevent this, we will train our algorithms on databases that differ substantially from the testing sets. Additionally, we will try to keep the size of the training set small. This approach is better than the traditional cross validation, because the latter does not guarantee that the testing set does not contain the same *types* of attacks that were used during the training (although the particular *samples* from both sets indeed differ). We will try to address the most pessimistic variant (the situation that one would most probably have to deal with in practice), i.e. the case when the particular attack is not similar to any of the training samples, for example because some unique microphone-speaker combination had been used by the impostor.

As we had focused on searching for the most general features and decided to rely on support vector machine as a classifier, the problem of correct interpretation of experimental results, obtained for different definitions of features, became particularly important. Competing solutions are labeled by the values of the two coefficients: the false acceptance rate (FAR), which represents the percentage of successful attacks, and the false rejection rate (FRR), representing the probability that a legitimate user would be treated as an attacker. Typically, the equal error rate (EER) is used for final comparisons. However, it is often difficult to reach EER when non-linear SVM is used. Therefore, we decided to keep the SVM parameters constant most of the time and treated security (low FAR) as more important than usability (low FRR).

Courtesy of the authors of [8], we were allowed to use the corpus that they prepared (we will call it the AGH set). It contains recordings of 21 (10 male and 11 female) participants of the experiment who uttered the same Polish phrase several times. There are 81 authentic recordings and 4 sets ($4 \times 81 = 324$ samples) of their spoofed versions. The quality of the authentic samples is probably too good to reflect real-life conditions.

The VL-Bio database, prepared by VoiceLab.AI, the company providing commercial speech recognition solutions, contains 160 authentic recordings of 5 persons (4 male, 1 female), captured with 4 different smartphones, together with 480 examples of illegal access attempts (the detailed description of the VL-Bio corpus can be found in [9]). In this setup neither specialized microphones nor high-quality speakers were used, so, intuitively, the attacks should be easier to detect than in the case of the AGH database. On the other hand, the genuine signals are highly variable because of the different characteristics of microphones built into the mobile devices. Additionally, languages and passphrases varied from attack to attack.

In order to study the generalization capabilities of antispoof features, four training subsets were extracted from the VL-Bio database. All of them contain voices of only two persons – Male1 and Female1. The configuration of recording and replaying devices, however, varies from set to set:

- T1 – genuine authorization attempts were collected using Apple smartphone, ZTE and Samsung served as eavesdropping devices,
- T2 – LG used by the legitimate user, Samsung and Apple by the impostor,
- T3 – Samsung used by the legitimate user, LG and ZTE by the impostor,
- T4 – ZTE used by the legitimate user, LG and Samsung by the impostor.

Notice, that each of the above training subsets contains recordings of one Polish and one English passphrase only. As a result, during the tests most of the passphrases will be different from those selected for training. This way we try to reduce the risk of overfitting to patterns present only in particular passphrases.

3 Basic Algorithm

The basic version of our playback attack detection method, which we call the LTPgram algorithm, has been described in [9]. The LTPgram was inspired by the Textrogram method [10] and follows the same general approach: voice samples are converted to images, which are afterwards described with the use of typical texture features. However, we introduced some important modifications to the original solution (Fig. 1):

1. we used regular spectrograms instead of mel-scaled cepstrograms,
2. we replaced Local Binary Patterns with Local Ternary Patterns (LTP) [13],
3. instead of calculating the histogram of texture descriptors for each line of the image separately, we used 11 uniformly distributed horizontal regions (each was 23 pixels high).

We have already shown [9], that the LTPgram significantly outperforms the Textrogram during the 10-fold cross-validation tests, reaching FAR = 0.31% and FRR = 5% on the AGH dataset, as well as FAR = 0.42% and FRR = 0% on the

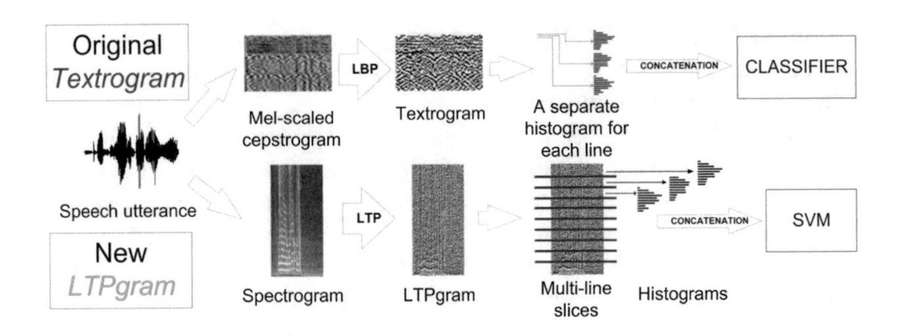

Fig. 1. The LTPgram algorithm compared to the original Textrogram method

VL database. The results might look like a proof of a major success of our modifications; unfortunately, the new round of tests did not confirm that. This time we used the T1 – T4 subsets as the training data, and the whole VL-Bio database as the testing set. We also trained the system using the whole AGH database and tested it on the entire VL-Bio dataset. Then the roles of the AGH and VL-Bio databases were exchanged. Moreover, we used two additional datasets: Tx, containing all the training samples collected for Male1 and Female1 (taken from the VL-Bio database), and AGH+, i.e. the AGH set supplemented by the *authentic* recordings of Male1 and Female1. Table 1 shows the new results. Initially, we tried to adjust the SVM parameters to obtain equal error rates (EER), however, in most of the cases it appeared to be impossible (OpenCV implementation of the SVM was used). Therefore, since we opted for good generalization, the γ parameter was set to 0.1 (a low value meaning that the influence of a single support vector reaches far), and ν to 0.4 (a high value indicating large tolerance to data variations, typically used when we suspect that the training set is missing some representative examples).

Table 1. Results of the tests of LTPgram method for training sets substantially different from the testing data

Experiment	Training set	Testing set	FRR (%)	FAR (%)
1	Tx	VL-Bio	0	0
2		AGH	22.22	35.19
3	VL		23.46	35.19
4	AGH	VL-Bio	100	0
5	T1		7.74	9.38
6	T2		48.39	0
7	T3		1.94	25.21
8	T4		0	31.25
9	AGH+		2.58	76.04

The results of experiment no. 1 were very good, despite the fact that the training set contained utterances collected from only two persons. However, all possible combinations of recording/playback devices, as well as all passphrases were represented in the training set, and this guaranteed the success. When the attacks were substantially different from the training samples (different devices, passphrases and voices were used), as in the case of cross-database experiments 2 and 4, the results appeared to be poor. In case no. 3 the number of voices was increased (5 persons instead of 2), but the error rates remained unchanged. The radical failure was observed when the training set contained only high-quality samples (case 4): all the authentic recordings from the testing set were rejected, probably because the genuine samples included in the training set were unrealistic (too good). The AGH set was created to prove the effectiveness of the method [8] that is supposed to detect attacks which are TOO SIMILAR to previous legal authorization attempts; consequently, it does not contain low-quality attacks, as they would be too different from the stored samples, and thus erroneously treated as authentic recordings. On the other hand, to increase the task difficulty, the positive samples are of high quality, which makes them similar to the attacks – and TO EACH OTHER. As a result, a system trained on the AGH database is very sensitive and treats even the smallest deviations from the training recordings as attacks. To compensate for this deficiency of positive samples (which, by the way, can be collected easily), we created the AGH+ set including some *authentic* recordings from the VL-Bio database. This changed the situation completely (case 9): false rejection rate dropped to the acceptable level of 2.58%, but the false acceptance rate raised from 0 to 76%. Such a result indicates that the most representative features of attacks from the AGH+ database were not present in the large part of the attacks included in the VL-Bio database. In other words, the features that we defined were not general enough, not common for diverse types of attacks. In further experiments we have paid particular attention to the cross-database tests 2 and 9, as these were the most challenging cases and results obtained for them are especially important. Still, we must remember about the high variance of the results – experiments 5–8 show how extremely different outcomes can be obtained when feature definitions are the same, but training set contents varies.

4 Optimization

In the first step of optimization we applied the silence removal algorithm together with a number of image filtering operations. Table 2 shows the results. If we assume that false acceptances are more dangerous than false rejections, the results may be interpreted as indicating that silence removal (SR) deteriorates the performance of the algorithm (test no. 2) or, in the best case, does not influence it (test no. 9). Silence elimination, however, should not be treated as a typical optimization: it is more a necessity that reduces the probability of overfitting to unimportant features, although even the silence recordings, when played back, may contain some patterns that distinguish them from the authentic signals. Nevertheless, in further experiments only cropped samples were used. The histogram equalization appeared to be a very effective operation. The application of other image processing methods, however, such as gamma correction or median filtering, did not improve the results.

Table 2. Results of the spectrogram preprocessing experiments; SR – silence removal, HE – histogram equalization

	Test case 2		Test case 9	
Preprocessing	FRR(%)	FAR (%)	FRR (%)	FAR (%)
SR	0	67.28	2.58	74.38
SR+HE	0	45.99	8.38	30.21

Table 3 shows that more complex algorithms, using longer (12 bit) LTP codes, extracted from larger neighborhoods, do not provide any improvements; extending the LTP histograms (by taking the non-uniform patterns into account) appears to be counterproductive. On the other hand, simpler solutions (based on 4-bit codes) significantly reduce the FAR level.

Table 3. Results of the LTP code length optimization experiments

		Test case 2		Test case 9	
Code length	Non-uniform patterns	FRR (%)	FAR (%)	FRR (%)	FAR (%)
8	+	0	41.67	6.45	32.50
12	–	0.69	58.33	4.52	48.96
	+	0	50.93	0	63.93
4	–	6.25	11.11	11.16	23.33
	+	6.25	11.28	12.90	21.25

In the next experiment we varied the number of horizontal stripes defining the regions for which the individual LTP histograms were calculated. Recall that the height of each spectrogram is always equal to 257, so more stripes means a longer feature vector, containing more histograms; in other words it means that the more detailed description of the spectrogram is processed. No matter if we used 4 or 8-bit LTP codes, the best results (i.e. the optimal balance between FAR and FRR) were obtained when the histograms were extracted from 13 regions (stripes of 19 pixels height). Table 4 lists the error rates. Interestingly, they were particularly low for test case 2 if 4-bit codes were applied (for test case 9 the changes of code length did not influence the results considerably). By using short 4-bit codes we are able to train the system on a small set of straightforward attack attempts, and then apply it effectively for the detection of much more sophisticated attacks. Surprisingly, in the reverse case, when the training set contains refined playback samples, indistinguishable from the genuine ones for human listeners, the simple attacks are difficult to detect. It seems that the we are dealing with a feature space in which the distribution of training samples is different from what human intuition suggests: the primitive attacks are closer to each other than the advanced ones. This is beneficial because it means that a small training set of simple playback recordings is sufficient to create a system that can cope even with sophisticated attacks.

Table 4. Results obtained for the optimal number of horizontal regions (13 stripes). Silence removal and histogram equalization was applied

	Test case 2		Test case 9	
Code length	FRR (%)	FAR (%)	FRR (%)	FAR (%)
8	0	25.00	8.38	17.92
4	1.23	4.94	10.32	19.38

In the subsequent experiment we made an attempt to optimize the distribution of the regions in which the LTP histograms are calculated. To facilitate this we used the well-known simulated annealing procedure. The number of horizontal stripes remained constant (13, as the previous experiment indicated) but their borders were allowed to move freely; the only constraint was that the minimum height of the region, i.e. 3 pixels. Note, that this way we only tried to separate the *training* samples as much as possible. The optimization was performed separately for 8-bit and 4-bit codes, individually for training set Tx (test case 2) and AGH+ (test case 9). It appeared that the results depended slightly on the training set but not on the length of the code. In both cases the algorithm significantly increased the resolution in the low frequency range in order to achieve better separation of the training samples (in addition, the highest frequency regions were selected, especially when samples coming from the AGH+ set were used for training). Unfortunately, the direct application of the obtained distributions lead to considerable overfitting. However, following the recommendation provided by the optimization process we reduced the number of regions from 13 to 9 by merging the stripes which represent the middle frequencies (regions 3 to 6). This way we achieved slightly better results (Table 5).

We have also tried to optimize the weights assigned to the horizontal regions. To calculate the weight of a given stripe the features of all training samples were extracted on the basis of this particular stripe only, and then the distance between the class means was computed. The higher the distance, the larger was the weight. In both test cases, 2 and 9, the distribution of weights appeared to be similar. Again, the particular importance of stripes representing the lowest frequencies was evident. To prevent overfitting, the following simplified set of weights was adopted: 0.1-0.1-0.1-0.1-0.1-0.2-0.2-0.8-1. The application of weights improved the results obtained for test case 2 (Table 5). Although the false rejection rate increased in the case of test no. 9, significant reduction of false acceptance level (from 16.25 to 2.71%) confirmed the positive influence of region weighting on the overall performance of the system.

Table 5. Results after merging middle frequency regions (3 to 6), and after application of the region weights

	Test case 2		Test case 9	
Optimization	FRR (%)	FAR (%)	FRR (%)	FAR (%)
Stripe merging	0	5.56	6.45	16.25
Weights	0	3.70	11.61	2.71

Despite numerous trials, neither the optimization of LTP thresholds assigned to subsequent stripes nor the selection of LTP codes provided positive results. This was probably caused by the fact that this time, contrary to our earlier experiments [9], we were using short 4-bit codes instead of standard 8-bit representation. Therefore, only 30 features per stripe were calculated, and we have possibly reached the limits of optimization: the shortening of the codes itself guaranteed even better results than the time-consuming techniques that we had used previously. The results may be interpreted in several ways. As we see, the modifications that we introduced provide measurable benefits: we have managed to increase the security level by significantly lowering the FAR value in both most difficult cases (from 35.19% to 3.70% and from 76.04% to 2.71%). At the same time in case no. 2 the percentage of false rejections was reduced from 22.22% to 0%, although in case no. 9 FRR increased from 2.58% to 11.61%. Certainly, such error rates are not acceptable for systems used in the real world; however, we have to consider the fact that the results pertain to a series of particularly demanding experiments: the amount of training samples was limited and, more importantly, they were collected in the conditions that differed considerably from the setup in which the testing attacks were prepared. In case no. 9 the FAR lowered by 75 pp at the cost of a moderate FRR increase, which seems to be a favorable tradeoff.

5 Conclusions

Unsurprisingly, we have not reached the ultimate goal of our research: we cannot state that we have defined the features that are present in every spectrogram representing a playback attack and can never be found in any genuine recording. The classifier still needs a proper, not accidental, set of examples to be able to delimit the border between the two classes. Nevertheless, thanks to the proposed modifications, the task seems to be a bit simpler, as the results obtained in case no. 2 indicate. It is now enough to collect examples of straightforward attacks performed with the use of a few smartphones and just two voices (one male and one female) to train the classifier that will be able to detect sophisticated attacks which involve high quality equipment and are aimed at persons whose voices are completely different. We should also remind, that even the base version of our method appeared to be far more effective [9] than the reference solution described in [10].

In our future research we would like to focus on new definitions of texture features, better suited to spectrogram images than the Local Ternary Patterns. Another interesting direction is the search for the more effective measure of distance between the newly designed feature vectors, which should probably be based on the hypothesis testing approach.

References

1. Bimbot F, Bonastre J-F, Fredouille C et al (2004) A tutorial on text-independent speaker verification. EURASIP J Adv Sig Process 2004:430–451
2. Sadjadi SO, Pelecanos J, Ganapathy S (2016) The IBM speaker recognition system: recent advances and error analysis. In: Interspeech 2016, San Francisco, pp 3633–3637
3. Boves L, Den Os E (1998) Speaker recognition in telecom applications. In: IEEE 4th workshop interactive voice technology for telecommunications applications, IVTTA 1998, Torino
4. Farrell KR, James DA, Ganong III WF, Carter JK (2013) Speaker verification methods and apparatus. Nuance Communications, Inc., US Patent No. US 8,386,263 B2
5. Wang Z-F, Wei G, He Q-H (2011) Channel pattern noise based playback attack detection algorithm for speaker recognition. In: International conference on machine learning and cybernetics, Guilin, vol 4, pp 1708–1713
6. Shiota S, Villavicencio F, Yamagishi J, Ono N, Echizen I, Matsui T (2015) Voice liveness detection algorithms based on pop noise caused by human breath for automatic speaker verification. In: Interspeech 2015, Dresden, pp 239–243
7. Shang W, Stevenson M (2008) A playback attack detector for speaker verification systems. In: 3rd international symposium on communications, control and signal processing, St. Julians, pp 1144–1149
8. Gałka J, Grzywacz M, Samborski R (2015) Playback attack detection for text-dependent speaker verification over telephone channels. Speech Commun 67:143–153
9. Smiatacz M (2018) Playback attack detection: the search for the ultimate set of antispoof features. In: Kurzynski M, Wozniak M, Burduk R (eds) Proceedings of the 10th international conference on computer recognition systems, CORES 2017. AISC, vol 578. Springer, Cham, pp 120–129
10. Janicki A, Alegre F, Evans N (2016) An assessment of automatic speaker verification vulnerabilities to replay spoofing attacks. Secur Commun Netw 9:3030–3044
11. Luo D, Wu H, Huang J (2015) Audio recapture detection using deep learning. In: IEEE China summit and international conference on signal and information processing, Chengdu, pp 478-482
12. Korshunov P, Gonçalves AR, Violato RPV, Simoes FO, Marcel S (2018) On the use of convolutional neural networks for speech presentation attack detection. In: IEEE 4th international conference on identity, security, and behavior analysis, Singapore
13. Tan X, Triggs B (2011) Enhanced local texture feature sets for face recognition under difficult lighting conditions. IEEE Trans Img Process 19:1635–1650

Image Smoothing Using ℓ^p Penalty for $0 \leq p \leq 1$ with Use of Alternating Minimization Algorithm

Jacek Klimaszewski(✉) and Marcin Korzeń

Faculty of Computer Science and Information Technology,
West Pomeranian University of Technology in Szczecin,
Żołnierska 49, 71-210 Szczecin, Poland
{jklimaszewski,mkorzen}@wi.zut.edu.pl

Abstract. Total Variation (TV) filtering is a very popular filtering technique. Recently, using variants of TV penalty term, many filtering methods were presented. The main objective of such approaches is to provide sparse solutions and retain some important details of an image, such as edges. The paper presents a possible solution for the penalty term consisting of p-th power of the differences of neighbouring signal values. The special cases are: $p = 1$, $0 < p < 1$ and $p = 0$. For that purpose, the alternating minimization algorithm proposed by Wang et al. 2008 with the Gauss-Seidel iterations is used. We propose a solution for the special case $p = \frac{1}{2}$ and for the generalized case $0 < p < 1$.

Keywords: Image filtering · TV filtering · Sparse “norms” · L0 smoothing

1 Introduction

Image smoothing considered in this paper can be viewed as a special case of the general optimization problem—minimization of a function L defined as:

$$L(\mathbf{w}) = \|\mathbf{x} - \mathbf{w}\|_2 + \lambda\|\mathbf{D}\mathbf{w}\|_p \tag{1}$$

where: $\mathbf{x} \in \mathbb{R}^n$ is the vectorized source image, $\mathbf{w} \in \mathbb{R}^n$ is the resulting image, $\mathbf{D}$ is some matrix with dimension $m \times n$, and $\| \cdot \|_p$ is some penalty function with properties similar to the norm of vector. A typical penalty function is the norm of ℓ^p space defined as:

$$\|\mathbf{x}\|_p = \sqrt[p]{{x_1}^p + \ldots + {x_n}^p}, \quad p > 1. \tag{2}$$

This function is convex and this causes that L is also convex, thus optimization of (1) is easy [1]. The paper concerns somewhat difficult cases:

$$p = 1: \qquad \|\mathbf{x}\|_1 = |x_1| + \ldots + |x_n|, \tag{3}$$

$$0 < p < 1: \qquad \|\mathbf{x}\|_p = \sqrt[p]{{x_1}^p + \ldots + {x_n}^p}, \tag{4}$$

$$p = 0: \qquad \|\mathbf{x}\|_0 = \#\{i\colon x_i \neq 0\}. \tag{5}$$

R. Burduk et al. (Eds.): CORES 2019, AISC 977, pp. 224–234, 2020.
https://doi.org/10.1007/978-3-030-19738-4_23

The penalty functions (3)–(5) are sometimes called "sparse norms" [6], however (4) and (5) are not norms in the strictly mathematical sense (they violate e.g. the triangle inequality or homogeneity condition). All three penalty functions are non-differentiable, additionally the function (1) with penalty (5) is also discontinuous. Those kinds of penalty functions are very useful in filtering, because they smooth connected areas and sharpen edges. This behaviour is in opposition to ℓ^2 norm penalty, which also smoothes connected areas, but unfortunately blurs edges. It should be kept in the mind that the considered problem is a bit different than the general problem of image filtering including deconvolution with a known model of the filter or noise, and also in such cases the sparse prior can be used [5].

The kind of filter depends on either the penalty function or the matrix $\mathbf{D}$. In this paper as a start point we use Total Variation Filtering (TVF), where the operator $\mathbf{D}$ for one dimensional signals is defined as:

$$\mathbf{D}\mathbf{w} = \left[w_2 - w_1, w_3 - w_2, \ldots, w_n - w_{n-1}\right]^T, \tag{6}$$

and the corresponding matrix is defined as follows:

$$\mathbf{D} = \begin{bmatrix} -1 & 1 & 0 & \ldots & 0 \\ 0 & -1 & 1 & \ldots & 0 \\ \vdots & \ddots & \ddots & \ddots & \vdots \\ 0 & \ldots & 0 & -1 & 1 \end{bmatrix}_{(n-1)\times n} \tag{7}$$

Using above notations we have:

$$TV(w) = \|\mathbf{D}\mathbf{w}\|_1. \tag{8}$$

In the two-dimensional case differences in columns have to be calculated additionally, therefore matrix $\mathbf{D}$ is a bit more complicated. Below there is a MATLAB™ instruction which constructs such a matrix with assumption that vectorization of the source image $\mathbf{x}$ was done in a column-major order:

```
D=[spdiags([-ones(m*(n-1),1) ones(m*(n-1),1)],[0 m],m*(n-1),m*n)
   kron(speye(n),diff(eye(m),1))];
```

where m and n are dimensions of the original image (height and width respectively). For a row-major vectorization, dimensions m and n must be swapped.

2 Solutions

For the total variation (TV) filtering, many techniques were proposed, such as: gradient-projection method [9], smooth approximation of TV criterion [13], iterative clipping algorithm [11], Bregman iterations [3]. In the paper [16] the iteratively reweighted total variation algorithm (IRTV) is described, and it is also adapted to the cases $0 < p < 1$. In this paper we utilize an approach called Alternating Minimization Algorithm described in [14] for TV filtering (the case $p = 1$), but the method is general and it can be used for the case L^0 ($p = 0$) [15], and (as we show below) it also can be used for some special cases of $p \in (0, 1)$. This section presents solutions to the aforementioned problems.

2.1 The Case p=0

The task is to find such $\mathbf{w}$ that minimizes objective function

$$L(\mathbf{w}) = \frac{1}{2} \cdot \|\mathbf{x} - \mathbf{w}\|_2^2 + \lambda \cdot \|\mathbf{D}\mathbf{w}\|_0. \tag{9}$$

This function cannot be optimized directly, because the penalty function $\|\cdot\|_0$ is a non-convex, non-differentiable and discontinuous function. To overcome this problem, a new variable $\mathbf{z}$ is introduced that corresponds to the $\mathbf{D}\mathbf{w}$ [15]. Hence the new objective function has a form

$$L(\mathbf{w}, \mathbf{z}) = \frac{1}{2} \cdot \|\mathbf{x} - \mathbf{w}\|_2^2 + \lambda \cdot \|\mathbf{z}\|_0 + \frac{\mu}{2}\|\mathbf{D}\mathbf{w} - \mathbf{z}\|_2^2, \tag{10}$$

where μ parameter penalizes deviation—as $\mu \to \infty$, $\mathbf{z} \to \mathbf{D}\mathbf{w}$. Now Eq. (10) is easier to solve as we may alternate between minimization of variables $\mathbf{w}$ and $\mathbf{z}$:

$$\mathbf{w}^{(k+1)} = \underset{\mathbf{w}}{\operatorname{argmin}} \; \frac{1}{2} \cdot \|\mathbf{x} - \mathbf{w}^{(k)}\|_2^2 + \frac{\mu}{2}\|\mathbf{D}\mathbf{w}^{(k)} - \mathbf{z}^{(k)}\|_2^2, \tag{11}$$

$$\mathbf{z}^{(k+1)} = \underset{\mathbf{z}}{\operatorname{argmin}} \; \lambda \cdot \|\mathbf{z}^{(k)}\|_0 + \frac{\mu}{2}\|\mathbf{D}\mathbf{w}^{(k+1)} - \mathbf{z}^{(k)}\|_2^2. \tag{12}$$

There are many ways of solving (11). Here we propose the use of Gauss-Seidel method. Differentiating (11) with respect to $\mathbf{w}$ yields:

$$\frac{\partial}{\partial \mathbf{w}}\left(\frac{1}{2} \cdot \|\mathbf{x} - \mathbf{w}\|_2^2 + \frac{\mu}{2}\|\mathbf{D}\mathbf{w} - \mathbf{z}\|_2^2\right) = \mathbf{w} - \mathbf{x} + \mu\mathbf{D}^T(\mathbf{D}\mathbf{w} - \mathbf{z}). \tag{13}$$

Setting $\frac{\partial}{\partial \mathbf{w}}$ to equal 0 and moving terms not involving $\mathbf{w}$ to the other side of the equation leads to the following system of linear equations:

$$(\mu\mathbf{D}^T\mathbf{D} + \mathbf{I})\mathbf{w} = \mathbf{x} + \mu\mathbf{D}^T\mathbf{z}, \tag{14}$$

where $\mathbf{I}$ stands for the identity matrix. Matrix $\mu\mathbf{D}^T\mathbf{D}+\mathbf{I}$ is a sparse, strictly diagonally dominant, tridiagonal block matrix. Due to diagonal dominance, Gauss-Seidel method converges to the solution [10].

Problem (12) is simple to solve using the following observation: the minimum of variable z_i lies either at 0—there function reaches value $\frac{\mu}{2}(\mathbf{D}_i\mathbf{w})^2$ ($\mathbf{D}_i$ stands for i-th row of matrix $\mathbf{D}$)—or at vertex of parabola, where it reaches value λ. To sum up:

$$z_i^{(k+1)} = \begin{cases} 0, & \frac{\mu}{2}(\mathbf{D}_i\mathbf{w})^2 < \lambda, \\ \mathbf{D}_i\mathbf{w}, & \frac{\mu}{2}(\mathbf{D}_i\mathbf{w})^2 > \lambda. \end{cases} \tag{15}$$

Minimization of $\mathbf{z}$ is performed element-wise. One may notice that there are 2 global minima when $\frac{\mu}{2}(\mathbf{D}_i\mathbf{w})^2 = \lambda$, so one of them has to be chosen.

To speed-up convergence, minimization of (10) has to be done for increasing value of μ [14]. We start from $\mu = 2^{-8}$ and each time multiply it by $\kappa = 2$ (or some other value from interval $(1, 2]$) until it exceeds $\mu_{\max} = 2^{24}$.

Presented solution is very similar to [15]. The difference is in solving Eq. (12), where instead of Gauss-Seidel iterations the Fast Fourier Transform is used.

2.2 The Case $0 < p < 1$

We consider here the problem of minimizing:

$$L(\mathbf{w}) = \frac{1}{2} \cdot \|\mathbf{x} - \mathbf{w}\|_2^2 + \lambda \cdot \|\mathbf{D}\mathbf{w}\|_p^p, \text{ for } 0 < p < 1. \tag{16}$$

A natural approach is to use a direct approach used in the previous section in the form (10). The problem of minimizing L is changed to a successive minimization of (10) with respect to variables $\mathbf{w}$ and $\mathbf{z}$. The $0 < p < 1$ case is harder to solve than the case of $p = 0$, because the minimum of the function:

$$L(\cdot, z_i) = \text{Const} + \lambda \cdot |z_i|^p + \frac{\mu}{2}(\mathbf{D}_i\mathbf{w} - z_i)^2 \tag{17}$$

cannot be obtained directly in general. However, we can consider some special cases. For example selecting $p = \frac{1}{2}$ we obtain:

$$L(\cdot, z_i) = \text{Const} + \lambda \cdot \sqrt{|z_i|} + \frac{\mu}{2}(\mathbf{D}_i\mathbf{w} - z_i)^2, \tag{18}$$

and after substitution $\eta_i = \sqrt{|z_i|}$ Eq. (18) is reduced to the polynomial of fourth degree, whose minima are found analytically. Let denote vertex of parabola $(\mathbf{D}_i\mathbf{w} - z_i)^2$ by $c_i = \mathbf{D}_i\mathbf{w}$. Because $\sqrt{|u|}$ is monotone on interval $[0, c_i]$, then local optimum of (18) should be in interval $[0, c_i)$. This case will be used in further sections.

Assume $z_i \geq 0$—then $\eta_i = \sqrt{z_i}$. Now we need to find a solution of

$$\operatorname*{argmin}_{\eta_i} \lambda \cdot \eta_i + \frac{\mu}{2}(c_i - \eta_i^2)^2. \tag{19}$$

Taking derivative with respect to η_i and setting it to equal zero yields depressed cubic equation:

$$\eta_i^3 - c_i\eta_i + \frac{\lambda}{2\mu} = 0. \tag{20}$$

Because (18) is continuous at zero and its derivative changes sign in the neighbourhood of zero (from $-$ to $+$), therefore $z_i = 0$ is a potential minimum. The second minimum is the root of (20). To find it, we use Cardano formula (see e.g. [8]). Critical value of λ, for which (18) has 2 global minima, is:

$$\lambda_{\text{critical}} = \mu \cdot \left(\sqrt{\tfrac{2}{3}c_i}\right)^3. \tag{21}$$

Hence we may write:

$$\eta_i^* = \begin{cases} 0, & \lambda > \lambda_{\text{critical}}, \\ \sqrt{\frac{4}{3}c_i} \cos\left(\frac{1}{3}\arccos\left(-\frac{\lambda}{4\mu\left(\sqrt{\frac{1}{3}c_i}\right)^3}\right)\right), & \lambda < \lambda_{\text{critical}}. \end{cases} \tag{22}$$

Because $\eta_i = \sqrt{z_i}$, we need to square η_i^* to obtain z_i. Negative case ($z_i < 0$) is handled similarly.

In general case $0 < p < 1$ to solve (17) with respect to z_i we use Newton's method. Details can be found in [4].

2.3 The Case p=1, TV Filtering

In total-variation filtering [9], absolute differences between neighbouring pixels are penalized, so objective function may be written as

$$L(\mathbf{w}) = \frac{1}{2} \cdot \|\mathbf{x} - \mathbf{w}\|_2^2 + \lambda \cdot \|\mathbf{D}\mathbf{w}\|_1. \tag{23}$$

The problem was studied before and some algorithms were devised to solve it efficiently [2,11,13]. For TV filtering we use Split Bregman method according to [3,17], this approach was proved to be more efficient than alternating minimization used in previous sections.

3 Experimental Results

All algorithms were implemented in C++ and programs were launched on Samsung RC520 laptop (CPU: Intel i5-2410M 2.3 GHz, 8 GB RAM) under Ubuntu OS. We set multiplier $\kappa = 2$ in all cases. All images were not normalized to within [0, 1]—instead a natural 8-bit colourspace was used. We skipped comparison with other conventional edge-preserving methods, because it was already done in [15].

3.1 Determining Stopping Criterion for the Gauss-Seidel Method

In the first experiment, we checked how the accuracy of the Gauss-Seidel method affects the resulting image. The method stops when condition

$$\max \left(\left| \frac{w_i^{(k+1)} - w_i^{(k)}}{w_i^{(k)}} \right| \right) < \delta \tag{24}$$

is met. Three values of tolerance were checked: 10^{-1}, 10^{-2} and 10^{-3}. Results are depicted in the Fig. 1. We may see that increase of δ makes result better, but time processing also increases. For $\delta = 10^{-1}$ it took ≈1.5 s, for $\delta = 10^{-2}$ it took ≈2.3 s and for $\delta = 10^{-3}$ it took ≈6 s. We concluded that $\delta = 10^{-2}$ is some compromise between computation time and achieved quality.

3.2 Denoising Phantom Example

In this section, we show the quality of the proposed approach on the artificial phantom image [12]. During the experiment we change three parameters of the

(a) original image (b) $\delta = 0.1$

(c) $\delta = 0.01$ (d) $\delta = 0.001$

Fig. 1. Impact of tolerance used in stop condition for Gauss-Seidel method. In all cases $\lambda = 200$.

algorithm: (1) exponent p, (2) regularization parameter λ, and (3) the number of iterations over μ. Conditions of the experiment are similar to [16]—the same level of noise equals 0.1, and PSNR for the noised image is about 20 dB. The results are shown in the Fig. 2. One can see that the our approach gives a slightly better PSNR value than [16] for larger values of p (38.21 dB for p = 0.75, 38.16 dB for p = 0.5), and a slightly worse for p = 0.25 (PSNR = 36.01). The cause of this behaviour is not clear, but we want to mention that our reference algorithm [15] gives for $p = 0$ even worse result—the best solution we found was about PSNR=33 dB.

3.3 Real Life Examples

In this part images available in [7] were used. Each channel of the RGB image was processed separately.

In Figs. 2, 3, 4 we present a comparison of different penalty functions ($\|\cdot\|_0$, $\|\cdot\|_{0.5}$, $\|\cdot\|_1$) on test images. Especially in Fig. 4 it can be seen how result depends on different λ parameter. Generally, it can be seen that $\|\cdot\|_1$ penalty slightly blurs edges and smoothes areas. The $\|\cdot\|_{0.5}$ penalty blurs less and sharpens more strongly the edges. Finally, the $\|\cdot\|_0$ smoothes areas and sharpens edges the most strongly.

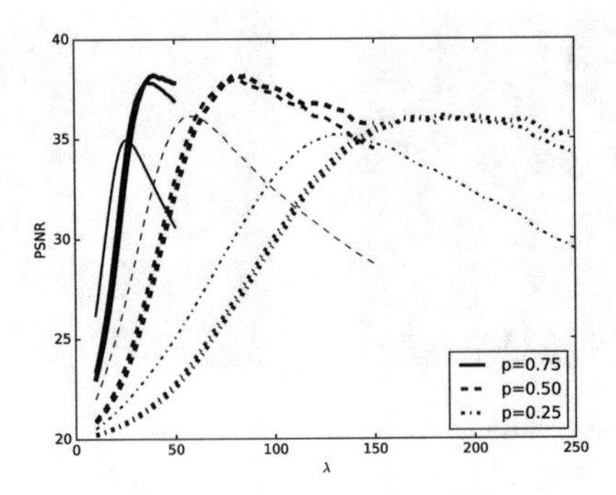

(a) The change of PSNR for parameters: p, λ, number of iterations over μ, for the phantom example. Thin line means 1 iteration, thicker: 5 iterations, the thickest: 10 iterations.

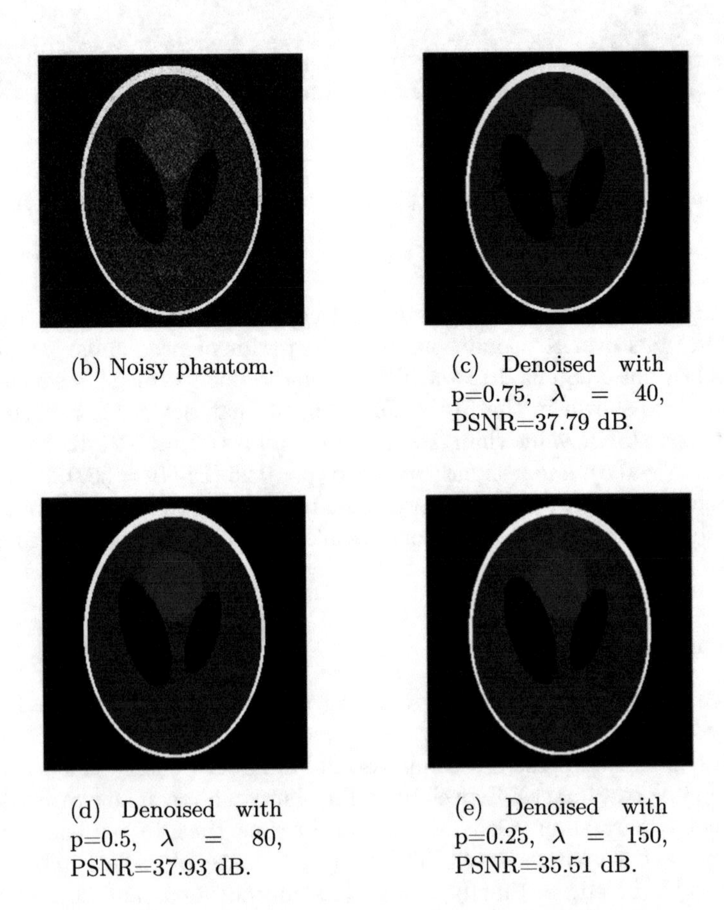

(b) Noisy phantom.

(c) Denoised with p=0.75, λ = 40, PSNR=37.79 dB.

(d) Denoised with p=0.5, λ = 80, PSNR=37.93 dB.

(e) Denoised with p=0.25, λ = 150, PSNR=35.51 dB.

Fig. 2. Results of denoising phantom example.

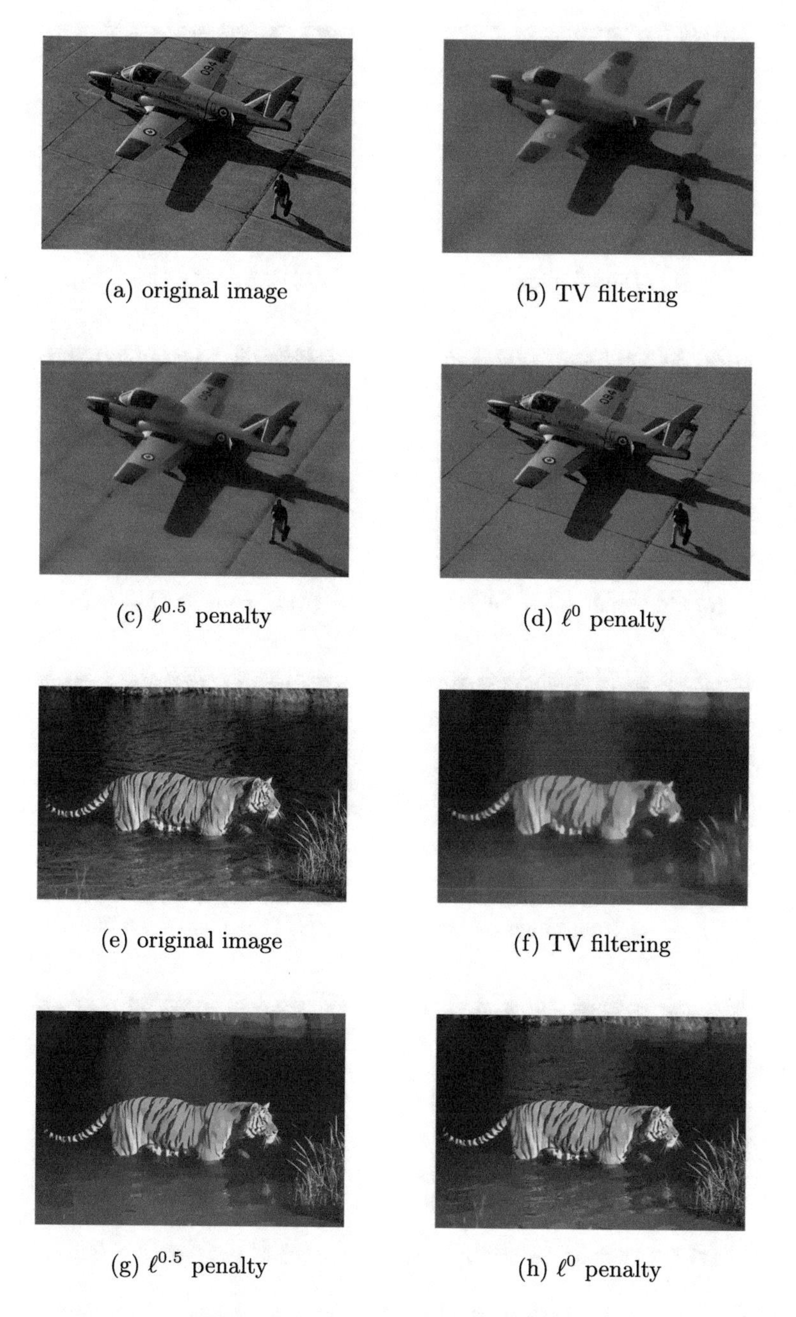

(a) original image (b) TV filtering

(c) $\ell^{0.5}$ penalty (d) ℓ^0 penalty

(e) original image (f) TV filtering

(g) $\ell^{0.5}$ penalty (h) ℓ^0 penalty

Fig. 3. Comparison of 3 regularizers on test images. In all cases $\lambda = 50$ and $\kappa = 2$.

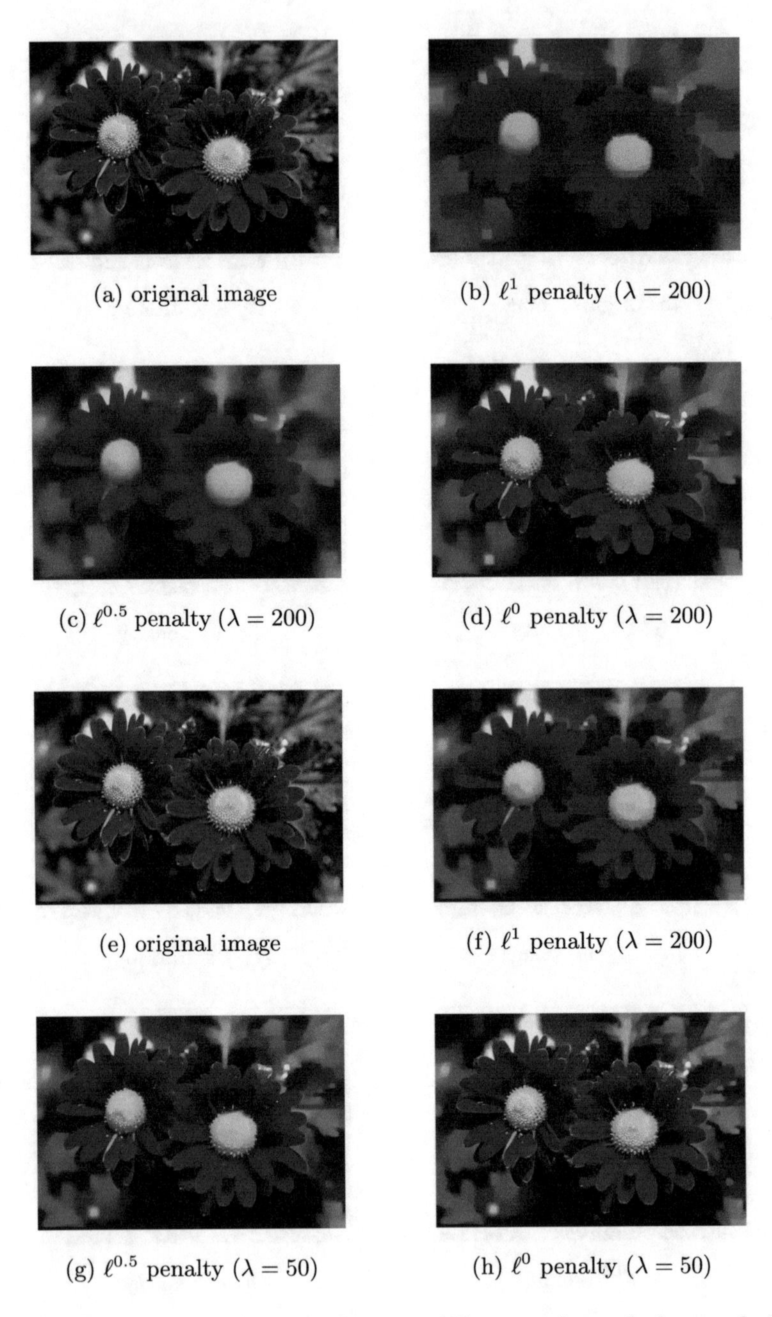

(a) original image

(b) ℓ^1 penalty ($\lambda = 200$)

(c) $\ell^{0.5}$ penalty ($\lambda = 200$)

(d) ℓ^0 penalty ($\lambda = 200$)

(e) original image

(f) ℓ^1 penalty ($\lambda = 200$)

(g) $\ell^{0.5}$ penalty ($\lambda = 50$)

(h) ℓ^0 penalty ($\lambda = 50$)

Fig. 4. Comparison of 3 methods for two different values of the regularization parameter.

4 Conclusions and Remarks

In the paper we have shown that the alternating minimization algorithm proposed by [14] can be also used for $\|\cdot\|_p$, $(0 < p < 1)$ penalty functions. However in only special cases (like for $p = \frac{1}{2}$, what yields to the cubic equation) we can obtain a closed form of the solution. In other cases to solve the optimization problem (17) one should use approximate methods like e.g. Newton's method.

The remarks about the convergence of algorithms are as follows. The method is sensitive on multiplier κ which is by default set to 2, however an artificial example (Fig. 2) requires a lower value.

It is difficult to compare with [15] because of a somewhat different λ parameter and stop condition, but in most cases our implementation is about 2 times slower than [15]. In comparison to the algorithm IRTV [16], our implementation is slightly better for larger values of p and slightly worse for smaller.

It seems more efficient to use Gauss-Seidel method instead of Jacobi, but they are not the only possible methods for solving the second subproblem (14).

References

1. Eilers PHC (2003) A perfect smoother. Anal Chem 75:3631–3636
2. Friedman J, Hastie T, Höfling H, Tibshirani R (2007) Pathwise coordinate optimization. Ann Appl Stat 1(2):302–332
3. Goldstein T, Osher S (2009) The split Bregman method for L1-regularized problems. SIAM J Imaging Sci 2(2):323–343
4. Klimaszewski J, Korzeń M (2016) Optimization of ℓ^p-regularized linear models via coordinate descent. Schedae Informaticae 25:61–72
5. Levin A, Fergus R, Durand F, Freeman WT (2007) Deconvolution using natural image priors. ACM Trans Graph 26(3):2
6. Liu B, Chen S, Qian M, Zhang C (2009) Sparse norm-regularized reconstructive coefficients learning. In: Proceedings of the 2009 ninth IEEE international conference on data mining, ICDM 2009, pp 854–859
7. Martin D, Fowlkes C, Tal D, Malik J (2001) A database of human segmented natural images and its application to evaluating segmentation algorithms and measuring ecological statistics. In: Proceedings 8th international conference on computer vision, vol 2, pp 416–423
8. Nickalls RWD (1993) A new approach to solving the cubic: Cardan's solution revealed. Math. Gaz. 77(480):354–359
9. Rudin LI, Osher S, Fatemi E (1992) Nonlinear total variation based noise removal algorithms. Phys D 60(1–4):259–268
10. Saad Y (2003) Iterative Methods for Sparse Linear Systems, 2nd edn. Society for Industrial and Applied Mathematics, Philadelphia
11. Selesnick IW, Bayram I (2009) Total Variation Filtering
12. Shepp LA, Logan BF (1974) The fourier reconstruction of a head section. IEEE Trans Nuclear Sci 21(3):21–43
13. Vogel CR, Oman ME (1996) Iterative methods for total variation denoising. SIAM J Sci Comput 17(1):227–238
14. Wang Y, Yang J, Yin W, Zhang Y (2008) A new alternating minimization algorithm for total variation image reconstruction. SIAM J Img Sci 1(3):248–272

15. Xu L, Lu C, Xu Y, Jia J (2011) Image smoothing via L0 gradient minimization. ACM Trans Graph 30(6):174:1–174:12
16. Yan J, Lu WS (2015) Image denoising by generalized total variation regularization and least squares fidelity. Multidimension Syst Sig Process 26(1):243–266
17. Ye GB, Xie X (2011) Split Bregman method for large scale fused Lasso. Comput Stat Data Anal 55(4):1552–1569

Deep-Based Openset Classification Technique and Its Application in Novel Food Categories Recognition

Jakramate Bootkrajang(✉), Jakarin Chawachat, and Eakkapap Trakulsanguan

Data Science Research Center, Department of Computer Science, Chiang Mai University, Chiang Mai 50200, Thailand
{jakramate.b,jakarin.c,eakkapap.t}@cmu.ac.th

Abstract. Being able to accurately recognise food categories from input images has many possibly useful applications such as content-based recipe searching or automatic intake calories tracking. Convolutional neural networks has been successfully applied in a number of food recognition tasks. Despite its impressive predictive performance on closed datasets, there is currently no standard mechanism for distinguishing unknown object classes from the known ones leading to invalid classification attempts even on non-food images. In this paper, we study a technique for detecting whether input images are beyond the scope of CNN's knowledge. The idea is to model the final activation vectors of data from the known classes using a data description method namely the support vector data description. We can then reject network's prediction if the activation vector of the query image is too different from the known ones as generalised by the model. Experimental results on a subset of UECFOOD100 datasets demonstrated that the proposed method was able to accurately classify instances from the known classes while also being able to satisfactorily reject the prediction of novel food image compared to two commonly used baselines.

Keywords: Openset classification · Food recognition · Deep learning

1 Introduction

Food categories recognition is one of the interesting topics among several visual recognition tasks. An accurate food recogniser can be employed in many real world applications such as intake calories estimation [12], dietary assessment [13] or image-based recipe searching [2]. Several attempts had been made to tackle the problem of food categories classification problem. The majority of previous research was based on Support Vector Machine using specially crafted visual features [5,16]. Some classical pattern recognition techniques such as k-nearest neighbour has also been employed for the task [7]. Recently, Convolutional Neural Networks (CNN) is gaining more popularity due to its impressive

R. Burduk et al. (Eds.): CORES 2019, AISC 977, pp. 235–245, 2020.
https://doi.org/10.1007/978-3-030-19738-4_24

performance in visual recognition tasks [6,18]. Unlike previous image classification approaches which rely on the quality of visual features extracted from the image, CNN learns good feature representation simultaneously with learning the classifier. The model has been adopted for food recognition task [10,11,20] and has been shown to outperform existing approaches.

Despite of its impressive predictive performance, CNN and in fact any classifier in general still has some limitation for real-world usage. In particular, classical supervised learning assumes query input comes from the same data distribution as the one used to train the model. In the context of food categories recognition, we implicitly assume that query food image be one of the known food categories. Unfortunately, this assumption does not always hold true in real-world recogniser deployment. Surely, in such case, the classification model still make a prediction even though the query image is not a food image or is a new type of food.

The aforementioned limitation motivates us to study the problem of detecting whether the prediction should be made for incoming query image. We would want our recognition model to be able to reject the prediction for input image which does not belong to the known classes while still gives accurate prediction for those that are within the classification scope. The problem of this kind is not new and it has been studied in the past under the name *novelty detection* [15] (and sometime interchangeably as outlier detection or anomaly detection). From machine learning perspective, novelty detection problem can be approached with one-class classification [14] where there are already many learning algorithms available. Recently, the problem is increasingly known in the machine learning community as an openset classification problem [9,17]. There is, however, one subtle difference between one-class classification and openset classification problem. In one-class classification, the task is to differentiate between a *single* target class and other possible inputs. Meanwhile, in openset classification, we want to differentiate *multiple* known classes from other possibilities. The challenge is that the set of known classes might not form an obvious single class and it is interesting if existing one-class classification technique will work in this more challenging case.

To the best of our knowledge, there were not many attempts to studying openset classification problem within the scope of convolutional neural network. A seemingly straightforward mechanism for detecting if an input is beyond the scope of training data is by means of *class posterior thresholding*, where a prediction would not be made if the posterior probability of the most probable class is below some predefined threshold. A more advanced approach could involve the modelling of the activation vector, e.g., the output of the final layer of the network. The approach is based on the observation that inputs that belong to the same object class should have similar activation vector. Therefore, if we can summarise and construct representative activation vector either globally or locally (i.e., having one representative for each of the data classes), we should be able to detect if the query image actually belongs to one of the known classes by comparing their activation vectors. The work in [1] took this route and modelled the representative using the mean activation vector computed over correctly classified examples in each class. The prediction is rejected if the distance of input's AV is too far from the mean AV. We shall also take this route but extend the data description model into a more complex one by using the Support Vector Data Description. Hopefully, the complex data boundary creates by the SVDD might be beneficial. Further, our work differ from [1] in that their analysis was based on activation values from the penultimate layer of the network while our approach models the final activation vectors directly. We believe that the relatively lower dimensional nature of the final activation vectors, which scales with the number of classes might better suit our modelling choice.

The rest of the paper is organised as followed. Background and the details of the proposed SVDD-based novel instance detection approach are presented in Sect. 2. Section 3 presents the empirical studies and discussion of the results. Section 4 presents the concluding remarks and outlines future research direction.

2 Background and Methods

Formally, constructing a classifier is a task of inferring a function $f : X \to Y$, which maps instances in a feature set X to an instance in label set Y using a subset of examples in the form of $(\mathbf{x}_i, y_i)_{i=1}^{N}$ pairs independently and identically (i.i.d) drawn from the joint distribution $D : X \times Y$. The goal of the learning is to be able to use the resulting classifier $f(\cdot)$ for assigning $y \in Y$ to an unseen query instance $\mathbf{x}_q$ from X with high accuracy. In classical setup, we implicitly assume that the query point is also from X. However, such assumption is quite unrealistic for real world classification model deployment where there is little guarantee that the query input will be one of the instance in X. And we would like to detect when this happens.

Our approach for detecting whether $\mathbf{x}_q$ is beyond the classification scope relies on the analysis of the final activation vector: the vector of output from the final layer of the neural network $f(\cdot)$. We denote the activation vector of an input image $\mathbf{x}_i$ by $f(\mathbf{x}_i) = \mathbf{v}_i = [v^1, \ldots, v^K]$. Usually, given a classification task of K target classes, there will be K output nodes. Accordingly the activation vector is a real-valued vector in $\Re^K$. Often, the output values from the final layer is normalised such that $\sum_{j=1}^{K} v^j = 1$.

2.1 Convolutional Neural Networks

Before proceeding, we would like to first outline the architectures of the deep neural networks employed. A convolutional neural network is a classifier which can be divided into two parts: the convolutional layers and the fully connected layers. The convolutional layers part is responsible for extracting visual features from the images while the fully connected layers takes visual features extracted by the convolutional layers and assigns class label to the image. Various CNN architectures published to date differ from each other primarily at the convolutional layers level. It has been empirically shown that abstract visual features (especially those found at the very first convolutional layers) are high level features and are shared among various kind of visual recognition tasks [21]. In practice, the weights of the convolutional layers can be transferred from some pretrained networks of which the weights were sufficiently learnt from massive visual datasets and should provide a good starting point for further fine tuning. In this work we employed four well-known deep convolutional architectures trained on ImageNet dataset, namely VGG16 and VGG19 [18], ResNet50 [6] and DenseNet121 [8]. The weights of the convolutional layers were transferred and were frozen during training and only the weights of the fully connected layer was trained via the standard back-propagation methodology. Our fully connected network is composed of an input layer with 256 nodes, a hidden layer with 64 nodes and an output layer with 20 nodes. The weights from input to hidden and hidden to output layer are subjected to 0.3 dropout rate. Activation functions in the input and hidden layers were the sigmoid while the softmax function is used in the output layer. We trained the fully-connected network using batch size of 16 and small learning rate of 10^{-5}.

2.2 Activation Vector Data Description Model

To differentiate between activation vectors belonging to the set of known classes and those from the unknowns, we study the idea of employing the Support Vector Data Description (SVDD) [19] well used in the area of one-class classification to model activation vectors from the known classes. Intuitively, SVDD starts with a hypersphere of radius R centred at $\mathbf{a}$. The objective is to find a hypersphere with minimum radius which also encloses all of the data. Formally, the objective of SVDD is to minimise the following loss function: $L(R, \mathbf{a}) = R^2$ and subject to $||\mathbf{v}_i - \mathbf{a}||^2 \leq R^2, \forall i$. The above formulation is rather rigid such that requiring all $\mathbf{v}_i$ to lie in the hypersphere could be difficult in real world usage where some outliers exist. To mitigate the problem, *slack variable* $\xi_i \geq 0$ could be introduced into the formulation, yielding an objective function in its primal form: $L_{primal}(R, \mathbf{a}) = R^2 + C\sum_i \xi_i$ subject to $||\mathbf{v}_i - \mathbf{a}||^2 \leq R^2 + \xi_i, \forall i$. Here C is the hyperparameter which controls the trade-off between the volume of the hypersphere and errors. The new formulation expresses the fact that *almost* all data (the activation vectors in our case) is required to fall within the hypersphere. Following [19], minimising the primal is equivalent to maximising its dual form:

$$L_{dual}(\alpha) = \sum_i \alpha_i \kappa(\mathbf{v}_i, \mathbf{v}_j) - \sum_{i,j} \alpha_i \alpha_j \kappa(\mathbf{v}_i, \mathbf{v}_j) \tag{1}$$

subject to $0 \leq \alpha_i \leq C$ and $\sum_i \alpha_i = 1$. Here, $\kappa(\cdot, \cdot)$ is a positive definite reproducing kernel function that enables the construction of non-linear data boundary. The activation vectors $\mathbf{v}_i$ where its corresponding $\alpha_i > 0$ are called the Support Vectors (SVs) for the description. In this work, we will work with the Radial Basis Function kernel (RBF) given by $\kappa(\mathbf{v}_i, \mathbf{v}_j) = \exp(||\mathbf{v}_i - \mathbf{v}_j||^2)/2\sigma^2$. We used $\sigma = 2^{-4}$ throughout the experiments. Accordingly, a test image $\mathbf{x}_q$ is considered novel if the distance of its $\mathbf{v}_q$ from the SVs, given by

$$\sum_{i \in SVs} \alpha_i \exp\left(\frac{||\mathbf{v}_q - \mathbf{v}_i||^2}{2\sigma^2}\right), \tag{2}$$

is greater than some predefined threshold ρ. Since the value of $\mathbf{v}$ can be very small e.g., less than 10^{-10}, and can cause some numerical instability, we propose to work instead with the logarithm of $\mathbf{v}$. We shall refer to the method described above as Activation Vector Data Description (AVDD) to emphasise the modelling of the activation vectors using SVDD. Algorithm 1 summarises the steps to construct the AVDD while the steps for detecting novel instance are outlined in Algorithm 2.

Algorithm 1. Construction of the Activation Vector Data Description

Input: Activation vectors of training data $(\mathbf{v}_i)_{i=1}^N$
1 Perform logarithmic transform $\tilde{\mathbf{v}} = \log \mathbf{v}$
2 Construct SVDD model using Eq. (1) based on the transformed $(\tilde{\mathbf{v}}_i)_{i=1}^N$
Output: Optimised $(\alpha_i)_{i=1}^N$

Algorithm 2. The openset detection step

Input: Activation vector of test data $(\mathbf{v}_q)$ and parameters of AVDD $(\alpha_i)_{i=1}^N$
1 Perform logarithmic transform $\tilde{\mathbf{v}}_q = \log \mathbf{v}_q$
2 Calculate distance of $\tilde{\mathbf{v}}_q$ from the support vectors $(\tilde{\mathbf{v}}_i)_{i \in SVs}$ using Eq.(2).
3 **if** *distance* $> \rho$ **then**
4 $\hat{y}_q =$ 'unknown'
5 **else**
6 $\hat{y}_q = \arg\max_j \tilde{v}_q^j$
7 **end**
Output: $\hat{y}_q$

3 Empirical Evaluations

We will now study the effectiveness of the proposed AVDD detection method in openset classification problem. The main question is how well the proposed method identify unknown input instance while also being able to recognise instances from target classes. We shall compare the detection performance of our proposed method with two commonly used baselines. The first baseline is the simplest mechanism for novelty detection. The scheme rejects the prediction of input $\mathbf{x}_q$ if the class posterior probability of the most probable class turns out to be less than some predefined threshold, e.g., $\max_j v_q^j < \theta$. We will refer to this method as Class Posterior Thresholding (CPT). The second baseline involves the calculation of Mean Activation Vectors (will be referred to hereafter as MAV method) for each class. The approach then rejects the prediction if the activation vector of the query $\mathbf{v}_q$ is too different from the mean activation vector of the predicted class, e.g., $dist(\mathbf{v}_q, \mu_{\hat{y}_q}) > \beta$. For simplicity, we considered standard Euclidean distance for similarity measurement.

3.1 Datasets and Protocol

The food images used in this study are from the UECFOOD100 dataset [11]. The original dataset contains visual images of 100 Japanese food categories. Region Of Interest (ROI) information is provided for every image. Our preprocessing steps involve extracting food images according to the ROIs and resizing the image to 224×224 pixels to match the input requirement of the VGG16 network.

We randomly sampled 20 food classes from the dataset for our experiment. We will refer to this set of data as *FOOD20* dataset. To evaluate the novelty detection performance, we set apart another 10 classes from UECFOOD100, called *OPEN-FOOD* and another 10 classes of general objects images from Imagenet dataset [3] which are irrelevant to food called *OPEN-OBJECT*. Table 1 summarises the datasets used in this study.

Table 1. The datasets used in this study are divided into three groups. *FOOD20* is used to train the recognition model. *OPEN-FOOD* is used to test the capability of the model in detecting unknown but related objects. *OPEN-OBJECT* is a set of unrelated objects.

Dataset	Class labels
FOOD20 #instances 3338 #classes 20	rice, eels on rice, pilaf, sushi, chicken rice, fried rice, toast, croissant, roll bread, hamburger, pizza, sandwiches, udon noodles, spaghetti, Japanese pancake, takoyaki, gratin, cutlet curry, potato salad
OPEN-FOOD #instances 1396 #classes 10	chicken-and-egg on rice, pork cutlet on rice, beef curry, tempura bowl, bibimbap, raisin bread, chip butty, beef noodle, tensin noodle, fried noodle
OPEN-OBJECT #instances 301 #classes 10	apple, bird, car, carrot, cat, dog, doll, fish, orange, plane

3.2 Results: Performance on Known Classes

We first want to establish a *closed set accuracy*. The accuracy is identical to the accuracy obtained in the idealised supervised learning scenario where testing data are from the same data distribution as that of the training data used to train the model. To do this, we randomly split *FOOD20* data into training and testing set using 90/10 percent ratio. We trained the models until they sufficiently converged on the training data. We then validated theirs performance on the remaining 10 percent of data and recorded the classification accuracies. We note that in this case the models are allowed to predict all of the testing examples without employing the novelty filtering mechanism. We repeated the aforementioned procedure for 10 repetitions in order to get reliable statistics. Table 2 reports the average top-1 and top-5 classification accuracies together with their standard errors of the four CNNs employed.

Table 2. Top-1 and Top-5 closed set accuracies of four convolutional neural networks employed in this study.

	VGG16	VGG19	ResNet50	DenseNet121
Top-1 accuracy	83.34 ± 1.48	82.68 ± 1.91	86.71 ± 1.45	76.66 ± 1.84
Top-5 accuracy	97.36 ± 0.07	96.61 ± 1.03	97.37 ± 1.21	94.61 ± 1.14

We observed that top-1 accuracies of all CNNs are well above 80% except for DenseNet121 which slightly lagged behind. In general, there seems to be some confusion among similar kind of food e.g., rice-based dishes. Meanwhile, the top-5 performances from all models are quite impressive with accuracies exceeding 97%. Although, the top-5 performances are acceptable, we believe that more sophisticated technique can surely be incorporated into the classification model to enhance the top-1 predictive performance and we plan to do so in the subsequent work. The results are also inline with the results reported in the original paper of the dataset [10]. This suggests that the CNN architectures used in this study are deemed suitable for the task as all were capable of learning the regularities in the data to some extent. Next, we shall turn to study the effect of novelty detection mechanism on the performance of the chosen CNN models.

3.3 Results: Novel Classes Detection

In this section we shall evaluate the proposed novelty detection mechanism. We would like to quantify the error that each of the comparing mechanisms makes during the detection process. There are two types of error: type 1 and type 2 error. Type 1 error occurs when the detector thinks that query image is from a novel class but in reality the query is from one of the known classes. Type 2 error occurs when the query image is indeed outside the classification scope but the detector thinks that it is not. A good way to summarise both errors graphically is by constructing a Receiver Operating Characteristic (ROC) curve [4]. We followed the same training protocol as described above but instead of predicting class labels during testing, we mixed the remaining 10 percent of *FOOD20* data held out for testing with data from *OPEN-FOOD* to get the first set of open data, and with *OPEN-OBJECT* to get the second open dataset. The task is then to tell whether images in the mixed testsets are the images from the held-out *FOOD20* or not. We repeated the experiment for 10 repetitions while recording True Positive Rate and False Positive Rate in each run. The average ROC curves of the three mechanisms combined with four respective CNNs are presented in Fig. 1.

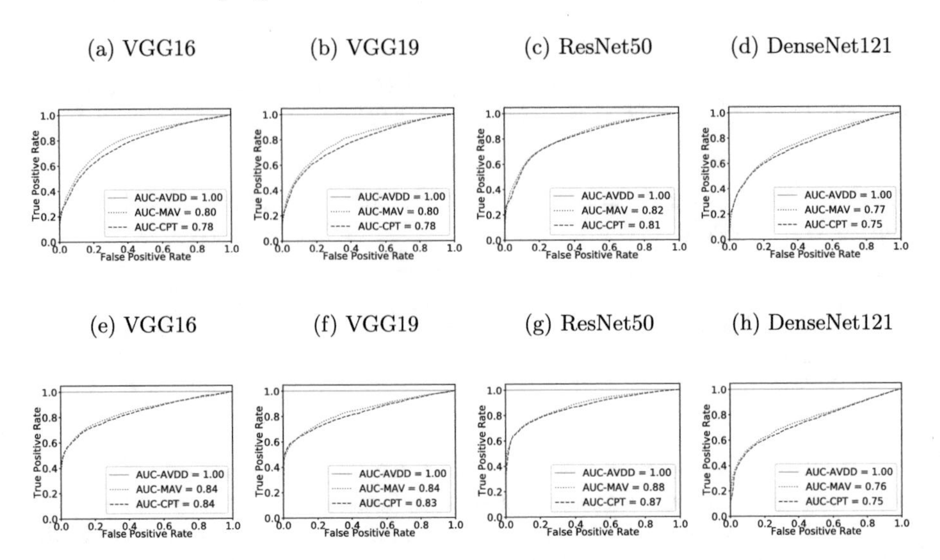

Fig. 1. The ROC curves for the three novelty detection mechanisms on OPEN-FOOD (top) and OPEN-OBJECT (bottom), together with their associated AUCs.

From the results, we notice that detecting unknown objects based on activation vectors, either by MAV approach or AVDD approach, is more effective than the standard CPT method. We also see that AVDD was better than both MAV and CPT methods by a large margin in both of the open datasets, partly thanks to its non-linear data boundary. We speculate that the similarities between some of the classes in *FOOD20* and *OPEN-FOOD*, e.g., chicken rice vs chicken-and-egg on rice, might contribute to the poor performance of CPT. Meanwhile, MAV method might not be delicate enough to differentiate between two different AVs that happened to have the same Euclidean distance from the mean. This suggests that the proposed mechanism is quite promising in detecting novel classes for various CNN-based food classification models in real-world usage.

The ROC curve summarises the detection performance at various thresholding values. Still, one question remains unanswered namely, how do we choose the cutoff threshold sensibly? It is somehow unrealistic to assume there exists open validation set for threshold selection as by definition the possibilities of instances in the openset are endless. For the posterior thresholding baseline, i.e., CPT in this study, the threshold reflects our requirement for classifier's confidence which varies from task to task. For less sensitive task we could, for example, aim for $\theta \approx 0.9$ while in more critical situation we might want to set θ a little bit higher. For MAV method, the determination of cutoff threshold is less straightforward as the distance measure lacks probabilistic semantics, and we think this is one of the difficulty associated with this kind of approach. Interestingly, for AVDD, we observed that the distances of AVs of novel points from the support vectors mostly concentrate at some value. This allows us to select good cutoff threshold using only a small set of open data.

We then investigated whether or not input instances from *OPEN-OBJECT* dataset can be used to facilitate the selection of ρ, the cutoff threshold. According to the concentration phenomenon mentioned earlier, we set ρ to be the distance of an AV of a random example from *OPEN-OBJECT* from the support vectors, minus some small number e.g., 0.0001. This is to compensate the tiny variance associated with the distances of AVs of other instances from *OPEN-OBJECT* dataset. The heuristic is sensible for two reasons. First we did not assume the availability of novel food images since if we have such data we could have already included it in the training set. Second, *OPEN-OBJECT* is publicly available and can be obtained quite easily without additional overhead. We adopted the same mechanism for setting β, the threshold for MAV. The cutoff thresholds of CPT was simply set to $\theta = 0.9$. We then evaluate the threshold selection heuristic by measuring an *open set accuracy* which is the ratio of instances, which were not caught by the detection mechanism and were also correctly classified, over the total number of test instances in *FOOD20*. Due to page limit, we present the open set performances of the three detection methods combined with VGG16 and ResNet50 using *FOOD20 + OPEN-FOOD* dataset in Table 3.

From the results, we see that all detection mechanisms incurred a slight drop in both top-1 and top-5 accuracies. This is expected though because some legitimate predictions might have been discarded by the detection mechanisms. Still, we observe that AVDD, among the three methods, was able to retain the top-1 and top-5 accuracies better while also being effective in detecting novel inputs. The results also validated the usefulness of the threshold selection heuristic and hinted that VGG16 + AVDD might be a good pair for the task.

Table 3. Top-1 and Top-5 open set accuracies of the convolutional neural network paired with each respective detection mechanisms on *FOOD20 + OPEN-FOOD* dataset. Closed set performances are included for reference.

Methods	Top-1	Top-5	*OPEN-FOOD* recognition rate
VGG16 + AVDD	**83.34 ± 1.48**	**97.36 ± 0.07**	**99.78 ± 0.07**
VGG16 + MAV	73.73 ± 5.55	79.76 ± 8.12	58.08 ± 16.17
VGG16 + CPT	66.46 ± 1.85	69.04 ± 2.31	72.06 ± 1.11
VGG16 (closed set)	83.34 ± 1.48	97.36 ± 0.07	N/A
ResNet50 + AVDD	82.51 ± 2.35	89.06 ± 2.71	**100.00 ± 0.00**
ResNet50 + MAV	76.54 ± 3.82	79.72 ± 4.74	63.62 ± 9.55
ResNet50 + CPT	83.05 ± 1.82	89.82 ± 1.47	36.48 ± 1.44
ResNet50 (closed set)	86.71 ± 1.45	97.37 ± 1.21	N/A

4 Conclusions

We studied a novel food categories recognition for convolutional neural networks. Our method relies on the construction of data description model by means of the support vector data description. The model rejects the prediction and alerts that the input is from unknown food class if the activation vector of the query image is too different from the model. The empirical study was setup to compare the effectiveness of the proposed method with the traditional baselines of class posterior thresholding and mean activation vector methods. The detection capability of the proposed method was shown to be promising. What remains unexplored in this work is how to further make use of the novel objects. One possibility is to combine a so-called self learning methodology to query similar images from image search engine while extracting label from the most probable image tags and use the new information to retrain the recognition model.

Acknowledgement. The research is supported by the Faculty of Science, Chiang Mai University.

References

1. Bendale A, Boult TE (2015) Towards open set deep networks. CoRR abs/1511.06233
2. Chen J, Ngo CW (2016) Deep-based ingredient recognition for cooking recipe retrieval. In: Proceedings of the 2016 ACM on multimedia conference, pp 32–41
3. Deng J, Dong W, Socher R, Li LJ, Li K, Fei-Fei L (2009) Imagenet: a large-scale hierarchical image database. In: Proceedings of IEEE CVPR, pp 248–255
4. Fawcett T (2006) An introduction to roc analysis. Pattern Recogn Lett 27(8):861–874
5. He H, Kong F, Tan J (2016) DietCam: multiview food recognition using a multikernel SVM. IEEE J Biomed Health Inform 20:848–855
6. He K, Zhang X, Ren S, Sun J (2016) Deep residual learning for image recognition. In: Proceedings of IEEE CVPR, pp 770–778
7. He Y, Xu C, Khanna N, Boushey CJ, Delp EJ (2014) Analysis of food images: features and classification. In: Proceedings of IEEE image processing, pp 2744–2748
8. Huang G, Liu Z, Van Der Maaten L, Weinberger KQ (2017) Densely connected convolutional networks. In: Proceedings of IEEE CVPR, pp 4700–4708
9. Jain LP, Scheirer WJ, Boult TE (2014) Multi-class open set recognition using probability of inclusion. In: ECCV. Springer, Cham, pp 393–409
10. Kawano Y, Yanai K (2014) FoodCam: a real-time food recognition system on a smartphone. Multimed Tools Appl 74:5263–5287
11. Matsuda Y, Hoashi H, Yanai K (2012) Recognition of multiple-food images by detecting candidate regions. In: Proceedings of IEEE ICME
12. Meyers A, Johnston N, Rathod V, Korattikara A, Gorban A, Silberman N, Guadarrama S, Papandreou G, Huang J, Murphy KP (2015) Im2Calories: towards an automated mobile vision food diary. In: Proceedings of IEEE international conference on computer vision, pp 1233–1241
13. Mezgec S, Koroušić Seljak B (2017) Nutrinet: adeep learning food and drink image recognition system for dietary assessment. Nutrients 9(7):657
14. Moya MM, Koch MW, Hostetler LD (1993) One-class classifier networks for target recognition applications. NASA STI/Recon Technical Report 93
15. Pimentel MA, Clifton DA, Clifton L, Tarassenko L (2014) A review of novelty detection. Sig Process 99:215–249

16. Pouladzadeh P, Villalobos G, Almaghrabi R, Shirmohammadi S (2012) A novel SVM based food recognition method for calorie measurement applications. In: Proceedings of IEEE ICME (workshops), pp 495–498
17. Scheirer WJ, de Rezende Rocha A, Sapkota A, Boult TE (2013) Toward open set recognition. IEEE TPAMI 35(7):1757–1772
18. Simonyan K, Zisserman A (2014) Very deep convolutional networks for large-scale image recognition. arXiv preprint arXiv:1409.1556
19. Tax DM, Duin RP (2004) Support vector data description. Mach Learn 54(1):45–66
20. Yanai K, Kawano Y (2015) Food image recognition using deep convolutional network with pre-training and fine-tuning. In: Proceedings of IEEE international conference on multimedia & expo workshops, pp 1–6
21. Yosinski J, Clune J, Bengio Y, Lipson H (2014) How transferable are features in deep neural networks? In: NIPS, pp 3320–3328

Singing Power Ratio Analysis in the Context of the Influence of Warm up on Singing Voice Quality

Edward Półrolniczak(✉)

Faculty of Computer Science and Information Technology,
West Pomeranian University of Technology, Szczecin, Żołnierska Street 49,
71-210 Szczecin, Poland
epolrolniczak@zut.edu.pl
http://www.zut.edu.pl/eng/home/news/current-news.html

Abstract. The article attempts to assess the impact of warm up on the singers voice. To estimate the impact, values of The Singing Power Ratio (SPR) were calculated for the singing samples. The hypothesis is that the warm up will have a positive effect on the parameters of the singing voice. The conclusion resulting from the statistical analysis is that the warm up has a positive effect on the quality of the singing voice but the strength of the impact depends on the personal abilities of the singers.

Keywords: Singing voice · Singing quality · Warm up · Singing Power Ratio · Singing assessment

1 Introduction

Musical performances are often an intense activity which should be preceded by proper voice warm up. Most of the voice production experts claim that the warm up is a must before singing. According to the literature, well performed warm up should have positive effect on the singing voice parameters. In fact, experts' opinions about whether it is necessary are divided. This is due to the small amount of research and publications that support or disprove the thesis that it is necessary to carry it out before singing. Some subjective effects reported by singers seem to be significant, but the physiological effects are largely unknown [1]. During singing warm up it is possible to see a tendency for increased blood flow in the muscles, lungs, lips or tongue. Singing causes an increase in muscle temperature, lowers blood pressure and reduces the viscosity of vocal folds, which contributes to a better quality of singing [10]. Singing may be helpful in removing excess mucus from vocal cords [2], as well as in reducing excessive muscle tone. No singing warm up can lead to vocal fatigue, hoarseness - or worst of all, nodules. On the other hand one can find in the literature the opinion that warm up is not a must [3] and carried out in an inappropriate way, without consulting a professional, may even harm. There are various procedures of the warming-up

R. Burduk et al. (Eds.): CORES 2019, AISC 977, pp. 246–254, 2020.
https://doi.org/10.1007/978-3-030-19738-4_25

which may differ [4,5] from each other depending on the person conducting the warm up. Usually the process of the warming-up consists of: (1) body relaxation exercises, (2) breathing exercises, and (3) voice productions and placement at different pitches, registers, and amplitude levels [6,7]. Taking into account the facts found in the literature, it can be concluded that warm up in singing may have a positive impact on the quality of the voice increasing its capabilities. In the same time it can prevent damage to the voice apparatus. If it is carried out by an experienced specialist and the exercises are tailored to the singer's needs it may positively influence the singer's voice quality. On the other hand, improper warm up may be dangerous for the voice. Regardless of the above, the influence of warm up on voice quality is usually evaluated by specialists and is based on the effects they can hear which does not guarantee a reliable assessment, stable and consistent results. This article attempts to objectify the assessment of the effect of warm up on the quality of voice. For this purpose, the SPR (Singing Power Ratio) values were estimated for each recorded sound sample and the collected data was subjected to an analysis considering the knowledge about quality of the singing voice.

2 Literature Review

As mentioned above in order to objectify the assessment of the effect of warm up on the quality of voice the SPR was estimated. In some studies SPR parameter is used to measure the singing voice quality. In the case of this article the SPR is used to find if warm up gives any effects on the voice quality. By reviewing the literature it is found that SPR and $LTAS$ (Long Term Average Spectrum) are applied in studying differences between normal and pathological voices [12] and in analysing various vocal pathologies [9,13]. $LTAS$ method has been also used to study individual and gender, age and language related differences [14,15]. Characteristics of vocal expression of emotions [16] and acoustic differences between specific voice qualities [18] have also been studied with this method. Moreover, $LTAS$ has been used to investigate singing voice [8], in particular to investigate the differences between voice categories and various singing styles [17].

There are few studies on the differences between singing skills in the group of untrained singers. One of such studies can be found in [21]. The authors have evaluated the abilities of precise reproduction of note intervals, as well as ability to control the pitch and timbre quality. In this context, an SPR, calculated over the $LTAS$ spectrum, has been used. According to the authors, the quality of singing, expressed by SPR, reflects the strengthening or attenuation in the vocal tract of the harmonics generated by the source of the sound. The values of this ratio were measured for untrained singers, divided on talented and not talented singers to objectively analyse differences in voice quality. The $LTAS$ has been also analysed in that study. It allowed to draw a conclusions on the degree of talent. Higher harmonics were more dominant in the spectrum of more talented singers. An increase of energy in the range of 2500–4000 Hz was observed.

SPR is based on the difference between the two maxima peaks found at two consecutive ranges in the $LTAS$. The first peak lies in the frequency range of

0–2000 Hz and the second one falls into the 2000–4000 Hz range. Mostly *SPR* value is calculated as the difference subtracting the value of the second peak from the first one.

3 Material for the Study

The room for recordings was organized in the building in which the Choir of the Maritime University of Szczecin is working. A didactic room was used for this purpose, the walls of which were lined with sound-absorbing material. In order to improve the acoustic parameters, a sound-absorbing 7 cm pyramids foam panel was used. The foam was stably fixed on a stand. It additionally allowed reducing reflections and reverberation. After adaptation, the conditions were considered sufficient to allow the obtained recordings to be used as research samples.

The recording equipment consisted of the Audio-Technica AT4050 condenser microphone, ZOOM H4n digital recorder and PreSonus HD7 headphones allowing for a proper listening session. The chosen microphone is a good choice to record vocal parts, piano, string instruments or as an overhead for recording drums or guitars. The excellent quality of the microphone makes it sufficient for implementation of the planned task. The equipment has an externally polarized capacitive transducer based on two gold-plated membranes and records correctly in each of the characteristics: cardioid, omnidirectional, octal. During recording cardioid characteristics were used.

At the recording sessions the Yamaha PSR-213 synthesizer was used, which served as a source of sounds that the singers were supposed to repeat. The recorded singer was facing the microphone at a distance of 40 cm. About 70 cm from the microphone, behind the back of the person being recorded, a pyramid foam absorbing the sound was placed. During the recording session, a person responsible for the course of the tasks was also present. Her duties included: playing the right sounds on the musical instrument and ensuring that the sensitivity of the microphone was at the optimal level - in order to avoid the occurrence of sound clipping while recording. The recorder settings used during the session are a resolution of 24 bits and a sampling rate of 48 kHz. Time, which was devoted to one person, is about 30 min on average. The sensitivity of the microphone remained at the same unchanged level throughout the study.

The recorded singers were members of the Choir of the Maritime University of Szczecin. The study was carried out on the group of seventeen choir singers. Each person was in good health. In order to ensure diversity and to get a full picture of the quality of voices, 4 altos, 6 sopranos, 6 basses and 1 tenor were chosen for that study. The Table 1 lists the types of the voices in the estimated group.

4 Research Background

After the singing samples were acquired the *SPR* values have been estimated for each recorded sample. The whole process is illustrated in Fig. 1.

Table 1. The types of the voices in the estimated group

Coded name	Symbol in the database	Voice type	Subtype
a01f	AK	Soprano	1
a02f	BD	Alto	1
a03f	BK	Alto	2
a04f	GP	Alto	2
a05f	KN	Soprano	2
a06f	KS	Soprano	2
a07f	MC	Soprano	2
a08f	MR	Soprano	2
a09f	MS	Soprano	2
a10f	NK	Alto	1
a11m	AB	Bass	1
a12m	GW	Bass	1
a13m	KK	Bass	1
a14m	ŁE	Bass	1
a15m	PB	Bass	2
a16m	PR	Bass	1
a17m	RW	Tenor	2

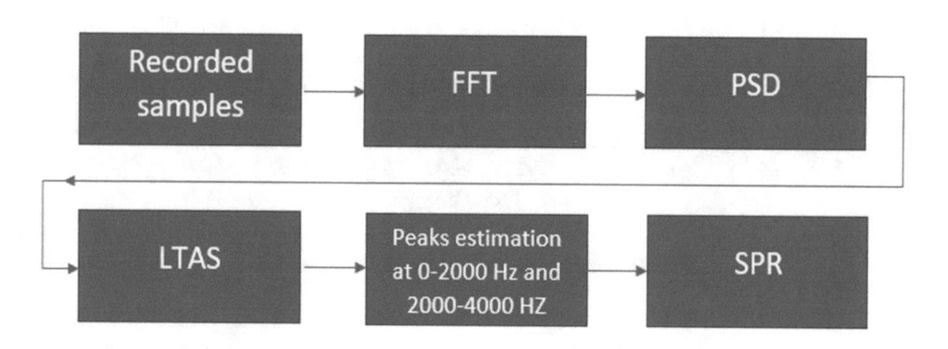

Fig. 1. Data extraction for the analysis

The base for SPR estimation may be $LTAS$. To estimate the $LTAS$ first the Power Spectra Distribution (PSD) needs to be calculated. Therefore the signal has to be divided into independent segments where each one contains N samples. Then, for each segment a Fourier spectrum is calculated. In the next step the PSD function for each Fourier spectrum has to be calculated (Eq. 1).

$$PSD(k) = \frac{|X(k)|^2}{Ndt} \tag{1}$$

where:

$X(k)$ - signal spectrum,
dt - signal duration,
N - quantity of signal samples.

Taking mean value of each PSD the $LTAS_{dBHz}$ (Eq. 2) is calculated which is normalized $LTAS$ function. (The $LTAS_{dBHz}$ better reflects the hearing curve and thus better corresponds to the hearing impressions of the human.)

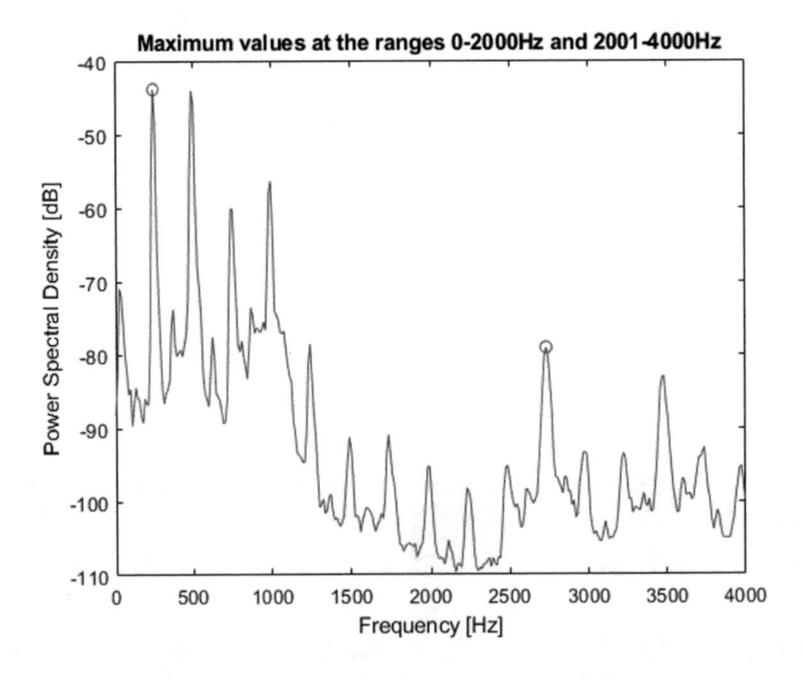

Fig. 2. Example PSD for the singer a17m

$$LTAS_{dBHz}(f) = 10log_{10}\frac{LTAS(f)}{{P_0}^2} \tag{2}$$

where:

P_0 - is the value of acoustic pressure understood as the threshold of human hearing at the frequency of 1 kHz, and the value of P_0 is $2 * 10^{(-5)}$ Pa. The unit of measure for $LTAS_{dB}$ is $\frac{dB}{Hz}$.

The SPR [21] is generally the ratio of the value of highest peak in the band 2–4 kHz to the value of the highest peak in the band 0–2 kHz (as presented in Fig. 2). As it was mentioned before, SPR is well known and interpreted as the parameter representing the singing voice quality - it shows how good singer is the analysed person. Lower SPR means better singer (better trained, more advanced in singing). It has to be also noticed that the SPR parameter may depend on the content of the analysed samples (for example different kinds of the sung vowels). In the case described in this article $LTAS$ was calculated from the average Power Spectral Density (PSD) obtained from a series of overlapping FFTs. The FFT was set to 4096, and the overlap size was assumed as 2048. The segments of signal were Hann-windowed. The average PSD were Gaussian-smoothed to 1/3-octave resolution. This was made by averaging FFT spectrum values calculated over consecutive Hamming windows. To estimate SPR the maximum peaks in the two ranges of $LTAS$ were found and the differences between those peaks were obtained at each case.

The values of peaks of SPR are analysed to prove their hypothetical dependencies on the warm up.

The abbreviations used in different publications may cause doubts - at some publications $LTAS$ peak values are divided by each other [19,20] and at other publications SPR is the difference in spectrum energy at the intervals analysed for the SPR [21].

5 The Analysis and the Results

After the singing samples were acquired the SPR values were estimated for each recorded sample. It was supposed that the analysis of the data will answer the posed question whether the warm up influences positively on the singing voice of the choral singers.

The study was carried out for two situations: 1 - singing in piano and 2 - singing in forte. In each scenario the samples were collected before and after warm up. Apart from the fact that the answer to the general question, whether the singing improves the quality of the singing voice was sought, the question was also whether influence of the warm up is stronger in the case of piano or forte singing.

In order to answer the questions the below figures and tables are presented. The values of mean SPR obtained for the singers are shown in Fig. 3.

As can be seen in Fig. 3 the values don't differ too much. This results from the fact that the level of advancement of singers in the studied group was similar. The figure presents mean SPR found for each singer while singing piano. Two results are presented for each case: SPR before and after warm up. It has been assumed here that improvement is noticed if the SPR value decreases after singing. This situation occurred at 9 from 17 subjects. Does it mean that warm up "works" in the intended way? It should be noted that in the remaining cases a deterioration of the SPR (SPR was higher than before warm up) value was observed. That means a deterioration in the quality of the voice. However, what is important, the average difference of SPR "piano before" minus "piano after" was twice as high in the case of improvement than deterioration.

To further investigate the results Fig. 4 is presented. The graph shows min values of SPR calculated over the samples sung in piano.

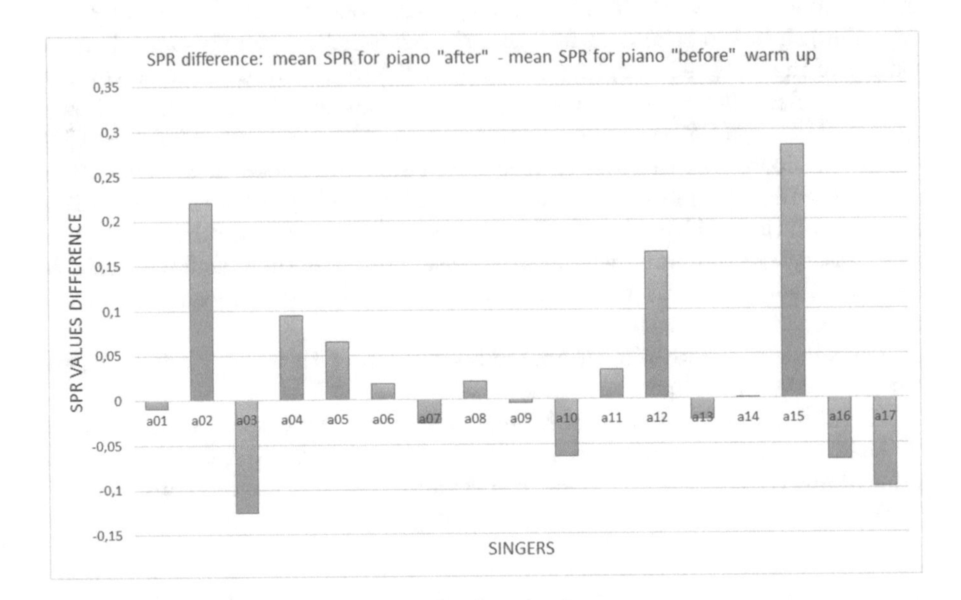

Fig. 3. Differences of mean SPR values for the choral singers while singing piano (after-before)

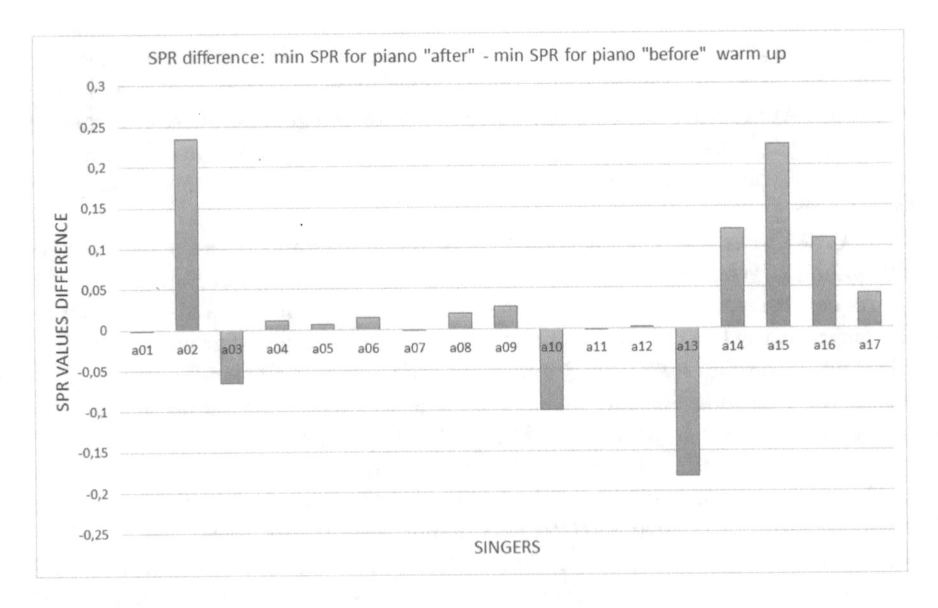

Fig. 4. Differences of minimum SPR values for the choral singers while singing piano (after-before)

Table 2. SPR difference values in the context of the warm up influence on the quality of singing voice

SPR_DIFF			
Mean		Minimum	
Piano	Forte	Piano	Forte
0,0282	−0,0123	0,0037	−0,0369
The number of singers for which progress has been observed			
9	9	11	10

Finally the table summarizing some analysis of SPR values is presented.

The Table 2 needs some description. The SPR_DIFF denotes difference between SPR values of samples recorded before and after warm up. Each SPR_DIFF was calculated for two scenarios: 1 - singing in piano, 2 - singing in forte. In each scenario some statistics have been calculated, among others mean, min and max values of SPR_DIFFs for each person (in each scenario). In the table there are presented: mean of SPR differences (SPR before - SPR after warm up) while singing in piano, mean of SPR differences (SPR before - SPR after warm up) while singing in forte, mean of minimum values of SPR differences (SPR before - SPR after warm up) while singing in piano, mean of minimum values of SPR differences (SPR before - SPR after warm up) while singing in piano. Those values present the scale of progress in context of warm up. The last line presents the number of people in which the positive change of the SPR_DIFF parameter was observed. Taking into account that the number of the investigated singers was 17, it can be seen (through minimum SPR_DIFF found for each person in piano and forte) that the progress was noticed for most of the singers.

6 Conclusion

The article was focused on the analysis of the singing voice with the use of SPR parameter. The goal was to confirm the hypothesis that a warm up for singing may influence positively on the voice quality. It was assumed that the hypothetical influence of the warm up on the voice of singers may be reflected in the SPR values. The analysis has shown that values of the SPR may be consistent with abilities of the singers. Because the SPR is well researched and recognized in the field of the assessment of the singing voice quality it may be assumed that the lack of quality improvement in some cases may be caused by personal voice production problems. The final conclusion is that, at the investigated group, the positive influence of the warm up on the quality of singing voice has been observed. The effect was a little bit stronger while singing in piano dynamics. This can be explained by the fact that the investigated group is a group of amateur singers, and as precise singing in piano is more demanding for them than singing in forte, the warm up may reveal stronger influence in this case.

References

1. Elliot N, Sundberg J, Gramming P (1995) What happens during vocal warm up? J Voice 9(1):37–44
2. Campbell JE (2008) A comparison of a hierarchical vocal function warm up regimen and a routine calisthenic warm up procedure in the choral ensemble rehearsal. University of Southern California
3. Sadolin C (2000) Complete vocal technique. Shout Publishing Copenhagen, Denmark
4. Raphael BN, Sataloff RT (2006) Increasing vocal effectiveness. In: Vocal health and pedagogy, volume II: advanced assessment and practice, vol 2, p 259. Plural Publishing
5. Freed SL, Raphael BN, Sataloff RT (2006) The role of the acting-voice trainer in medical care of professional voice users. In: Vocal health and pedagogy, volume II: advanced assessment and practice, vol 2, p 291. Plural Publishing
6. Phillips KH (1994) Twelve crucial minutes for voice building. Choral. Teach. Music 2(3):40 ERIC
7. Stegman SF (2003) Choral warm ups: preparation to sing, listen, and learn. Music Educators J 89(3):37–58
8. Półrolniczak E, Kramarczyk M (2017) Vocal tract resonance analysis using LTAS in the context of the singer's level of advancement. In: Hard and soft computing for artificial intelligence, multimedia and security, pp 249–257
9. Hammarberg B, Fritzell B, Gaufin J, Sundberg J, Wedin L (1980) Perceptual and acoustic correlates of abnormal voice qualities. Acta oto-laryngologica 90(1–6):441–451
10. Półrolniczak E, Kramarczyk M (2015) Computer analysis of the noise component in the singing voice for assessing the quality of singing. Przeglad Elektrotechniczny 91:79–83
11. Williams J (2011) Warm ups: what exactly are we trying to achieve
12. Formby C, Monsen RB (1982) Long-term average speech spectra for normal and hearing-impaired adolescents. J Acoust Soc Am 71(1):196–202
13. Wendler J, Doherty ET, Hollien H (1980) Voice classification by means of long-term speech spectra. Folia Phoniatrica et Logopaedica 32(1):51–60
14. Pavlovic CV, Rossi M, Espesser R (1990) Statistical distributions of speech for various languages. J Acoust Soc Am 88(S1):S176–S176
15. Byrne D, Dillon H, Tran K, Arlinger S, Wilbraham K, Cox R, Hagerman B, Hetu R, Kei J, Lui C et al (1994) An international comparison of long-term average speech spectra. J Acoust Soc Am 96(4):2108–2120
16. Pittam J, Gallois C, Callan V (1990) The long-term spectrum and perceived emotion. Speech Commun 9(3):177–187
17. Rossing TD, Sundberg J, Ternstrom S (1986) Acoustic comparison of voice use in solo and choir singing. J Acoust Soc Am 79(6):1975–1981
18. Pittam J (1987) Discrimination of five voice qualities and prediction to perceptual ratings. Phonetica 44(1):38–49
19. Omori K, Kacker A, Carroll LM, Riley WD, Blaugrund SM (1996) Singing power ratio: quantitative evaluation of singing voice quality. J Voice 10(3):228–235
20. Lundy DS, Roy S, Casiano RR, Xue JW, Evans J (2000) Acoustic analysis of the singing and speaking voice in singing students. J Voice 14(4):490–493
21. Watts C, Barnes-Burroughs K, Estis J, Blanton D (2006) The singing power ratio as an objective measure of singing voice quality in untrained talented and nontalented singers. J Voice 20(1):82–88

Evaluation of a Feature Set with Word Embeddings to Improve Named Entity Recognition on Tweets

Onur Büyüktopaç and Tankut Acarman(✉)

Computer Engineering Department,
Galatasaray University,
Ortaköy, 34349 İstanbul, Turkey
onurbuyuktopac@gmail.com, tacarman@gsu.edu.tr
http://gsu.edu.tr

Abstract. In this paper, we present the Named Entity Recognition system and we evaluate baseline classifiers. We use tweets as informal and noisy texts including emoticons, abbreviations, which significantly degrade the performance of classifiers. We present the dataset format, the feature set, we evaluate and test each classifier subject to different combinations of features. Finally, we discover the most representative set of features. Our experimental results show that the presented system is reached at 72% level in precision, 69% in recall and 69% in F1 (micro average), respectively.

Keywords: Named Entity Recognition · Information extraction · Twitter · Word embedding · Classification · Machine learning · Cosine similarity

1 Introduction

Named Entity Recognition (NER) identifies and categorizes textual contents such as person, thing, organization, location, product, event and character. While traditional NER approaches produce successful results when working with well structured texts, the prediction scores are much lower on unstructured microblog texts like Twitter. These type of texts contain emoticons, abbreviations, grammar mistakes and mixed languages. In this study, we present an improved approach using syntactic, semantic and domain specific features while augmenting the dataset used in [9]. For this purpose we extend the dataset, use feature scaling and evaluate all feature combinations. For benchmarking purposes, we use publicly available dataset provided by Named Entity rEcognition and Linking (NEEL) 2016 Challenge, [8]. A feature based approach with Word2Vec in addition to syntactic and domain specific features is elaborated in [9]. This study is an extension of [9] while focusing on the most representative feature set analysis and the improvement of the classifier's performance. In [5] a feature based system combining existing NER systems and domain specific Part-Of-Speech (POS)

R. Burduk et al. (Eds.): CORES 2019, AISC 977, pp. 255–263, 2020.
https://doi.org/10.1007/978-3-030-19738-4_26

tagger is presented. Candidate name generation and classical NER systems such as Stanford NER, MITIE, twitter_nlp and TwitIE are used in [7]. An adapted Kanopy system for Twitter domain is implemented in [10].

2 Data and Features

2.1 Data

We use the dataset from NEEL 2016 Challenge [8], which consists of TAB separated tweet identifier, starting index of the word, ending index of the word, link to DBpedia resource or NIL, confidence score and NER type. Dataset is provided with Google Apps Script (.gs) format as shown below:

[Tweet id] [Word start index] [Word end index] [Word dbpedia link]
[Confidence score] [NER type]

An illustrative example can be given as follows:

674869443671941120 93 101
http://dbpedia.org/resource/Egyptians 1 Thing

To create the feature vector of a word in the dataset, we analyze the tweet involving the word. But the dataset contains only labeled word information and tweet id. Finding tweets from the tweet id is a challenging task because either some tweets were already deleted or they are private. Therefore, we augment the dataset size with the whole tweet set provided by an open source project from GitHub, Webpack Bundle Analyzer [4].

We expand the training dataset from 1645 to 4073 unique tweets. Multi-class classification problem is considered. The training data is labeled by the NER types as given in Table 1. "Person" is the most labeled type in the dataset and in terms of occurrence it is followed by "Location", "Organization", "Product", "Thing","Event" and "Character". "Character" label has relatively few occurrences in the dataset.

The testing dataset is constituted by 296 unique tweets, which are labeled by the NER types. Occurrence order of each label is changed with respect to the training dataset. While the most labeled type is "Product", it is followed in terms of occurrence by "Person"', "Organization", "Character", "Thing" and "Event", respectively.

Our approach requires a vector representation to evaluate the "cos" feature. In consequence, we reduce the training data from 8665 to 7009 and the testing data from 1022 to 921 labeled words. Each NER type occurrence is also given in Table 1.

Table 1. NER type occurrences in the reduced training and testing dataset

Class	Occurrences in original training dataset	Occurrences in studied training dataset	Occurrences in original testing dataset	Occurrences in studied testing dataset
Person	2485	2197	337	331
Thing	570	576	49	42
Organization	1641	1338	158	154
Location	1868	1526	43	34
Product	1196	990	354	306
Event	482	344	24	19
Character	63	38	57	35

Finally, the collected tweets are harmonized and merged with the NEEL dataset and stored in a format where each sample contains the following information: tweet identifier, tweet, start index of the word, end index of the word, NER type and the word.

[Tweet id] [Tweet] [Word start index] [Word end index] [NER type] [Word]

As an example:

674869443671941120 RT @EntheosShines: Just As Some Parents Have A Favorite Child, Obama Has Favorites (sign at Egyptian Airport) @chirofrenzy @PatVPeters htt... | 93 101 Thing Egyptian

2.2 Feature Set

We label words from tweets with NER types. Each feature represents a numeric value in the feature vector, which is an input to the classifier.

'#'. Word starts with '#'. Checking the word whether it starts with hashtag '#'. Twitter uses hashtags to categorize tweet topics. This information leads us to find that the word with a hashtag can be a highly probable valuable entity.

'@'. Word starts with '@'. Checking the word whether it starts with "at" sign '@'. Twitter uses at sign to mention someone. It means the word starts with an "at" sign is a person or an organization.

'Title'. Word starts with capital letter. Checking the word whether the first letter is started with uppercase and followed by lowercase. This feature reduces the probability of the word being a proper noun.

'All capital'. Word is all capital. Checking the word whether all letters are uppercase. This type of words is probably used to highlight something. Highlighted words have a higher probability being an entity than words written with lowercase letters.

'POS'. Stanford POS Tagger. A Part-Of-Speech Tagger (POS Tagger) is a piece of software that reads text in some language and assigns parts of speech to each word such as noun, verb, adjective, etc., [11].

'Next POS'. Next word POS tagger. Labeling POS tag of the next word.

'Position'. Ranking of the word position with respect to the total word count of tweet.

$$Rank = \frac{(Wordposition)}{(Totalwordcount)} \quad (1)$$

'Cos'. Cosine Similarity. Calculating the cosine similarity with the corpus of 400 million tweets [6] and the word. First, we calculate an average vector for each NER type. Then, every average vector is compared with the word and each average vector is represented as an additional feature.

3 Tests and Evaluation

3.1 Implementation

We implement our algorithm using Python programming language with "genism" library [2] in order to execute word embedding through "word2vec" and "sklearn" library [3] for feature scaling and classifying trained and tested results. We use the following 9 supervised classification algorithms from "scikit-learn" library; Logistic Regression, Support Vector Machines (SVM), k-Nearest Neighbors (k-NN), Gaussian Naive Bayes (Gaussian NB), Bernoulli Naive Bayes (Bernoulli NB), Extra Tree, Decision Tree, Random Forest, and Gradient Boosting. For tokenizing and the Part-of-Speech (POS) tagging we use "nltk" library [1]. The library contains an adapter of Stanford POS Tagger, which is originally written in Java.

3.2 Test and Results

We evaluate 9 baseline classifiers subject to different combinations of 8 features. Firstly, we examine the highest statistical metric values. Secondly, we evaluate features and classifiers to discover the most representative feature set. We use feature scaling to scale measurements in the given range. The k-NN classifier when k = 5 by using the feature "**cos**", "**all capital**" and "#" achieves the highest statistical metric values; precision of 0.72, recall of 0.69 and F1 micro average of 0.69. k-NN classifier predicts 639 true NER type correctly from 921 ground truth NER type, confusion matrix is given in Table 2.

Table 2. Confusion matrix of k-NN with "cos", "all capital" and "#" features for 7 NER Types

Class	Person	Thing	Organization	Location	Product	Event	Character
Person	234	1	19	64	12	0	1
Thing	2	30	4	0	6	0	0
Organization	55	9	64	5	21	0	0
Location	7	1	1	24	1	0	0
Product	8	3	22	3	269	1	0
Event	0	3	1	0	7	8	0
Character	10	0	8	1	3	3	10

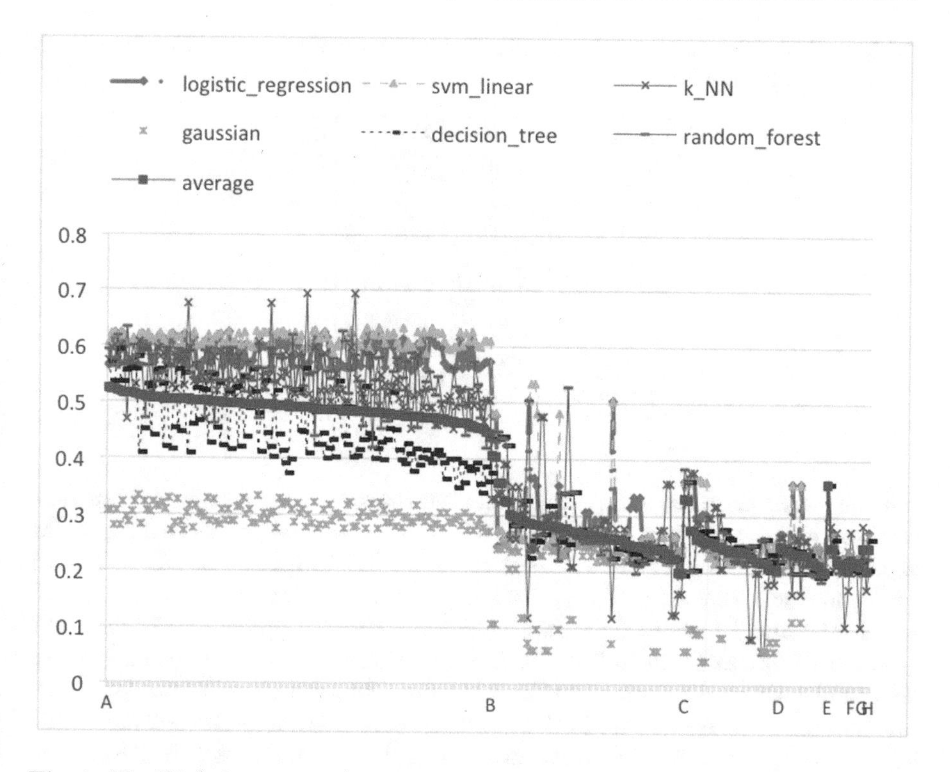

Fig. 1. The F1 (micro average) statistical metric value responses of the classifiers are compared with respect to the feature sets. Horizontal axis A, B, C, D, E, F, G, H denotes "cos", "title", "all capital", "POS", "next POS", "@", "#" and"position", respectively.

In Fig. 1, testing results are compared and F1 (micro average) statistical metric values are plotted for 9 classifiers subject to the combination of 8 features. The Cosine Similarity feature "**cos**", which is denoted by A in Fig. 1 is a representative feature for all classifiers. Then, the title feature is evaluated and classifiers have reached at lower level in F1 value. The SVM classifier's F1

Table 3. Confusion matrix of SVM linear kernel with "cos", "all capital", "POS" "@" and "#" features for 7 NER Types

Class	Person	Thing	Organization	Location	Product	Event	Character
Person	300	3	9	0	7	11	1
Thing	0	35	1	0	0	6	0
Organization	54	10	66	1	4	19	0
Location	5	2	3	22	1	1	0
Product	3	4	7	1	130	161	0
Event	0	1	0	0	1	17	0
Character	7	0	1	2	7	3	15

Table 4. F1 (minimum average) scores subject to the presence and the absence of each feature excluding the "cos" feature

Feature	Presence	Absence	Ratio
"#"	0.249722	0.260481	−1%
"@"	0.251428	0.258747	−1%
"title"	0.272437	0.237406	4%
"all capital"	0.267512	0.242409	3%
"POS"	0.263657	0.246325	2%
"Position"	0.24689	0.263358	−2%
"next POS"	0.25533	0.254783	0%

value is higher at average in comparison with the k-NN classifier, both classifiers reach at higher metric value by using the same combination of features. SVM algorithm assures higher value subject to "**cos**", "**all capital**", "**POS**" "@" and "#" and linear kernel algorithm reaches at level in precision about 0.80, recall about 0.64 and F1 about 0.63. Linear kernel algorithm predicts 585 true NER type out of 921 as given in Table 3. Overall, different combinations of features give similar average F1 statistical metric values for each classifier except Gaussian and Bernoulli algorithm. In the confusion matrix given in Tables 2 and 3, the high values appear between two classes of organization and person because some organizations have person names including organizations funded for donation or foundation.

The "cos" feature has an impact to the classifier algorithm performance and particularly to enhancement of the F1 (micro average) statistical metric value. To discover the most representative features for all classifiers, we focus on features improving the statistical metric values by simply removing and adding it into the feature set for cross checking. When the "cos" feature is removed, we calculate the average F1 statistical metric value subject to the existence or absence of each

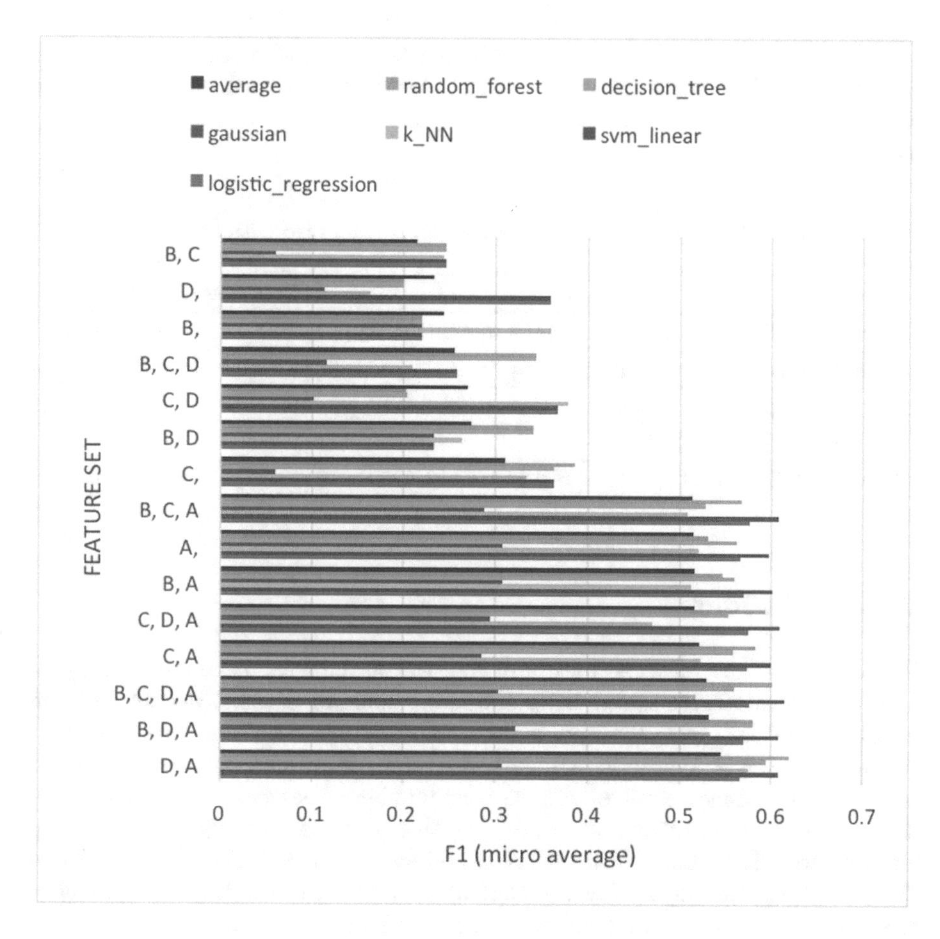

Fig. 2. Selected feature sets over the average of all classifiers. Vertical axis A, B, C, D represents in the following order "cos", "title", "all capital" and "POS".

individual feature. In Table 4, the average F1 value change gives the impact of each individual feature's impact to the classification performance.

On the one hand, according to the average F1 value subject to the presence or the absence of the given feature in Table 4, features having an impact to the improvement of classifier's performance is "title" followed by "all capital" and "POS", respectively. On the other hand, "#", "@", and "position" affects negatively the average of F1 metric values while the feature "next POS" does not contribute to the classifier's performance. We observe the most representative features as "cos", "title", "all capital" and "POS" for all classifiers. Then, in Fig. 2 we investigate different combinations of representative features and we discover the feature set of "cos" and "POS" has an impact to the classifier performance. When these two features are used, classifiers reach at higher level in F1 metric value.

In Table 5 the presented NER approach is compared with respect to the literature benchmarking the same dataset. Methods [5,7] presented during the NEEL 2016 workshop, and the method presented by [9] evaluates these three methods while testing the same dataset.

Table 5. The performance of our approach is compared with other studies

Study	Precision	Recall	F1
Our approach (k-NN, 2 features + Cosine Similarity)	0.72	0.69	0.693
A feature based approach performing Stanford NER, [5]	0.729	0.626	0.674
Logistic Regression, 5 features + Cosine Similarity, [9]	0.71	0.56	0.58
TwitIE (CRF Model), [10]	0.435	0.459	0.447
Stanford NER, MITIE, twitter_nlp and TwitIE, [7]	0.587	0.287	0.386

4 Conclusions

Because of the unstructured nature of the tweets, the features, which are representative on more structured texts like newspapers or articles, do not have an impact to the classifiers. To discover the most representative feature set, we investigate and analyze syntactic, semantic and domain specific features extracted from a richer dataset. First we increase the number of descriptive features in the feature set for a better classification. For this purpose, in addition to existing features, we add two new features. We also increase the volume of dataset to be able to extract the impact level of each individual feature to the level in statistical metric value of each classifier. We execute all combinations of features over all classifiers and we discover the feature set, which gives the highest F1 (micro average) score. We also observe the highest F1 (micro average) score assured by a given classifier varies largely since all features are not representative and the weight of a particular feature needs to be reduces, i.e., removing this particular feature in our case. Then, we analyze the impact of the feature combinations over classifiers and we observe the most representative features being "cos", "title", "all capital" and "POS", which assures classifiers reaching at acceptable statistical metric values at average. Finally, the "cos" and "POS" feature set has an impact to the NER classifiers.

Acknowledgements. The authors gratefully acknowledge the support of Galatasaray University, scientific research support program under grant #18.401.002.

References

1. A platform for building Python programs to work with human language data. https://www.nltk.org/. Accessed 8 Mar 2019
2. A Python library for topic modelling, document indexing and similarity retrieval with large corpora. https://pypi.python.org/pypi/gensim. Accessed 8 Mar 2019
3. A Python module for machine learning. http://sklearn.org/stable/index.html. Accessed 8 Mar 2019
4. Webpack Bundle Analyzer. https://github.com/webpack-contrib/webpack-bundle-analyzer. Accessed 8 Mar 2019
5. Ghosh S, Maitra P, Das D (2016) Feature based approach to named entity recognition and linking for tweets. In: #Microposts
6. Godin F, Vandersmissen B, De Neve W, Van de Walle R (2015) Multimedia lab @ ACL WNUT NER shared task: named entity recognition for twitter microposts using distributed word representations. In: Proceedings of the workshop on noisy user-generated text, pp 146–153. Association for Computational Linguistics. https://doi.org/10.18653/v1/W15-4322. http://aclweb.org/anthology/W15-4322
7. Greenfield K, Caceres RS, Coury M, Geyer K, Gwon Y, Matterer J, Mensch AC, Sahin CS, Simek O (2016) A reverse approach to named entity extraction and linking in microposts. In: #Microposts
8. Rizzo G, van Erp M, Plu J, Troncy R (2016) Making sense of microposts (#microposts2016) Named Entity rEcognition and Linking (NEEL) challenge. In: Proceedings of the 6th workshop on 'Making Sense of Microposts', pp 50–59
9. Taşpınar M, Ganiz MC, Acarman T (2017) A feature based simple machine learning approach with word embeddings to named entity recognition on tweets. In: Frasincar F, Ittoo A, Nguyen LM, Métais E (eds) Natural Language Processing and Information Systems. Springer International Publishing, Cham, pp 254–259
10. Torres-Tramón P, Hromic H, Walsh B, Heravi BR, Hayes C (2016) Kanopy4tweets: entity extraction and linking for twitter. In: #Microposts
11. Toutanova K, Klein D, Manning CD, Singer Y (2003) Feature-rich part-of-speech tagging with a cyclic dependency network. In: Proceedings of the 2003 conference of the North American chapter of the association for computational linguistics on human language technology - Volume 1, NAACL 2003, pp 173–180. Association for Computational Linguistics, Stroudsburg. https://doi.org/10.3115/1073445.1073478

Introducing Action Planning to the Anticipatory Classifier System ACS2

Olgierd Unold(✉), Edyta Rogula, and Norbert Kozłowski

Department of Computer Engineering,
Wrocław University of Science and Technology,
Wyb. Wyspiańskiego 27, 50-370 Wrocław, Poland
{olgierd.unold,norbert.kozlowski}@pwr.edu.pl

Abstract. This paper introduces and tests Action Planning mechanism in the Anticipatory Classifier System ACS2. Action Planning implies goal-directed learning and bidirectional search to strengthen reliable classifiers. It is shown that it can speed up the process of gaining knowledge about the learned environment. Experiments were performed over three environments (Hand-Eye, Maze and Taxi) extended with custom goal-generator functions.

Keywords: Learning Classifier System · Anticipatory Classifier System · Action Planning · OpenAI Gym

1 Introduction

Anticipations play an important role in our lives. They enable us to pursue both short- and long-term goals. By anticipating the consequences of our actions we are able to choose the best possible action to achieve them.

Anticipatory Classifier System (ACS) is a variant of Learning Classifier System (LCS) extending the classical human-interpretable, rule-based model with the psychological theory of anticipations. Every situation is accompanied by consequences after performing each possible behaviour. ACS is well-tested in both single- and multi-step interaction processes like knowledge discovery [11] or controlling mobile robot's arm [10]. Action Planning is a mechanism that uses known reliable classifiers to set a path of actions from the current state to some appointed goal. Previous studies [2,10] with the first anticipatory learning classifier system ACS [9] showed that ACS can use action planning to improve an animat environment learning process.

The work was supported by statutory grant of the Wrocław University of Science and Technology, Poland.

R. Burduk et al. (Eds.): CORES 2019, AISC 977, pp. 264–275, 2020.
https://doi.org/10.1007/978-3-030-19738-4_27

This work describes an integration between Anticipation Learning and the Action Planning (AP) in an enhanced version of ACS - ACS2 [5]. AP was implemented as an extension to PyALCS library[1] and testing environments were adapted to be OpenAI Gym [1] compliant.

The main contributions of the present research are as follows:

1. The Action Planning mechanism was integrated into the ACS2 model.
2. New goal-generators were proposed.
3. The Action Planning mechanism was tested in two new environments (Taxi, Maze).
4. The exhaustive experiments were performed and an influence of Action Planning onto the overall knowledge of the environment was analysed.
5. The framework for replicating experiments by using open access PyALCS library and OpenAI Gym environment was proposed.

This paper is divided into five main sections. Section 2 presents a general concept of ACS2. In Sect. 3 the AP mechanism is described including goal-generator and bidirectional search. Later on, in Sect. 4 three environments are used to evaluate the algorithm performance. Final conclusions and problems spotted so far are listed in Sect. 5.

2 Anticipatory Learning Classifier Systems (ACS2)

Learning Classifier Systems are rule-based machine learning algorithms that were created on the grounds of evolutionary biology and artificial intelligence. Their definition isn't strict and many other systems can be based on it. Below is the brief description of Anticipatory Learning Classifier Systems.

Anticipatory Learning Classifier System (ACS) was originally proposed by Stolzmann [9]. It is an LCS concept extended with psychological mechanism of the anticipative behavioural control which was introduced in cognitive psychology by Hoffmann [7]. In ACS as opposed to LCS, the knowledge is not represented by "condition-action" pairs, but by "condition-action-expectation" triples. Therefore, ACS can learn successfully even without an environmental reward and execute additional mechanisms (like Action Planning). Classifier structure consists of a rule and it's metadata - like a mark M (holding states where the classifier did wrong), quality q and anticipated reward r. By convention, classifiers with quality bigger than a specified threshold ($\theta_r > \theta$) are called reliable. Condition (C) describes an environmental state in which the classifier can be used. Action (A) is an action which can be performed in this state. Expectation (E) represents the anticipated state after executing action (A) in (C). Condition's and expectation's alphabet contain not only the characters specific for the environment, but also a '#' symbol. In condition string it's called

[1] https://github.com/ParrotPrediction/pyalcs.

the "*don't care*" and means that specified allele matches every state (increasing generality of a classifier). For the effect string it's called a *"pass-through"* symbol and states that certain allele remains unchanged after executing the action (Fig. 1).

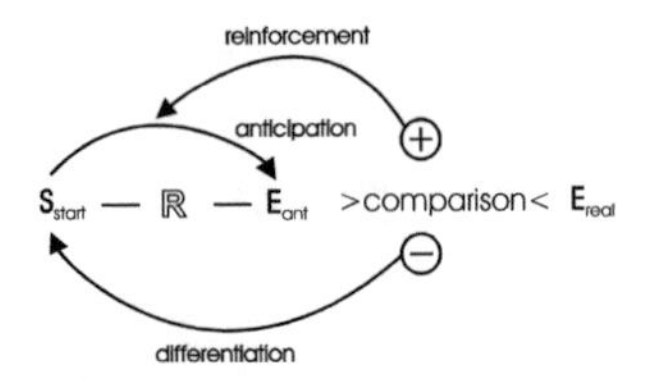

Fig. 1. The theory of anticipatory behavioral control. Figure adapted from [3, p. 4].

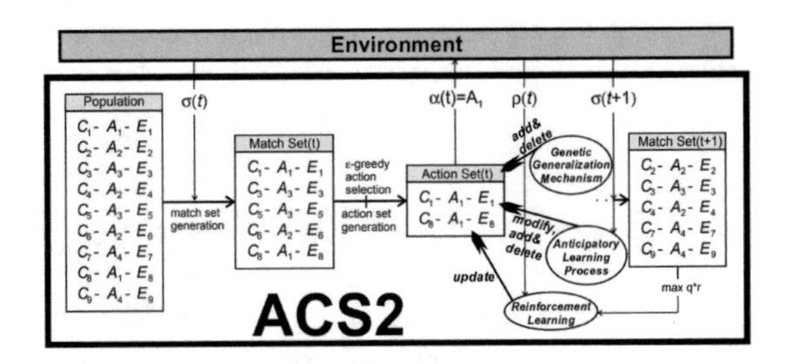

Fig. 2. A behavioral act in ACS2 with reinforcement learning and anticipatory learning process application. Figure adapted from [4].

Behavioral act (see Fig. 2) in ACS starts with generating a match set which is created by using the current snapshot of the environment (agent's perception of it). From a set of all classifiers (called a population [P]) only the classifiers with condition string matching the perception are chosen, forming a match set [M]. Next, the action is selected by using certain strategy. It can be picked randomly, the action from the most promising classifier (highest fitness) can be chosen or other more sophisticated techniques can be used. Then, the action set [A] is created. It contains all of the classifiers from the match set which have the chosen action in their action part. Afterward, the action is executed in the environment and all classifiers from the action set are modified based on the new perception from the environment (*Anticipatory Learning Process*) and the received reward (*Reinforcement Learning*). The next match set is created using the new perception and the process of learning is continued. Sometimes, over-specializations of classifiers can occur so the *Genetic Generalization Mechanism* is applied. Is selects classifiers from the action set [A] proportionally to the fitness value. Then the condition parts of classifiers are generalized by a mutation operator and crossed over. The whole algorithm is described in detail in [5].

3 Action Planning

The Action Planning is an extension to the ACS and can be described with the Algorithm 1.

Algorithm 1. Action Planning

```
1  while true do
2  |   goal ← get from the environment's goal-generator.
3  |   if goal does not exist then
4  |   |   end
5  |   else
6  |   |   using bidirectional search for reliable classifiers, find a path to the goal
7  |   |   if a path is found then
8  |   |   |   perform all classifiers that lead to the goal
9  |   |   end
10 |   end
11 end
```

The main goal of the AP mechanism is to precipitate agent's knowledge acquisition. It uses an external goal-generator which is a part of the environment and returns desirable states. Next, the agent searches for the shortest known path from the current state to the goal state using only reliable classifiers. If there aren't enough reliable classifiers to find such path, the AP process ends and the agent comes back to the exploration phase. Otherwise, the agent executes found actions, enhancing the classifiers (like in the exploration phase). When the goal is reached the whole cycle repeats.

3.1 Goal-Generator

In order to make use of the AP mechanism, agent needs to know what is the desired goal to reach. Therefore each evaluated environment must be enhanced with *goal-generator* feature. It's purpose is to notify the agent about the next desired state to accomplish.

Because the main purpose of the AP is to speed up reasoning about the environment, it might be also beneficial to utilize the goal-generator to suggest less frequent or less probable states (ex. gripping a box) for exploration (even in dynamic way).

For example in Hand-Eye environment [10] described wider in Sect. 4.1 (where some logical sequence of steps is required to reach the main goal) it might look like:

1. Move gripper over the box.
2. Grip the box.
3. Move gripper to another place.
4. Release the box.
5. Move gripper to another place.

Goal-generator states are also useful when in the same situation two actions can be done (for example moving the box to another place or releasing it) - then the states determine, which action should be preformed in certain situation.

Since the goal-generator is a part of the environment and not the AP mechanism, it has more knowledge about the environment than it is given in its perception. Therefore, goal-generator has a chance to generate goals which are important to the environment but the agent doesn't know the reasoning behind that.

3.2 Bidirectional Search

Bidirectional search is a searching mechanism that works in two ways - from the current state to the goal state (forward) and from the goal state to the current state (backward). It is a breadth-first search that doesn't include states which already occurred.

The next forward state is assigned by using the anticipation part of the reliable classifier. It is analogical for the next backward state but the condition and anticipation parts are switched.

Bidirectional search is effective in most cases but only in environments where perception is explicit, meaning one perception can describe only one state. In other cases bidirectional search (or any other search) can return incorrect results because identical perceptions would describe different environment states.

Bidirectional search always finds the shortest known path using the reliable classifiers. When there are not enough reliable classifiers, the path might be longer or it can be not found at all. In the latter case the AP mechanism ends.

4 Experiments

AP was implemented as a part of PyALCS Python library. All testing environments[2] were created in full compliance with the OpenAI Gym [1] interface. Doing so facilitates other researchers an easier comparison and benchmarking by using a common standard.

4.1 Environments

Experiments with AP mechanism were performed on three environments - Hand-Eye [10], Maze [8] and Taxi[3] [6]. The two last ones were expanded with goal-generator functions.

Hand-Eye is a discrete environment with three main elements: a square plain, a box and a robot arm (*Hand*). The camera (*Eye*) is observing all of these elements from above. The robot has also a sensor feeling the box underneath it. In this environment perception consist of discretized view on the whole plain and one

[2] https://github.com/ParrotPrediction/openai-envs.
[3] https://gym.openai.com/envs/Taxi-v2/.

sensor informing whether there is a box under the robot arm. Robot can perform one of the six actions - moving in four adjacent directions (north, south, east, west) plus gripping and releasing the box.

This environment is parametrized by the size of the squared plain. For example $n = 3$ means the plain is 3×3. It also doesn't have a specific goal with a reward. The reward is always equal to zero, making it harder for the agent to learn.

Maze. In this environment the agent explores a two dimensional maze trying to reach a goal. It can move in eight directions (north, north-east, east, south-east, south, south-west, west and north-west) and perceives only eight neighbouring cells (not the whole map). The environment difficulty can be balanced by choosing different maze configuration. Depending on that it's also possible that some non-deterministic perceptions will be presented to the agent.

The proposed goal-generator consists of two steps:

1. Reach the state one step away from the main goal
2. Obtain the main goal.

Taxi. Problem similar to Hand-Eye - the agent (taxi, hand) is moving an object (passenger, box) from one place to another. The difference is that in the Taxi environment the places (source and destination) are set (each one is chosen randomly from one of four locations) and perception consist of one element only - a number encoding current state of the environment. In addition to that, the Taxi environment has -1 reward for every action, -10 reward for an illegal *"Put down"* and *"Pick up"* actions and $+20$ reward for successfully delivering the passenger to his destination. To achieve the maximum reward, the agent needs to execute the following steps:

1. Move to the passenger location.
2. Pick up the passenger.
3. Move to the destination location.
4. Put down the passenger.

A goal generator created for this environment is suggesting the correct order of actions and increases the probability of obtaining maximum possible reward.

4.2 Results

All experiments were performed in the explore mode with 10 exploit trials performed at the end of the experiment, as described in Algorithm 2. Metrics were collected for each trial. To compare the results, the experiments were performed with AP on and off. Action Planning Frequency f determines in which trial the AP phase should be applied.

Algorithm 2. Experimental protocol

Data: n - number of experiments,
e - name of the environment,
$n_{explore}$ - number of trials in explore mode,
θ_r - reliability threshold,
b - whether we should apply AP,
f - Action Planning frequency
Result: Mean metrics of the experiments

1 $i \leftarrow 0$
2 $env \leftarrow$ create environment e
3 build agent configuration using θ_r, b and f
4 mean metrics explore $\leftarrow$ []
5 mean metrics exploit $\leftarrow$ []
6 **while** $i < n$ **do**
7   explore metrics $\leftarrow$ results of $n_{explore}$ trials in explore mode in env
8   exploit metrics $\leftarrow$ results of 10 trials in exploit mode in env
9   mean metrics explore $\leftarrow$ count mean of explore metrics
10   mean metrics exploit $\leftarrow$ count mean of exploit metrics
11   i++
12 **end**
13 save mean metrics to file

Hand-Eye. Parameters: $\beta = 0.05$, $\gamma = 0.95$, $\theta_r = 0.9$, $\theta_i = 0.1$, $\epsilon = 1.0$, $\theta_{GA} = 100$, $m_u = 0.3$, $\chi = 0.8$, Action Planning Frequency $= 50$.

The most interesting thing in testing the AP mechanism is the overall knowledge about the environment. First all possible transitions that cause a change in the environment were determined and later it was checked how many of these moves can be predicted using the reliable classifiers.

The ACS agent is not deterministic, so the experiments were repeated 100 (for 2×2 and 3×3 environments), 50 (4×4 environment) or 10 (5×5 environment) times and then averaged.

As shown in Fig. 3 the agent at first quickly gains knowledge about the environment - for example in 20 trials for 5×5 environment it gains 85% knowledge. After this point learning becomes really slow. Using AP speeds up the learning process.

The results show that using AP makes the process of learning the environment also shorter in all tested sizes of the environment. It demonstrates that benefits are bigger for larger environments.

Comparing the achieved results with [10] AP in the PyALCS library achieves nearly as good results. The differences may occur due to different parameter settings (which were not revealed in the original work).

Maze. Parameters: $\beta = 0.05$, $\gamma = 0.95$, $\theta_r = 0.9$, $\theta_i = 0.1$, $\epsilon = 1.0$, $\theta_{GA} = 100$, $m_u = 0.3$, $\chi = 0.8$, Action Planning frequency $= 30$.

The knowledge measured in Maze environment is calculated in the same way as in the Hand-Eye example above. All experiments were performed 10 times and then averaged.

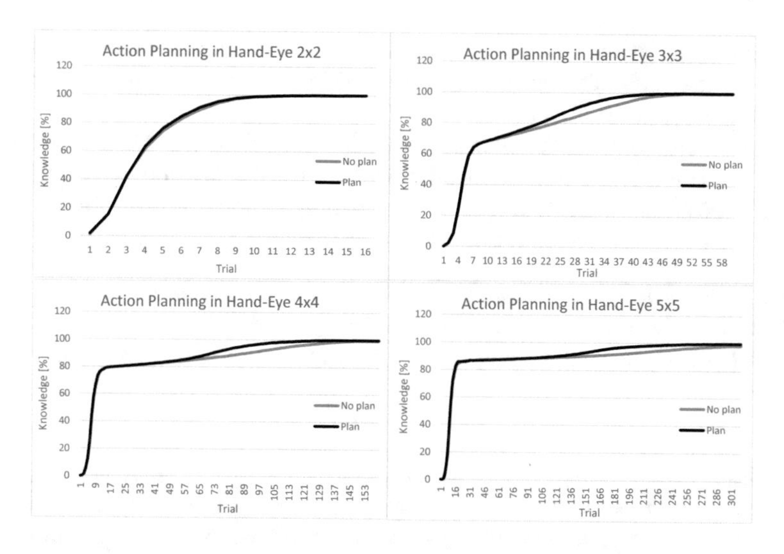

Fig. 3. Action Planning in Hand-Eye environment of various sizes.

There have been no prior experiments with AP on Maze environment before so there was a need to establish not only a goal-generator function but also the AP frequency parameter. It was done by performing preliminary experiments on the Maze4-v0 environment.

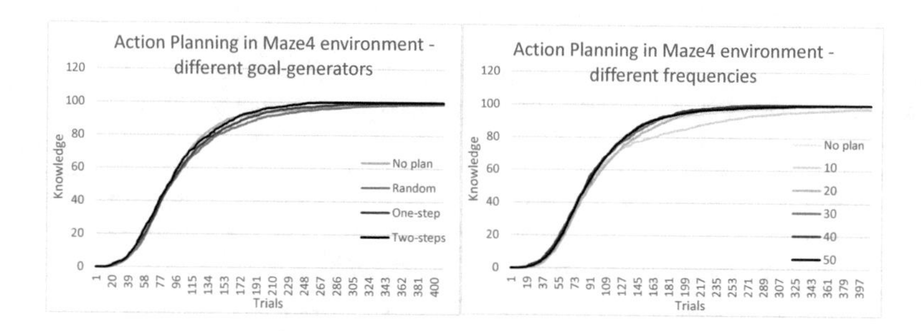

Fig. 4. Action Planning in Maze4-v0 environment - different goal-generators and frequencies.

Firstly, three goal-generator functions were tested: completely random function, one-step function leading straight to the goal and two-steps function described before (in Sect. 4.1). Random function only worsen the results, one-step function made it slightly better but the best ones were obtained with the two-steps function (Fig. 4). Therefore, in the succeeding experiments, only the two-steps function was used.

Secondly, the AP frequency had to be determined. Because in each Maze trial the agent can perform up to 50 steps before it's terminated, the following frequencies were evaluated: 10, 20, 30, 40 and 50. The best results were achieved with the value of 30. This frequency was used in all succeeding experiments. The results can be seen in Fig. 4.

The effect of AP has significantly smaller impact in the Maze environment than in the Hand-Eye environment. Sometimes it's application even worsens the results. The possible cause might be the wrong strategy chosen for generating desired states. Below three distinct Maze environments are presented:

- **MazeF3-v0** in which using AP makes the results much worse,
- **Maze5-v0** in which using AP makes nearly no difference,
- **Woods14-v0** in which using AP makes the results better (Fig. 5).

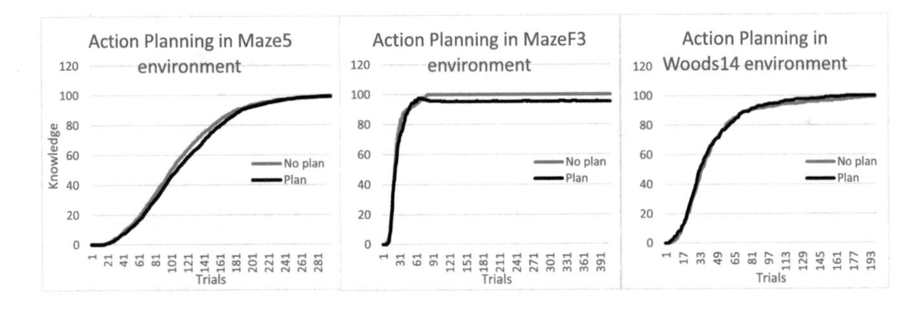

Fig. 5. Action Planning in different Maze environments.

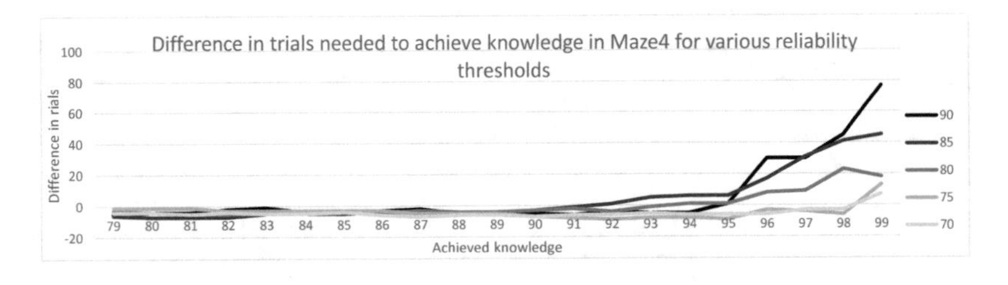

Fig. 6. Difference in trials needed to gain knowledge between using AP and not using AP for various reliability thresholds in the Maze4-v0 environment.

Additional experiments were performed to check whether changing the reliability threshold θ_r would help AP learn the knowledge faster. It indeed gained the knowledge faster but it was a consequence of the way the knowledge is measured - using the reliable classifiers. Lower threshold of reliability means there are more reliable classifiers, so the obtained knowledge is automatically higher. Figure 6 shows the difference when using AP for testing the reliability threshold. The difference in trials needed to gain knowledge was counted using Eq. 1. The positive difference (meaning AP is better than no planning) is smaller for lower reliability threshold. This might mean that the gained knowledge is bigger but the AP mechanism doesn't make a big difference at all.

$$\Delta_t = t_p - t_n \tag{1}$$

where:

Δ_t - difference in trials needed to achieve knowledge,
t_p - trials needed to gain knowledge when using AP,
t_n - trials needed to gain knowledge when not using AP.

Taxi. Parameters: $\beta = 0.05$, $\gamma = 0.95$, $\theta_r = 0.9$, $\theta_i = 0.1$, $\epsilon = 1.0$, $\theta_{GA} = 100$, $m_u = 0.3$, $\chi = 0.8$, Action Planning frequency $= 30$.

In this environment, there were two metrics taken into consideration - knowledge (to compare it with other environments) and the cumulative reward (sum of all rewards from one trial, to compare it with other agents). Only one goal-generator was tested due to the results of previous experiments. In Taxi environment, a goal-generator was created analogously, as described in Sect. 4.1.

Two kinds of experiments were performed - first one to confirm that the AP frequency of 30 is better then the frequency of 50 for the Taxi environment. That was confirmed and results are presented in Fig. 7. It is worth noting, that for Cumulative Reward plot (Fig. 7) the moving average was used and in some trials the cumulative reward even above zero was obtained.

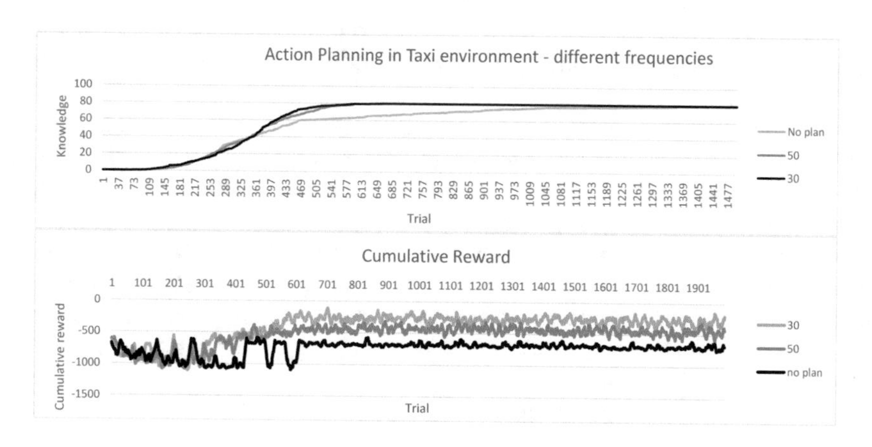

Fig. 7. AP in Taxi environment - different frequencies

Experiments checking if the AP is dependant of reliability threshold were performed. In the Taxi environment, this effect (higher reliability threshold making greater impact) was even greater than in the Maze environment, see Fig. 8. The cause of this effect is yet unknown. For the achieved knowledge, it is better to utilize AP. For example, the knowledge of 80% is achieved in 655 trials with AP

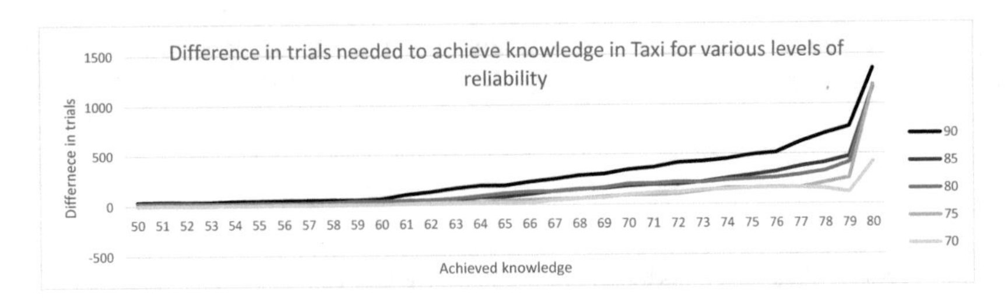

Fig. 8. Difference in trials needed to achieve knowledge between using AP and not using AP for various levels of reliability in the Taxi environment.

and it is not achieved at all in 2000 trials without it. The compared results are even better than in the Hand-Eye environment.

As for the cumulative reward in the explore phase it is a lot better to use the AP - sometimes the cumulative reward is positive which is difficult to reach in this environment (means that the passenger was delivered in less than 20 steps). In the exploit phase though there is no difference if the AP mechanism was used or not - the agent always chooses "safe" actions and the cumulative reward equals −200 for 200 moves, which means it didn't perform illegal actions but it also didn't succeed in delivering the passenger.

5 Conclusions

AP mechanism facilities the process of gaining knowledge about the environment. Some flaws were spotted, but in general the impact was positive on all tested environments.

The main downside of the mechanism is the goal-generator which is essential for it to work. Most (if not all) of the environments compatible with the OpenAI Gym doesn't have a goal-generator. It's the researcher duty to create an "environment-tailored" goal-generator.

Another flaw of the AP mechanism, which probably cannot be fixed, is the problem of using it in non-deterministic environments (without explicit perceptions). In this case the bidirectional search will most probably fail while searching for the path from the current state to the goal state, but it won't know whether the path is correct or not.

Other aspects requiring further investigation:

1. **Goal-generators.** For each tested environment, there is a need to create goal-generator. The presented ones are good, but there is obviously a possibility to create better ones.
2. **Action Planning frequency** turned out to be the best equal to 30 in Taxi and Maze environment, but it wasn't properly tested in the Hand-Eye environment. Besides, the effect of bigger frequencies (ex. 60) is yet unknown.
3. **Reliability threshold.** The AP mechanism performed better (made a bigger difference) when the reliability threshold was higher. It probably needs more research to understand the reasons.

References

1. Brockman G, Cheung V, Pettersson L, Schneider J, Schulman J, Tang J, Zaremba W (2016) OpenAI Gym. ArXiv e-prints
2. Butz M, Stolzmann W (1999) Action-planning in anticipatory classifier systems. In: Proceedings of the 1999 genetic and evolutionary computation conference workshop program, pp 242–249
3. Butz MV (2002) Anticipatory learning classifier systems, vol 4. Springer, New York
4. Butz MV, Goldberg DE (2003) Generalized state values in an anticipatory learning classifier system. In: Anticipatory behavior in adaptive learning systems. Springer, pp 282–301
5. Butz MV, Stolzmann W (2002) An algorithmic description of ACS2. In: Lanzi PL, Stolzmann W, Wilson SW (eds) Advances in learning classifier systems. Springer, Heidelberg, pp 211–229
6. Dietterich TG (2000) Hierarchical reinforcement learning with the MAXQ value function decomposition. J Artif Intell Res 13:227–303
7. Hoffmann J (2016) Vorhersage und Erkenntnis. Universität Würzburg
8. Kozlowski N, Unold O (2018) Integrating anticipatory classifier systems with OpenAI gym. In: Proceedings of the genetic and evolutionary computation conference companion. ACM, pp 1410–1417
9. Stolzmann W (1997) Antizipative classifier systems. Ph.D. thesis. Fachbereich Mathematik/Informatik, University of Osnabrück
10. Stolzmann W, Butz M (1999) Latent learning and action planning in robots with anticipatory classifier systems. In: International workshop on learning classifier systems. Springer, pp 301–317
11. Unold O, Tuszyński K (2008) Mining knowledge from data using anticipatory classifier system. Knowl-Based Syst 21(5):363–370

Supervised Classification Box Algorithm Based on Graph Partitioning

Ventzeslav Valev[1], Nicola Yanev[1], Adam Krzyżak[2](✉), and Karima Ben Suliman[2]

[1] Institute of Mathematics and Informatics, Bulgarian Academy of Sciences, 1113 Sofia, Bulgaria
{valev,choby}@math.bas.bg

[2] Department of Computer Science and Software Engineering, Concordia University, Montreal, QC H3G 1M8, Canada
krzyzak@cs.concordia.ca, karima.b.soliman@gmail.com

Abstract. In this paper we introduce the supervised classification algorithm called Box algorithm based on feature space partitioning. The construction of Box algorithm is closely linked to the solution of computational geometry problem involving heuristic maximal clique cover problem satisfying the k-nearest neighbor rule. We first apply a heuristic algorithm to partition a graph into a minimal number of maximal cliques and subsequently the cliques are merged by means of the k-nearest neighbor rule. The main advantage of the new approach is decomposition of the l-class problem ($l > 2$) into l single-class optimization problems. The performance of the Box algorithm is demonstrated to be significantly better than SVM in computer experiments involving real Monk's dataset from UCI depository and simulated normal data.

Keywords: Supervised classification · Graph partitioning · Maximal clique cover problem · k-nearest neighbor rule · Box algorithm

1 Introduction

We consider the supervised classification problem in which a pattern is assigned to one of a finite number of classes. The goal of supervised classification is to learn classifier $f(x)$ that maps features $x \in X$ to a discrete label (color), $y \in \{1, 2, \ldots, l\}$ using the training data (x_i, y_i). Our proposal is to approximate f by partitioning the feature space into uni-colored box-like regions. The optimization problem of finding the minimal number of such regions is reduced to the well-known problem of minimum clique cover of a properly constructed graph. The solution results in feature space partitioning. We provide a brief survey of relevant results.

In [4] sequential feature partitioning is applied to regression estimation problems and in [3] to the problem of diagnosis of gastric carcinoma using a classifier that maximizes the overall quality. In [9] the feature space partitioning is

R. Burduk et al. (Eds.): CORES 2019, AISC 977, pp. 276–285, 2020.
https://doi.org/10.1007/978-3-030-19738-4_28

accomplished by means of the principal component analysis. The feature space partitioning was applied to palm-print recognition [7], to A analysis of image data [8] and for subspace pattern classification [6]. Cheng et al. [2] proposed a unified framework for partition based on subspace projection techniques. This approach was applied to palm-print recognition [2], it was adapted for image data [19] and for pattern classification [22].

In [13] the solution of the classification problem is achieved using parallel feature partitioning of the initial n-dimensional space. This problem is reduced to an integer optimization problem, which leads to the construction of minimal covering. The learning phase consists of geometrical construction of the decision regions for classes in n-dimensional feature space.

A different approach based on multiple class partitioning has been applied to the efficient nearest neighbor search in multidimensional spaces, see [1] and [11]. The first paper used multidimensional binary search tree (or k-d tree) and the second one used projection combined with the novel data structure. To carry out nearest neighbor search efficiently in high dimensional spaces paper [1] uses multidimensional binary search tree (or k-d tree), while [11] introduces projection combined with the novel data structure.

Many important intractable problems are easily reducible to minimum number of the Maximum Clique Problem (MCP), where the Maximal Clique is the largest subset of vertices such that each vertex is connected to every other vertex in the subset. It has been applied to clustering in [19]. In the literature much attention has been devoted to developing efficient heuristic approaches for MCP for which no formal guarantees of performance exist. These approaches are nevertheless useful in practical applications. In [22] a flexible annealing chaotic neural network has been introduced, which on graphs from the Center for Discrete Mathematics and Theoretical Computer Science (DIMACS) has achieved optimal or near-optimal solution. In [20] the proposed learning algorithm of the Hopfield neural network has two phases: the Hopfield network updating phase and the gradient-ascent learning phase. In [12] annealing procedure is applied in order to avoid local optima. Another algorithm for MCP on arbitrary undirected graph is described in [10]. It first divides vertices into color classes and then uses them to prune branches of the maximum clique search tree. In [5] a Markov chain Monte Carlo algorithm is introduced, to study some NP-complete problems and MCP in particular.

In contrast to the sequential methods, in [14] and in [15] all features are considered in parallel. In [14] the parallel feature partitioning approach is used for transforming real, categorical, and fuzzy features into binary ones, if this transformation exists. Combinatorial and decision-tree approaches for solving supervised classification problems are discussed in [16].

Another approach to the supervised classification problem is described in [17], where it is solved by reducing it to the solution of an optimization problem for partitioning of the graph on the minimal number of maximal cliques. This approach is similar to the one-versus-all SVM with a Gaussian radial basis function kernel, however unlike in the previous case no assumptions are made about statistical distributions of classes. An alternative approach is discussed in [21], where classification rule becomes a hyperplane which misclassifies the fewest number of patterns in the training set. Initial results concerning the proposed approach have been presented in [18].

The rest of the paper is organized as follows. The supervised classification as *G-cut* problem is described in Sect. 2, class cover problem in Sect. 3 and a minimal clique cover Box algorithm in Sect. 4. Box algorithm classifier is presented in Sect. 5 and its relation to k-NN rule and decision trees is described in Sect. 6. Results of experiments are presented in Sect. 7 and conclusions are drawn in Sect. 8.

2 Supervised Classification as a *G*-cut Problem

We can formulate the supervised classification problem as a *G-cut problem*. The feature space partitioning problem can be regarded as an n-dimensional cutting stock problem and is thus equivalent to making, say k_1 guillotine cuts orthogonal to the x_1 axis, then all $k_1 + 1$ hyperparallelepipeds are cut into k_2 parts by cuts orthogonal to the x_2 axis, etc. Let us call such cuts "axes-driven-cuts". Thus, if only axes-driven-cuts are allowed, the classification problem by parallel feature space partitioning could be stated as follows.

***G*-cut Problem.** *Divide an n-dimensional hyperparallelepiped into a minimal number of hyperparallelepipeds, so that each of them contains either patterns belonging to only one of the classes or is empty.*

Since the classes are separable according to their class label, the *G*-cut problem is solvable. This problem was first formulated and solved in [13] using parallel feature partitioning. The solution was obtained by partitioning the feature space into a minimal number of nonintersecting regions by solving an integer-valued optimization problem, which leads to the construction of minimal covering. The learning phase consists of geometrical construction of the decision regions for classes in n-dimensional feature space.

3 Class Cover Problem by Colored Boxes

Recall that the patterns $\mathbf{x} = (x_1, x_2, \ldots, x_n)$ are points in R^n and $\mathbf{x} \in M$, where M is the training set. In the sequel, the hyperparallelepiped $P = \{X = (x_1, x_2, \ldots, x_n), X \in I_1 \times I_2 \times \cdots \times I_n\}$, where I_i is a closed interval, will be referred to as a box. Suppose that the set K_c of patterns belonging to class c are painted in color c. For any compact $S \subset R^n$, let us denote by $P(S)$ the smallest (in volume) box containing the set S, i.e. $I_i = [l_i, u_i]$, where $l_i = \min x_i, \mathbf{x} \in S$ and $u_i = \max x_i, \mathbf{x} \in S$. A box $P^c(*)$ is called painted in color c, if it contains at least one pattern $\mathbf{x} \in M$ and all patterns in the box are of the same color c, i.e. $P^c(*) \cap M \neq \varnothing$ and $P^c(*) \cap M \subset K_c$. Under these notations, we obtain the following Master Problem (MP):

MP: Cover all points in M with a minimal number of painted boxes.

Note that in the classification phase, a pattern $\mathbf{x}$ is assigned to a class c, if $\mathbf{x}$ falls in some $P^c(*)$. It is not necessary to require non-intersecting property for equally painted boxes. Suppose now that $P(c) = \{P^c(S_1), P^c(S_2), \ldots, P^c(S_{t_c})\}$ (minimal set of boxes of color c, covering all c colored points) is an optimal solution to the following problem:

$MP(c)$: *Find the minimal cover of the points painted in color c by painted boxes.* Then, one can easily prove that $\cup P(c)$ (minimal cover) is an optimal solution to MP. Thus MP is decomposable in $MP(c), c = 1, 2, \ldots, l$. In [17] the $MP(c)$ problem has been considered as a problem of partitioning the vertex set of a graph into a minimal number of maximal cliques. In general, this problem is NP-complete, but as it will be shown below the instances originating from the boxes approach are polynomially solvable.

4 A Minimal Clique Cover Box Algorithm

To introduce the Box algorithm (BA) we need to introduce additional notation. Consider again the master problem $MP(c)$. Let $B = \{\mathbf{x} : l_i \leq x_i \leq u_i,\ i = 1, \ldots, n\}$. If $u_i - l_i > 0,\ i = 1, \ldots, n$ then we call the box B a **full dimensional box**. Suppose that two sets X_b and X_r of training patterns (points in the hypercube $F \in R^n$) are given and suppose that they are colored in blue and red, respectively.

We will call the box B **colored** iff it only contains points of the same color. A pair of points $\mathbf{y} = (y_1, y_2, \ldots, y_n)$ and $\mathbf{z} = (z_1, z_2, \ldots, z_n)$ generates B if $l_i = \min\{y_i, z_i\}$ and $u_i = \max\{y_i, z_i\},\ i = 1, \ldots, n$.

Problem A: *Find a coverage of $X_b \cup X_r$ with the minimal number of colored full dimensional boxes, such that X_b and X_r are covered by blue and red boxes, respectively.*

Define a graph $G_X = (V, E)$, $V = X$, $E = \{e = (v_i, v_j)\}$ and let e be a colored box generator. An edge e is colored green if it is a full dimensional box generator.

Let now $e = (a, b)$ and $f = (c, d)$ be green and let B_e and B_f are the corresponding full dimensional boxes. An operation $e \oplus f$ is color preserving if the full dimensional box C, $C = B_e \oplus B_f$, $l_i = \min\{a_i, b_i, c_i, d_i\}$, $u_i = \max\{a_i, b_i, c_i, d_i\}$ is colored. An edge e **dominates** f (say $e > f$) if $B_e \supset B_f$.

Obviously, there is one-to-one correspondence between full dimensional boxes and the green edges. The dominance relation on the set of full dimensional boxes (say $B_e > B_f$) could be easily established. When the full dimensional box C is colored then it dominates B_e and B_f and the appropriate application of $\oplus$ operation allows generation of maximal colored cliques.

We call a clique **colored** if it contains green edges. The points contained in the full dimensional box C form the minimum clique cover, i.e., the vertex set (points in C) is partitioned in cliques and the number of cliques is minimal.

Now we can reformulate the Problem A as follows (here we only cover one class, say X_b, since the coverage of other class X_r is constructed in the same manner).

Problem A: *Cover the graph G_{X_b} with the minimum number of colored cliques.*

The Box algorithm for solving Problem A is constructed as follows.

Step 1. *(Build the graph) Create the partial subgraph of G_{X_b} from the list **GE** of all green edges.*
Step 2. *(Clique enlargement) Create a graph $GG_{X_b} = (V_{GG}, E_{GG})$, where $V_{GG} = \{v \in \boldsymbol{EGE}\}$ and $E_{GG} = \{(e, f), B_e \oplus B_f\}$ is colored. Call **try-to-extend**.*
Step 3. *(Save the cliques (full dimensional boxes)) If **EGE** is the list of all extended boxes then discard from **GE** all e not included in **EGE**. Save the set **EGE** ∪ **GE**. If all nodes are covered then stop else goto* **Exceptions**.
Try-to-extend: In all connected components of GG_X *find c-clique cover (cliques of size less or equal to c).*
Exceptions. *This function will be called if the set X is not coverable by the full dimensional boxes only. This case could be resolved by the algorithm above applied on the reduced X by covering it with lower dimensional boxes. Extreme instances when all nodes of G_X are singletons (nodes with degree one) will require rotation of the set X and are not discussed here.*

5 Classification Rule Based on Box Algorithm

Cliques-to-Painted Boxes. Let S be any clique in the optimal solution of $MP(c)$. The box painted in color c that corresponds to this clique is defined by $P(S) = \{\mathbf{x} = (x_1, x_2, \ldots, x_n), \mathbf{x} \in I_1 \times I_2 \times \cdots \times I_n\}$, where $I_i = [\min \bar{x_i}, \max \bar{x_i}]$. The points $\mathbf{x}$ correspond to the vertices in S. Geometrically, by converting cliques to boxes, one could obtain overlapping boxes of the same color. The union of such boxes is not a box, but in the classification phase the point being classified is trivially resolved as belonging to the union of boxes instead of a single box. If a pattern $\mathbf{x}$ from the test dataset falls in a single colored box or in the union of boxes with the same color the element $\mathbf{x}$ is assigned to the class that corresponds to this color.

It is possible that we encounter a case of overlapping boxes painted in different colors, with empty cross-section. Such empty cross-section is easily determined and corresponds to an empty box in the definition of the G-cut problem. The colored boxes constructed in this way are used to make decisions about class membership of classified patterns.

If a pattern $\mathbf{x}$ from the test dataset falls in an empty (uncolored) box then the pattern $\mathbf{x}$ is not classified. Another possible classification rule is that the pattern $\mathbf{x}$ can be assigned to a class with color that corresponds to the majority of adjacent colored boxes.

6 Relation of Box Algorithm to the Nearest Neighbor Rule and Tree Classifiers

6.1 Relation of Box Algorithm to the Nearest Neighbor Rule

A reasonable classification rule, known as a nearest neighbor rule, classifies pattern $\mathbf{x}$ as red (blue, etc.) if $argmin_{\mathbf{x}' \in X_b \cup X_r} \rho(\mathbf{x}', \mathbf{x}) = \mathbf{y}^*$ and $\mathbf{y}^*$ is red (blue, etc.) or equivalently when $\mathbf{x}$ falls in the appropriate region of the Voronoi diagram.

Let's shed more light on the relation of this rule and its k-NN generalization to BA. To this end introduce a function (map) f mapping a set X to $Y = \{1, 2, \ldots, l\}$. (In one dimensional case such function is called stepwise function). Without losing generality we can take X to be a box (rectangle in 2-d) and Y to be a finite set of colors, say $Y = \{red, blue, \ldots\}$. Thus to know f is equivalent to know the coloring of X based on the colors of a given colored set of training points $\{(\mathbf{x}_1, y_1), \ldots, (\mathbf{x}_m, y_m)\}$. Then the nearest neighbor rule is constructed as follows. For each point $\mathbf{x}_i$ colored in red (blue, etc.) paint in red (blue, etc.) all points in X that are closest to $\mathbf{x}_i$ (these are regions in the Voronoi diagram).

k-NN rule is obtained as follows. For each k-element subset A of T, having majority of red (blue, etc.) points, paint all points in X whose k closest points are A (not all A provide paintings).

If box $B = l_i \leq x_i \leq u_i \; i = 1, \ldots, n$ contains training patterns and ρ is the Manhattan distance, then for the pattern $\mathbf{y}$ the distance is equal to $\rho(\mathbf{y}, B) = \sum \max(0, l_i - y_i) + max(0, y_i - u_i)$. Now the idea of previously defined boxes becomes clear. We first approximate the above mentioned painted areas (not known in advance) by painted boxes (perfect candidates for Manhattan distance) and then classify patterns according to point-to-box distance rule.

BA is similar to NN if the Voronoi points are taken to be boxes. Let us note that instead of boxes we can consider the convex hulls of patterns. Applying now the nearest neighbor rule we get better classification, but this approach is computationally intractable either for convex hulls constructions or for computing the point-to-set distances. Now the $MP(c)$ problem can be formulated as an heuristic good clique cover problem satisfying the nearest neighbor rule.

6.2 Relation of Box Algorithm to Decision Trees

(i) The leaves of the decision trees (DT) are painted rectangles (boxes) obtained by preliminary given sets of intervals and priority (for univariate trees) for creating successors of a given node. Internal nodes are rectangles containing a mixture of colored points.
(ii) BA create painted boxes (under Manhattan distance) that are convex hull of given sets of unicolored sets of points.

For certain instances (i) and (ii) could create the same coloring if the set of intervals and priority are properly chosen. In general you can say that BA create DT with a root node and a list of successors (colored boxes, containing unicolored points, whose coordinates satisfy the test on the arc for box membership.) We could go further by saying that BA aims to create trees with minimum number of leaves and thus it attempts to reduce the generalization errors.

7 Experimental Results

In this section we compare the performance of our box algorithm and SVM classifier for synthetic data generated from 3-variate normal distributions and for real Monk's Problems data from UCI Machine Learning Repository. Simulations were carried out using MATLAB16b on Intel(R) Pentium CPU B960 @ 2.20 GHz with 4.00 GB RAM. Box algorithm is based on similarity and it is normal and reasonable to compare it with algorithms from the same class like SVM. This explains why Monk's problems are poorly classified by box and SVM but perfectly by neural networks. This is not because classes are separable as points in 6 dimensional space, but because they are separable by logical clauses.

7.1 Normal Attributes

The samples for a binary classification problem are generated for three cases and with 3-dimensional normal distributions with mean vectors and covariance matrices given in Table 1 below.

Table 1. Parameter settings for normal distributions

Case	Covariance matrices		Mean vectors	
1	I	I	0	$0.5\mathbf{e}$
2	I	$2I$	0	$0.6\mathbf{e}$
3	I	$4I$	0	$0.8\mathbf{e}$

where $\mathbf{e} = (1, 1, 1)^T$. For each distribution 100 samples are generated and Table 2 below, where bolded items represent the best results.

Table 2. Confusion matrices in percentage ratio for Box algorithm and SVM classifier for normal data

	Box algorithm		SVM classifier	
	Red Points	Blue Points	Red Points	Blue Points
	First normal distribution			
Red Points	**0.98**	0.02	0.72	0.28
Blue Points	0.02	**0.98**	0.45	0.55
	Second normal distribution			
Red Points	**0.97**	0.03	0.81	0.19
Blue Points	0.01	**0.99**	0.43	0.57
	Third normal distribution			
Red Points	**0.96**	0.04	0.89	0.11
Blue Points	0.01	**0.99**	0.27	0.73

In Table 2 we use SVM with the standard Gaussian kernel. Observe that the Box algorithm significantly outperforms SVM classifier in terms of true positive and true negative rates in all cases.

7.2 Nominal Attributes

In this section we present experimental results on three Monk's Database problems from UCI Machine Learning Repository. Each problem consists of training and testing data samples with the same 6 nominal attributes. Data sizes are as follows: Monk1 - 124, Monk2 - 169, Monk3 - 122 (train) and Monk1 - 432, Monk2 - 432, Monk3 - 432 (test), respectively. In Table 3 we used SVM classifier with the standard Gaussian kernel.

Table 3. Confusion matrices in percentage ratio for box algorithm and SVM classifier for Monk's data

	Box algorithm		SVM classifier	
	Red Points	Blue Points	Red Points	Blue Points
	Monk1			
Red Points	**1.00**	0.00	0.88	0.12
Blue Points	0.00	**1.00**	0.08	0.92
	Monk2			
Red Points	**1.00**	0.00	0.85	0.15
Blue Points	0.00	**1.00**	0.11	0.89
	Monk3			
Red Points	**1.00**	0.00	0.97	0.03
Blue Points	0.00	**1.00**	0.14	0.86

It can be observed in Table 3 through Table 5 that the Box algorithm significantly outperforms SVM classifier in all cases for normal distributions and for Monk's data in terms of true positive and true negative rates and in terms of accuracy, sensitivity, specificity and precision. Consequently, it can be concluded from the experimental results presented in this section that the Box algorithm is superior to SVM classifier.

Table 4. Accuracy and sensitivity of SVM classifier and the Box algorithm for normal data and Monk's

	Normal distributions			Monk's		
	1	2	3	1	2	3
	Accuracy					
SVM classifier	0.64	0.69	0.81	0.90	0.86	0.91
Box algorithm	**0.98**	**0.98**	**0.98**	**1.00**	**1.00**	**1.00**
	Sensitivity					
SVM classifier	0.55	0.57	0.73	0.92	0.89	0.86
Box algorithm	**0.98**	**0.99**	**0.99**	**1.00**	**1.00**	**1.00**

Table 5. Specificity and precision of SVM classifier and the box algorithm for Monks and normal data

	Normal distributions			Monk's		
	1	2	3	1	2	3
	Specificity					
SVM classifier	0.72	0.81	0.89	0.88	0.84	0.97
Box algorithm	**0.98**	**0.97**	**0.96**	**1.00**	**1.00**	**1.00**
	Precision					
SVM classifier	0.66	0.75	0.87	0.89	0.74	0.97
Box algorithm	**0.98**	**0.97**	**0.96**	**1.00**	**1.00**	**1.00**

8 Conclusions

We introduced a new geometrical approach for solving the supervised classification problem based on graph optimization approach for the well-known problem of partitioning a graph into a minimum number of maximal cliques which are subsequently merged using the k-NN rule. Equivalently, the supervised classification problem is solved by means of a heuristic good clique cover problem satisfying the nearest neighbor rule. The main advantage of the new approach which optimally utilizes the geometrical structure of the training set is decomposition of the l-class problem into l single-class optimization problems. One can see that the Box algorithm performs better than SVM in case of normally distributed data and for Monk's data from UCI depository.

An advantage of the proposed method is that we can directly work with the original data. The proposed theoretical approach seems quite promising as it can be applied to the general case of a supervised classification problem when classes are not linearly separable. We would like to emphasize that we have proposed a new classifier which is more general than the nonlinear classifier. An important advantage of this approach is the decomposition of a problem involving l classes into l optimization problems involving a single class. Using the proposed approach, the geometrical structure of the training set is utilized in the best possible way.

As a future work we are planning to compare the computational efficiency of the proposed algorithm with the classical classification techniques such as decision trees, ensembles of trees, and random forest.

Acknowledgments. Research of N. Yanev was partially supported by the French-Bulgarian contract "RILA", 01/4, 2018. Research of A. Krzyżak was supported by the Natural Sciences and Engineering Research Council under Grant RGPIN-2015-06412. K. Ben Suliman research was supported by the Libyan Government.

References

1. Bentley J (1975) Multidimensional binary search trees used for associative searching. Commun ACM 18:509–517
2. Cheng H, Vua K, Huaa K (2009) Subspace projection: a unified framework for a class of partition-based dimension reduction techniques. Inf Sci 179:1234–1248
3. Güvenir H, Emeksiz N, Ikizler N, Ormeci N (2004) Diagnosis of gastric carcinoma by classification on feature projections. Artif Intell Med 23:231–240
4. Güvenir H, Sirin I (1966) Classification by feature partitioning. Mach Learn 23:47–67
5. Iovanella A, Scoppola B, Scoppola E (2007) Some spin glass ideas applied to the clique problem. J Stat Phys 126:895–915
6. Kumar K, Negi A (2007) A feature partitioning approach to subspace classification. In: Proceedings of the TENCON IEEE conference, pp 1–4
7. Kumar K, Negi A (2007) A novel approach to eigenpalm features using feature-partitioning framework. In: Proceedings of the international conference on machine vision applications, Tokyo, pp 29–32
8. Kumar K, Negi A (2008) Novel approaches to principal component analysis of image data based on feature partitioning framework. Pattern Recogn Lett 29:254–264
9. Kumar K, Negi A (2008) SubXPCA and a generalized feature partitioning approach to principal component analysis. Pattern Recogn 41:1398–1409
10. Kumlander D (2005) Problems of optimization: an exact algorithm for finding a maximum clique optimized for dense graphs. In: Proceedings of the Estonian academy of sciences, physics, mathematics, vol 54, pp 79–86
11. Nene S, Nayar S (1997) A simple algorithm for nearest neighbor search in high dimensions. IEEE Trans Pattern Anal Mach Intell 19:989–1003
12. Pelillo M, Torsello A (2006) Payoff-monotonic game dynamics and the maximum clique problem. Neural Comput 18:1215–1258
13. Valev V (2004) Supervised pattern recognition by parallel feature partitioning. Pattern Recogn 37:463–467
14. Valev V (2004) Supervised pattern recognition with heterogeneous features. Int J Mach Graph Vis 13:345–353
15. Valev V (2011) Machine learning of syndromes for different types of features. In: Proceedings of the international conference on high performance computing and simulation, pp 504–509
16. Valev V (2014) From binary features to non-reducible descriptors in supervised pattern recognition problems. Pattern Recogn Lett 45:106–114
17. Valev V, Yanev N (2012) Classification using graph partitioning. In: Proceedings of the 21st international conference on pattern recognition. IEEE Xplore, Tsukuba, Japan, pp 1261–1264
18. Valev V, Yanev N, Krzyżak A (2016) A new geometrical approach for solving the supervised pattern recognition problem. In: Proceedings of the 23rd international conference on pattern recognition. IEEE Xplore, Cancun, Mexico, pp 1648–1652
19. Wang H, Obremski T, Alidaee B, Kochenberger G (2008) Clique partitioning for clustering. Commun Stat - Simul Comput 37:1–13
20. Wang R, Tang Z, Cao Q (2003) An efficient approximation algorithm for finding a maximum clique using hopfield network learning. Neural Comput 15:1605–1619
21. Yanev N, Balev S (1999) A combinatorial approach to the classification problem. Eur J Oper Res 115:339–350
22. Yang G, Tang Z, Zhang Z, Zhu Y (2007) A flexible annealing chaotic neural network to maximum clique problem. Int J Neural Syst 17:183–192

Algorithm of Multidimensional Analysis of Main Features of PCA with Blurry Observation of Facility Features Detection of Carcinoma Cells Multiple Myeloma

Mariusz Topolski(✉)

Department of Systems and Computer Networks, Wroclaw University of Technology, Wybrzeze Wyspianskiego 27, 50-370 Wroclaw, Poland
mariusz.topolski@pwr.edu.pl

Abstract. The article is mainly focused on the description of the PCA main component algorithm with fuzzy observation of object features. The author focuses on the application of this method to the reduction of similar traits, aiming at the diagnosis of not only the existence of cancerous rhinoceros, but also the assessment of the direction of its exposure and the degree of aggressiveness. An advantage of the developed algorithm is the ability to combine basic patient results with morphology or advanced indicators of tumor markers with microscopic or X-ray image. The occurrence is a possibility of the development of a group of people. Thanks to the fuzzy observation of the object. For this purpose, the author's points are the correlations. This gives a positive answer, where redundancy can indicate the same class. The analysis of the principal components does not exhaust the subject of research. Multidimensional exploration techniques that can improve the computerized medical diagnostics. The proposed solution can be used for unbalanced data. In the course of further research, it is possible to use the data imputation method. The method developed for selection of the traits of the subject is an original approach. It allows us to choose traits when data are either unconfirmed or incomplete. When creating statistical models for various medical institutions, the author encountered many classification problems. For doctors, the use of traditional statistical methods did not give satisfactory results. Very often the selection of traits was not very precise and could raise doubts as to its accuracy. The use of algorithms of fuzzy logic in PCA can be the foundation for a more accurate trait selection by combining different quantitative data. In the case of selecting cancer traits, it allows us to combine image data with test data. That way, it is possible to get additional information about the direction of cancer cell growth.

Keywords: Fuzzy sets · Main components analysis · Recognition · Multiple myeloma

R. Burduk et al. (Eds.): CORES 2019, AISC 977, pp. 286–294, 2020.
https://doi.org/10.1007/978-3-030-19738-4_29

1 Introduction

Modern computer science includes computer methods supporting human work, by proposing solutions in conditions of imprecise, incomplete data. Systems supporting human activity and making various decisions are called expert systems. These systems, unlike traditional methods, have overt algorithms that solve various problems using advanced methods of inference. Examples of such methods are fuzzy logic, neural networks, genetic algorithms, possibilistic, probabilistic, mathematical accounting theory and selected issues of mathematical statistics and hybrid models. Modern data analysis methods are being developed more and more often in modern research. Particularly noteworthy are the algorithms of a hierarchical set of single-row classifiers for unbalanced data [2]. The authors focus on the use of the above classifiers in the analysis of patients with hypertension. The same authors used the mergers of neighbors classifiers [3] to classify large data sets. The recognition systems also use fuzzy sets models that allow a better description of information uncertainty [7].

Modern methods of multidimensional data analysis are advanced analytical and graphic methods. They are widely used in many areas of the economy, including: medicine, science, logistics, management. In order to solve many classification problems, tools for modeling linear and non-linear interdependencies are used. Different statistical classifiers operate on continuous and qualitative predictors that take into account interactions that may occur between them. In terms of advanced statistical methods, hierarchical models are used.

The development of advanced computational methods gives more and more opportunities to solve various complex problems. When recognizing complex structures such as faces, simple algorithms do not give satisfactory results. It turns out that combining different methods of classifiers gives better results, because their merger, ie the submission of classifiers, allows to better describe the recognized problem. It is important to look for simple classifiers with uncomplicated computational complexity or simplicity of construction. The advantage is the ability to distinguish the set of features and classes more. We can also assume that such a metaclassifier will be less susceptible to the singularity of data and will have greater resistance to overfitting. One such algorithm is AdaBoost based on the so-called Adaptive strengthening. An algorithm for this property that in subsequent interactions trains k-weaker classifiers on a set of translations with weights [4].

In the analysis of the multidimensional space of features, we strive to reduce it. The possible co-occurrence of comparable symptoms is that the quality of classification for n-traits is comparable in the case of replacement or one of the most discriminating classes, or their common value, e.g. the average of scales, and when scales are roses is the average of the standardized scale Z. A very popular method of reducing traits is the analysis of the main PCA components. Different grouping techniques can give a good grouping of similar features [1,5,6,8]. Under certain circumstances, despite a good distinction, certain factors are similar to each other. An example of this is the structure of multiple myeloma, where, in addition to the testis nucleus characterized by higher density, there is a capsule, which also indicates the proliferation of tumor cells [NiLi] [Kyle]. Due to the uncertainty of the data considered, the author focuses on the group of classifiers opricked on fuzzy sets and multivariate analysis of the PCA main components. Fuzzy sets are widely used in medical diagnostics and advanced techniques of object recognition and classification [7].

2 A Mathematical Model of PCA Analysis with Fuzzy Observation of Object Features

The first stage of creating a PCA model is a fuzzy observation of the features of the object. The object is in state Y, which belongs to the interval $y \in \mathbf{R}+$. The object itself is not subject to direct observation, but it is the result of observing the value of features describing it at a given moment. Let $\boldsymbol{x} = [x_1, x_2, \ldots, x_d] \in \mathcal{X}$ be a d-dimensional vector of variables (characteristics) that have been measured. $x \in \mathbf{X}$ These variables $\tilde{T_i}$ are subject to fuzzy sets, where:

$$\tilde{T_i} = \left\{ \tilde{T_1}, \tilde{T_2}, \tilde{T_3}, ..., \tilde{T_d} \right\} \tag{1}$$

We call the set of fuzzy T in a certain (non-empty) space X, which we write as $T \subseteq X$, a set of pairs:

$$\tilde{T} = \{x, \mu(x); x \in X\}, \tag{2}$$

where: $\mu(x) = 1$ is a function of belonging to a fuzzy set $\tilde{T}$, where 3 cases can be distinguished:

- $\mu(x) = 1$ means full belonging to a fuzzy set, $\tilde{T}$ i.e $x \in \tilde{T}$,
- $\mu(x) = 0$ means no belonging to a fuzzy set, $\tilde{T}$ i.e $x \in \tilde{T}$,
- $0 < \mu(x) < 1$ means partial belonging to a fuzzy set $\tilde{T}$.

The classification task will be based on the fact that the PCA component classifier using the fuzzy obstruction of object features will estimate the eigenvalues $\lambda_1^S, \lambda_2^S, ..., \lambda_n^S$ of the S covariance matrix. Important in the case of the analyzed model is that the data is represented by the N set of fuzzy sets of measured features (2). Distributed matrix is determined by the formula:

$$S = \sum_{i=1}^{N} \left(\mu(x_i) - \mu(\bar{x})(\mu(x_i) - \mu(\bar{x})^T \right), \tag{3}$$

where $\mu(x_i)$ are the fuzzy values of the features, i $= 1, 2, ..., N$, and $\mu(\bar{x})$ and this is the fuzzy mean value of the features. The next step is to calculate the eigenvectors and the eigenvalues of the matrix S. This can be done by means of the spectral distribution of the matrix S:

$$S = A \wedge A^T, \tag{4}$$

where A is the matrix of eigenvectors, while $\wedge$ is a diagonal matrix, on the diagonal of which are the fuzzy eigenvalues of the matrix S:

$$S : \tilde{\lambda}_d^S, d = 1, 2, ..., n. \tag{5}$$

To obtain sharp values of the own value vector, the diffusifier was applied according to the average values of the centers:

$$\lambda_d^S = \frac{\sum_{l=1}^{M} c_1 \tilde{\lambda}_d^S}{\sum_{l=1}^{M} \tilde{\lambda}_d^S} \tag{6}$$

where c_l is the center (dispersion) of the fuzzy set. The next step of the algorithm is to organize the fuzzy eigenvalues of the covariance matrix S while maintaining the decreasing order:

$$\lambda_1^S \geq \lambda_2^S \geq ... \geq \lambda_d^S \geq 0 \tag{7}$$

Then a new subset of features is determined in the new space, which is unbuttoned by the orthogonal main components. New variable groups are determined on the basis of maximizing fuzzy observation of object features with simultaneous minimization of information loss. Next, an analysis of the cumulative variance of components is carried out, which is based on the percentage index of variation of the exited components. We can calculate this percentage from the formula:

$$V \doteq 100 \sum_{d=1}^{k} \frac{\sum_{l=1}^{M} c_l \tilde{\lambda}_d^S}{\sum_{l=1}^{M} \tilde{\lambda}_d^S} \left(\sum_{d=1}^{n} \frac{\sum_{l=1}^{d} c_l \tilde{\lambda}_d^S}{\sum_{l=1}^{d} \tilde{\lambda}_d^S} \right)^{-1} \tag{8}$$

In the PCA analysis, we assume that the fuzzy value $\tilde{\lambda}_d^S$ is higher, the feature has a high informative value in the discrimination of different objects from different classes. In the end, the reduction of the dimension comes down to the fact that we only select from the main components those that have V (7) variability not less than, for example 85%.

The final stage of the multidimensional PCA analysis is to project the fuzzy vet observations of the characteristics x_i $(i = 1, 2, ..., d)$ into a new coordinate system that is stretched by the matrix's own vectors. This transformation derives from the Karhunen-Loeva transform [8].

$$x^{'} = \frac{A_{nxk}^T \sum_{l=1}^{M} c_l \mu(x)}{\sum_{l=1}^{M} \mu(x)} \tag{9}$$

In the formula (9), the sharpened values of the fuzzy observation of the object's features are multiplied by the matrix A_{nxk}^T which contains the k columns of the matrix of the eigenvectors A, which correspond to k the largest eigenvalue. Then, into the results of the coordinate system rotation, the n dimension vectors of the fuzzy sets $\mu(x)$ are converted into the x' feature vectors.

An important feature of this method is that any pair of the new variable system is mutually uncorrelated. Feature vectors that are selected in accordance with the PCA method can be used to classify objects.

The proposed method is presented on the basis of the block diagram in Fig. 1

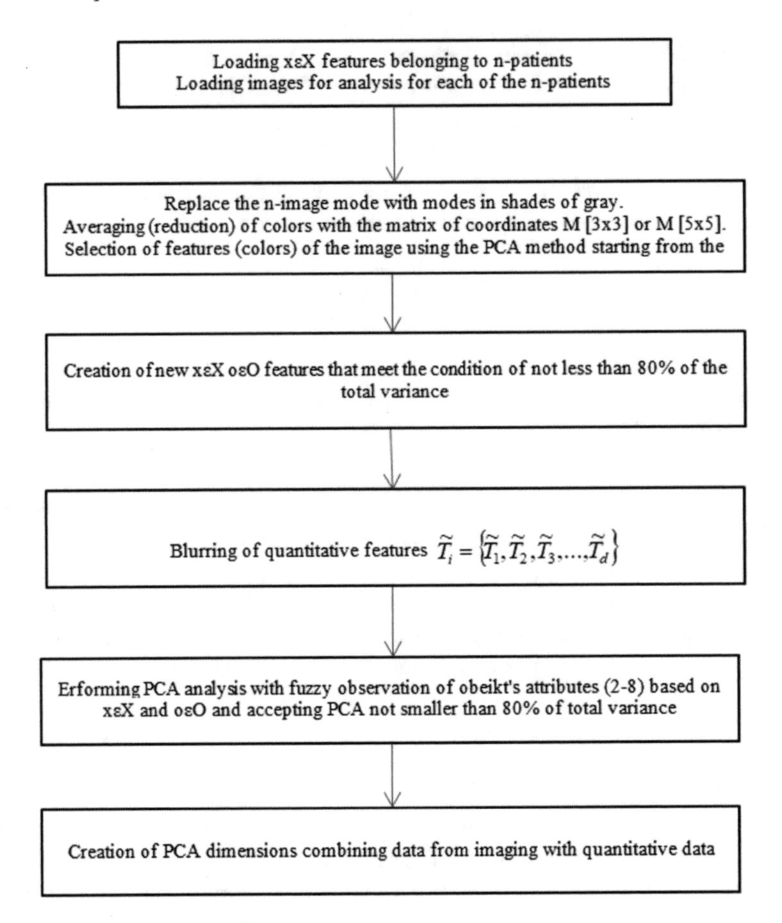

Fig. 1. Results of the main experiment.

3 Practical Examples of Using a Fuzzy PCA Model

The practical case under consideration includes the diagnosis of multiple myeloma, which is a malignant tumor of plasma cells. The purpose of the feature reduction is to detect similar fragments of the image to determine as accurately as possible the picture of cancer cells and their scattering state. Multiple myeloma is primarily located in the bone marrow. This disease is genetically heterogeneous. The image of multiple myeloma is shown in Fig. 2a. Myeloma diagnosis has taken into account data from the Xray image of the skeleton of multiple myeloma cells, i.e.

The position of pixels with the color values of the image in shades of gray are treated as separate factors. The image is searched by a 3×3 pixel matrix, averaging the color in this matrix. Such a picture has a cluster structure, where the number of clusters is the number of variables describing the image. The basic data is:

- blood counts,
- beta-microglobulin,
- beta-2-microglobulin,
- albumin,
- Biernacki's reaction,
- calcium concentration,
- creatinine concentration,
- results of protein tests,
- electrophoresis of serum proteins,
- testing of concentrated urine.

The results of specialist tests are:

- antibody concentration,
- assessment of the amount of monoclonal protein in urine or serum,
- specialized research of blood proteins in the form of immunofixation,
- examination of bone marrow using aspiration biopsy,
- determination of the concentration of free chains in serum.

Figure 2 presents the classification by means of PCA of the radiological image of multiple myeloma with the results of the characteristics resulting from the research.

Figure 2 shows two reduction methods, one classic PCA and the other author using fuzzy observation of object features. The PCA method with the fuzzy object of conservation features in addition to the cancer cell proliferation, which is very important in medical diagnosis. Let us analyze according to the algorithm. First, and scatter plot was taken to determine the number of PCA factors.

Table 1. Eigenvalues separated by the principal components method

	Method	Own value	Percent of total variance	Accumulated eigenvalue	Accumulated percentage
1	PCA1	4,86	58,77	4,86	58,77
2	PCA2	1,75	24,94	6,61	83,71
3	FPCA1	8,69	63,33	8,69	63,33
4	FPCA2	2,72	28,33	11,41	91,66
5	FPCA3	1,28	2,12	12,69	93,78

Based on Table 1, note that PCA1 in both methods determines the nucleus of the cancer cell. In the classical method, it explains 58,77% of the spearheads and FUZZY PCA 63.33%. From Fig. 1a, we can see that there are visible changes around the nucleus. With blurred observation of features, this area is isolated as a separate PCA2 in FUZZY PCA and all neoplastic changes are explained in 91.66%. The rest of the variance is a white background or no classification. In

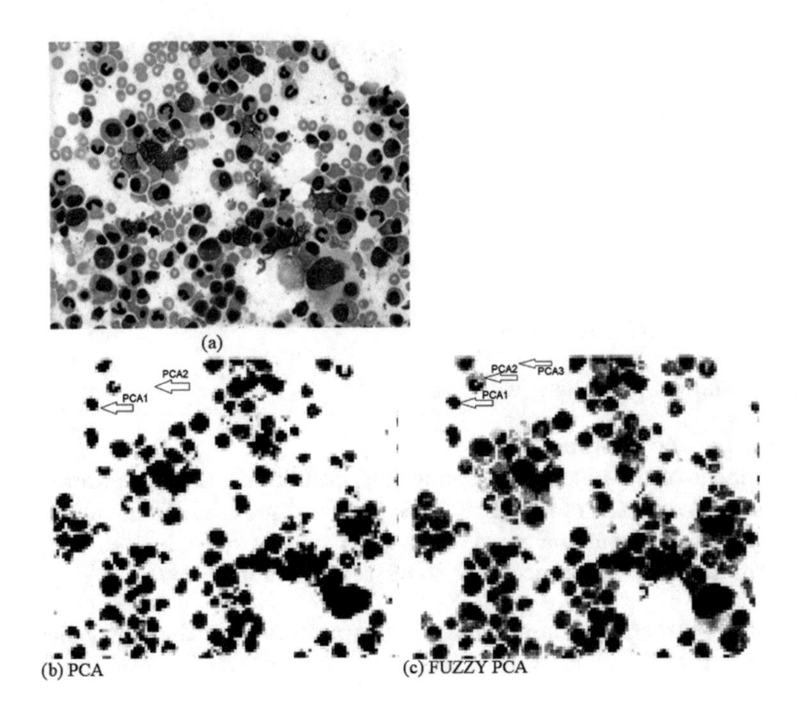

Fig. 2. PCA classification (a) original photograph of multiple myeloma of plasma cells, (b) reduction of PCA features, (c) reduction of PCA features with fuzzy feature obstruction. Source: https://doi.org/10.1016/j.celrep.2017.08.062.

Fig. 1a, we can see other fragments of cells, however, due to their low intrinsic value, they have been omitted. In the FUZZY PCA method, the myeloma tumor envelope explains 28.33% variability variance. Let's check correlations between PCA factors to determine the strength of PCA factors. The correlation results are presented in Table 2.

The PCA factors of cancer cells in both methods correlate very weakly with the background, which shows their independence. In the case of the FUZZY PCA algorithm, there is a very strong positive correlation ($r = 0.788$, $p = 0.001$) between PCA 1 and PCA 2. This suggests that the link between them may

Table 2. Pearson correlation coefficients between PCA factors

	Method	PCA1	PCA2	FPCA1	FPCA2	FPCA3
1	PCA1	-	r = 0,211	-	-	-
2	PCA2	r = 0,211	-	-	-	-
3	FPCA1	-	-	-	r = 0788	r = 0,233
4	FPCA2	-	-	r = 0,788	-	r = 0,178
5	FPCA3	-	-	r = 0,233	r = 0,178	-

indicate the same class that is shown in Fig. 2a to be multiple myeloma, and actually its expansion in the proliferation of plasma cells. It should be noted that PCA 1 and PCA 2 in FUZZY PCA have a strong charge (linkage) with the concentration of antibodies, with the over-normalization of monoclonal protein in urine or serum, with the presence of blood proteins in the form of immunofixation and free serum chain concentration.

4 Conclusion

The main problem of the research was the development of the main components algorithm with fuzzy observation of the object's features in the diagnosis of multiple myeloma. Multiple myeloma around the testicle creates infiltration, which may indicate further cancerous lesions. Full information on cancerous change is an important factor in correct diagnosis and further treatment. The PCA main components method allows the reduction of similar features, and thus the unification of changes in the X-ray image of bones or microscopic examinations. The PCA algorithm combines data from biochemistry and tumor markers with imaging data. As a result, the assessment of changes is more precise and allows to draw important conclusions. In the reduction task, the author used a rhythmic observation of the features of the object, which made it possible to improve the quality of the classification. The shell of the myeloma nucleus has the character of a filthy glow, and can be seen as a non-fragmented section of the bone slice. This information uncertainty can be described well by fuzzy sets. Thanks to this, it is possible to discover the deeper changes caused by myeloma. Quite important information is that variables must have a quantitative character. In the case of qualitative variables, Bayes classifiers based on the MCA method (multidimensional correspondence analysis) can be measured. This is a method comparable in the relegate to PCA. In order to improve the classification of similar data, you can search various methods of observing the features of the object. The results are promising and indicate the continuity of further searches. The method developed allows us to combine data from images of cancer cells with quantitative data from biochemical tests. As a result of this method, a more accurate description of the variables that discriminate against cancer cells the most is possible. Knowing the variables that are most important in description of cancerous changes allows us to suggest better treatment. By using fuzzy logic observation of the traits of the subject, at least one additional dimension can be distinguished, which may concern the differentiation of tumor cells. The use of fuzzy logic PCA method allows the evaluation of tumor cell proliferation. In further research, the author would like to create a fusion of the PCA and MCA methods. This way, it would be possible to combine quantitative and qualitative traits.

References

1. Webb AR, Copsey KD (2011) Statistical pattern recognition. Wiley. https://doi.org/10.1109/tpami.2015.2491929
2. Krawczyk B, Woźniak M (2014) Hypertension type classification using hierarchical ensemble of one-class classifiers for imbalanced data. Springer, pp 341–349. https://doi.org/10.1109/tpami.2015.2491929
3. Krawczyk B, Woźniak M (2015) Combining nearest neighbour classifiers based on small subsamples for big data analytics. CYBCONF, pp 24–26. https://doi.org/10.1109/tpami.2015.2491929
4. Burduk R (2017) The adaboost algorithm with linear modification of the weights. In: 9th international conference, Bydgoszcz, Poland. https://doi.org/10.1109/tpami.2015.2491929
5. Han J, Kamber M, Pei J (2012) Concept and techniques, 3rd edn. Elsevier. https://doi.org/10.1109/tpami.2015.2491929
6. Turk MA, Pentland AP (1991) Face recognition using eigenface. In: Conference on computer vision and pattern recognition, pp 586–591. https://doi.org/10.1109/tpami.2015.2491929
7. Trajdos P, Kurzynski M (2015) An extension of multi-label binary relevance models based on randomized reference classifier and local fuzzy confusion matrix. In: 16th international conference. https://doi.org/10.1109/tpami.2015.2491929
8. Sobczak W, Malina W (1985) Metody selekcji i redukcji informacji. WNT. https://doi.org/10.1109/tpami.2015.2491929

A New Benchmark Collection for Driver Fatigue Research Based on Thermal, Depth Map and Visible Light Imagery

Krzysztof Małecki(✉), Paweł Forczmański, Adam Nowosielski, Anton Smoliński, and Daniel Ozga

West Pomeranian University of Technology, Żołnierska 52, 71-210 Szczecin, Poland
{kmalecki,pforczmanski,anowosielski,ansmolinski}@wi.zut.edu.pl, od20417@zut.edu.pl

Abstract. The article has two goals. The first one is to present an original benchmark database for testing methods and algorithms for driver fatigue detection. Blinking eyes – opening and closing, squinting eyes, rubbing eyes, yawning, lowering the head and shaking the head are considered. The database includes recordings acquired from a thermal, depth map and visible light cameras. The imaging environment mimicked the conditions characteristic for driver's place of work. The second goal is to present a part of collected data. As an example of driver fatigue the eye rubbing motion was selected and the detection was made using contemporary TensorFlow-based detector, known to be accurate when working in visible lighting conditions. The results of driver's drowsiness detection in thermal and depth map imagery are compared with the detector's efficiency in the visible spectrum.

Keywords: Cvlab benchmark · Fatigue detection · Thermovision · Visible spectrum · Depth map · TensorFlow

1 Introduction

Many studies indicate human being as the main factor responsible for road accidents (e.g. 92% [24]). Such a large number of accidents are caused mainly by the condition of road [22,23], limited visibility of the road and its surroundings [12] but also by driver fatigue [5,16]. Today, in fact, many manufacturers equip their vehicles with the on-board Advanced Driver Assistance Systems (ADAS) to provide some assistance to the driver [10,19] and to make the travel more safe.

There are many factors that affect the condition and behaviour of drivers and motor vehicle operators. The attention of researchers is focused on detection of their undesirable psychophysical state on the basis of subjective, physiological, behavioural and vehicle-related factors. The analysis and evaluation the psychophysical condition of the driver is mainly based on observed external features and biomedical signals, i.e. face image and vital signs (pulse and blood pressure) [2,20,21]. The questionnaire technique is also used [17].

R. Burduk et al. (Eds.): CORES 2019, AISC 977, pp. 295–304, 2020.
https://doi.org/10.1007/978-3-030-19738-4_30

Vision-based solutions provide an excellent mean for fatigue detection and developing machine vision techniques allow continuous observation of the driver. Tired drivers show some observable behaviour in head movement, movements of eyelids, or in general the way they look [14]. A multimodal platform to identify driver fatigue and interference detection is presented in [4]. It captures audio and video data, heart rate, steering wheel and pedals positions. The authors present the results of fatigue detection reaching 98.4%. There are also dedicated hardware solutions and software solutions based on mobile devices like smartphones and tablets [11,13,25].

Vision systems for driver monitoring prove to be an effective tool supporting the safety of traffic participants and preliminary research works confirm this fact [9]. It should be remembered, that devices based on traditional imaging technique can work only in good lighting conditions, namely during a day. Therefore, modern researches are focused on other capturing devices, working in different lighting spectra (thermal and depth map) [3,4]. The problem of face detection in thermal imagery using several general-purpose object detectors known to be accurate when working in visible lighting conditions was investigated in [8]. Applications of cascading classifiers for human face detection in thermal images was described in [7].

The above analysis shows that many of current works are focused on the problem of recognizing driver's fatigue. Unfortunately, all of the authors use their own benchmarks. In the paper we focus our attention on the problem of obtaining multispectral video data on the fatigue of motor vehicle operators. The authors' achievement in the aspect of creating a common test database useful for testing methods and algorithms for recognizing signs of fatigue in people is presented. Additionally, a part of the developed benchmark is used to create a method oriented at detecting one of the example of driver drowsiness, i.e. rubbing the eyes.

The article is organized as follows. The following section presents the proposed approach; it includes the scheme of conducted research, the physical stand developed for obtaining multispectral data, the characteristics of the obtained test material and the most important information about developed database containing annotated videos. Then, there is a short note about TensorFlow and experimental results based on TensorFlow Object Detection API on sequences coming from above-mentioned database.

2 Proposed Approach

Figure 1 presents the scheme of conducted research. First, the stand for video data acquisition has been prepared. Then, more than 60 people were invited to the recording session, where video data has been acquired using visible, thermal and depth map cameras. People have followed the instructions by making motions symbolizing the fatigue of driver: closing eyes, rubbing eyes, yawning, tilting the head, etc. A benchmark has been prepared, available under web address: cvlab.zut.edu.pl. To present an example of recorded data, machine learning models have been created and a classifier has been trained to recognize human drowsiness in thermograms, depth maps and visible light videos. Using TensorFlow computational framework a detection of rubbing eyes motion have been presented. Finally, the results have been discussed, indicating the weak points of the approach.

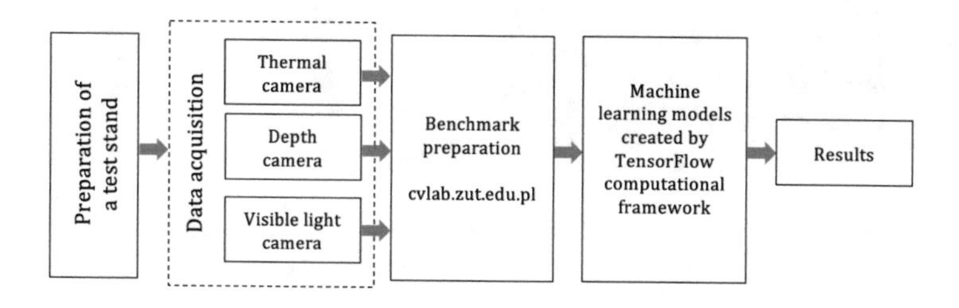

Fig. 1. The research scheme.

The experimental stand for data acquisition of human fatigue is presented in Fig. 2. The stand has been developed with the aim of multispectral data acquisition for the assessment of driver's fatigue [18]. It consist of five monitors, of which four imitate the vehicle windows showing real driving situations. There is a steering wheel and three cameras of different kind to capture driver's behaviours in selected spectral bands (visual image, depth map, and thermal image).

2.1 Characteristics of the Obtained Test Material

The video materials have been recorded in parallel with the use of three types of cameras. The scenes have been registered using SONY HDR CX-550 video camera, IR camera FLIR SC325 and depth map Intel SR300 camera (working in visible lighting and infrared band). The first one is a standard Full HD video camera, with a resolution of 12 MPix, operating in visible spectrum light. The other is a thermal imaging camera, equipped with 16-bit sensor of 320×240 pixels. It offers 25×18.8 degree FOV, works at 60 Hz and is interfaced by Ethernet [6]. The last one is a full 1080p color camera working with 30 frames per second (FPS), and offers an optimal distance between 0.2 m and 1.5 m for best depth perception.

2.2 Characteristics of the Benchmark

The shared database includes records of 50 persons with different characteristics (women, men, people with and without beards, with and without glasses, young ones and older). Sample original frames from the database are presented in Fig. 3. The size of the frame from cameras is different, depending on the camera characteristics. Selected frames presenting exemplary behaviour of participants are presented in Fig. 4.

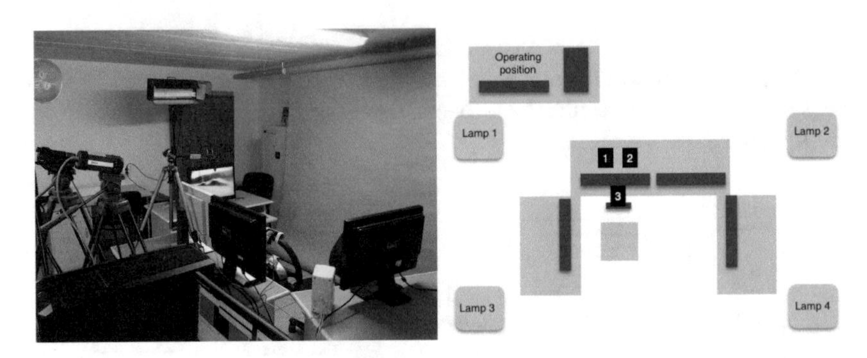

Fig. 2. Developed stand to acquire video sequences (squares marked with numbers 1, 2, 3 indicate the deployment of cameras, 1 – visible light camera, 2 – thermal camera, 3 – depth map camera).

Each person was recorded in a time interval of 3 to 35 min using three cameras simultaneously (thermal, visible-light and depth map). As the result, there are 3 independent videos for each person, one in each spectral band. In addition, if the person wore glasses, the registration was doubled and took into account both cases (with and without glasses).

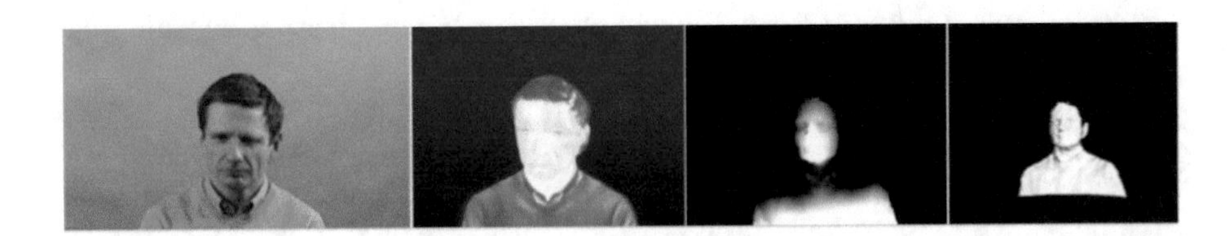

Fig. 3. Sample frames from the developed database, from the left: visible spectrum, thermal, depth map and cloud of points.

The registration procedure consists of the following stages. Firstly, the person being recorded was asked to sign an agreement to use the gathered material free of charge for fatigue recognition research. Then the person was asked to sit and listen to the instructions. At that time the person did not make any moves. After becoming acquainted with the procedure, the actual process of acquisition started. Three independent sections could be separated here. The first section constitutes a base material. It has a form of 20-s driving simulation with the ordinary blinking of the eyes and a natural observations of the road situation seen from the driver's perspective. The presentation has been displayed on the front and side monitors corresponding to the windshield and side windows of the vehicle.

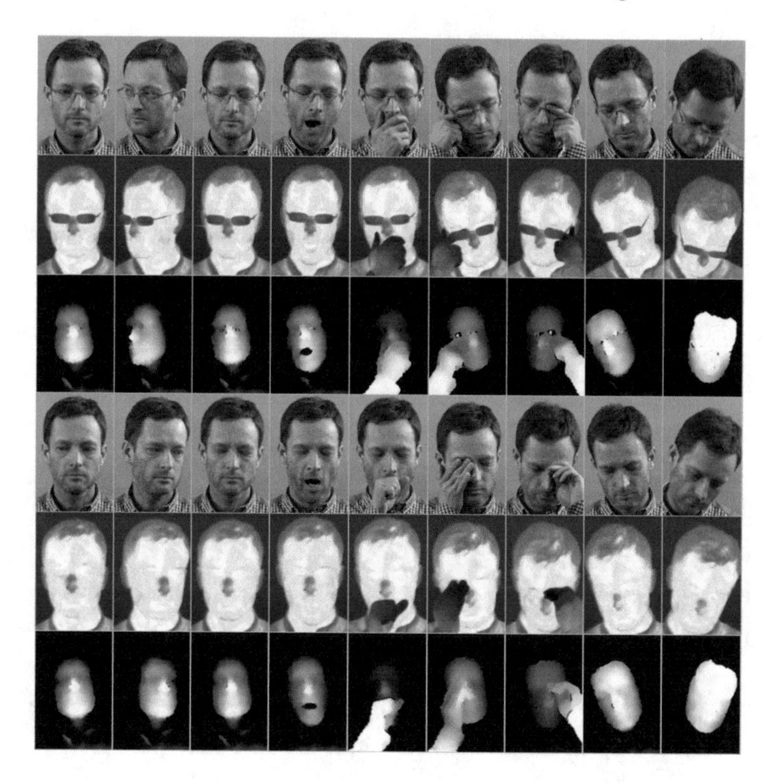

Fig. 4. Selected frames from the developed database for a person recorded with (top three rows) and without glasses (bottom three rows), presenting exemplary behaviour of a participant.

Then, there was a smooth transition to the Sect. 2, during which the person being recorded was asked to perform a series of four quick blinks with the eyes. After a moment the participant was asked to simulate the fight against drowsiness by rubbing and opening eyes for the next 10 s. Subsequently, the person was asked to squint the eyes and close them for 2 s. After opening the eyes and five seconds of normal driving, a transition to Sect. 3 followed. At that time, the person was asked to simulate yawning without covering the mouth, with varying length and frequency, for 10 s. Next, the participant was asked to simulate yawning with mouth covering, also with different length and frequency. Then the person being recorded was asked to rub the eyes several times with 10 s intervals.

The last section of the acquisition was related to the head movements and contained: head lowering with simultaneous eyes closing, simulation of counteracting to falling asleep by aggressive shaking of the head and return to the starting position. The last stage was associated with lowering the head in any direction with simultaneous closing of the eyes for 5 s. It was repeated several times.

2.3 TensorFlow

The TensorFlow framework [1] has been used for the classification of actions performed by drivers. Each time, an image coming from a respective sensor has been captured, which was later taken as an input for the detector. Each modality is associated with an individual detector, hence three unique networks have to be trained. All of them have the same architecture, yet the input data are different.

Input image size: 300×300 (all images were resized to that dimensions). Images were augmented by horizontal flipping and random cropping. The net was based on Single Shot MultiBox Detector (SSD) Mobilenet v1 architecture (Fig. 5) with two classes (rubbing/not-rubbing).

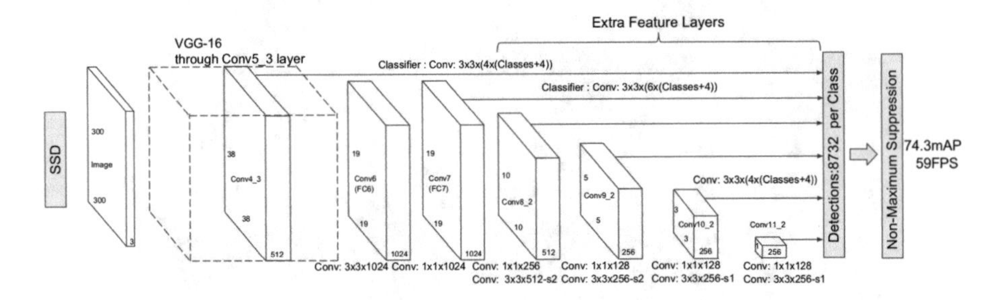

Fig. 5. The model of Single Shot MultiBox Detector (SSD) [15].

3 Experimental Results

The evaluation protocol is as follows. We manually marked the ground truth (the frames with people rubbing their eyes) in the validation video stream containing approximately 400–600 frames extracted from original benchmark data. We took 40 different videos and for each of them we took form 10 to 15 significantly different frames. For visible spectrum there was 500, for thermal –397 and for depth map –592 frames. They contained neutral poses as well as rubbing eyes occurrences (our model consisted of two classes).

The training time for each of the developed models was respectively: 23 h (3151 steps), 26.5 h (3071 steps) and 20 h (2740 steps), using Intel Core i7-6700HQ with 32 GB RAM under Windows 10 Education (x64) operating system.

Training models have been completed with the loss factor equal to 1.14 (visible spectrum), 1.27 (thermal imagery) and 1.12 (depth map). The exemplary frames, containing detected acts of rubbing eyes in visible spectrum, are presented in Fig. 6 and not rubbing eyes in Fig. 7, respectively.

The detection of the rubbing eye, for the visible spectrum was with 100% occurancy while the detection of the class not rubbing eye was only 88.9% effective.

Fig. 6. The exemplary frames, containing detected acts of rubbing eyes in visible spectrum.

Fig. 7. The exemplary frames, containing the class of not rubbing eyes in visible spectrum.

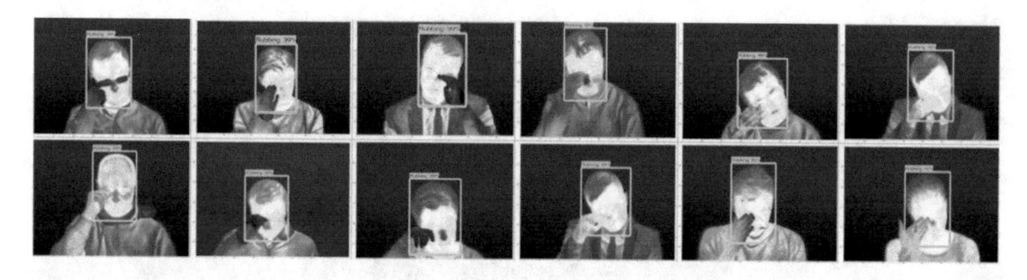

Fig. 8. The exemplary frames, containing the detection of rubbing eyes in thermal imagery.

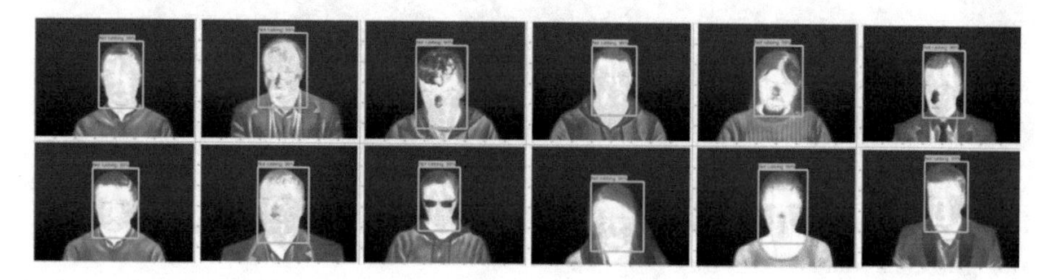

Fig. 9. The exemplary frames, containing the class of not rubbing eyes in thermal imagery.

The exemplary frames, containing detected acts of rubbing eyes in thermal imagery, are presented in Fig. 8 and not rubbing eyes in Fig. 9. The detection of the rubbing eye, for thermal imagery was with 89% occurancy while the detection of the class not rubbing eye was equal to 94.4% effective. However, in several

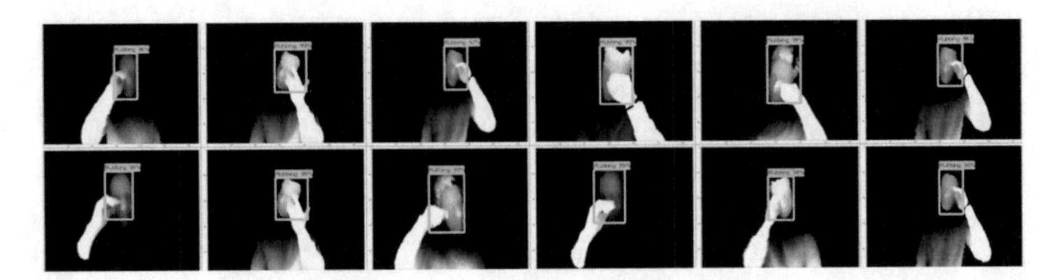

Fig. 10. The exemplary frames, containing the detection of rubbing eyes in depth map.

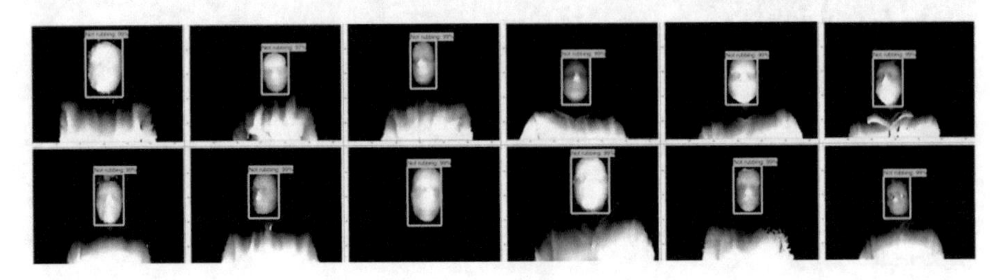

Fig. 11. The exemplary frames, containing the class of not rubbing eyes in depth map.

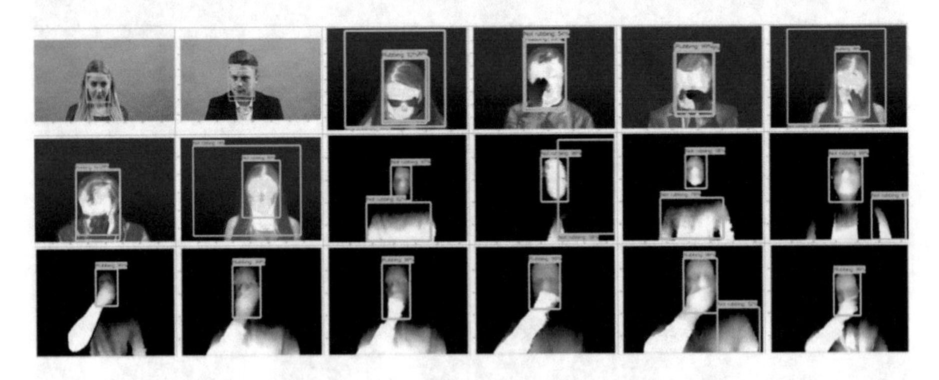

Fig. 12. Selected frames with wrong detection for all analysed spectra.

cases, along with the correct recognition of rubbing eye, the second class was recognized at the same time (Fig. 12).

The exemplary frames, containing detected acts of rubbing eyes in depth map, are presented in Fig. 10 and not rubbing eyes in Fig. 11. The detection of the rubbing eye was with 88.9% occurancy while the detection of the class not rubbing eye was 77.8% effective. However, in several cases, along with the correct recognition of not rubbing eye, this class was recognized more then ones at the same frame (Fig. 12). Additionally, there were a few cases with covering the yawning and they were classified as rubbing eye (Fig. 12).

4 Summary

The article presents a developed database (available at cvlab.zut.edu.pl) for testing methods and algorithms for drivers' fatigue detection. Characterization of the benchmark was made, the method of data collection was presented and a developed stand was presented, which served this purpose. As a research thread, part of the data was used and based on TensorFlow library for recognizing images in the visible spectrum, eye rubbing was detected for three different spectra. The results are illustrated in numerous examples. The detection of the rubbing eye, for visible spectrum, was with 100% occurancy while for depth map was 88.9%. The worst results TensorFlow achieved with the detection of the "not rubbing eye" class in the depth map band. The effectiveness was 77.8%.

References

1. Abadi M, Barham P, Chen J, Chen Z, Davis A, Dean J, Devin M, Ghemawat S, Irving G, Isard M et al (2016) Tensorflow: a system for large-scale machine learning. In: OSDI, vol 16, pp 265–283
2. Bortkiewicz A, Gadzicka E, Siedlecka J, Kosobudzki M, Dania M, Szymczak W, Jóźwiak Z, Szyjkowska A, Viebig P, Pas-Wyroślak A et al (2019) Analysis of bus drivers reaction to simulated traffic collision situations – eye-tracking studies. Int J Occup Med Environ Health 32(2):161–174. https://doi.org/10.13075/ijomeh.1896.01305
3. Chang H, Koschan A, Abidi M, Kong SG, Won CH (2008) Multispectral visible and infrared imaging for face recognition. In: IEEE computer society conference on computer vision and pattern recognition workshops, CVPRW 2008. IEEE, pp 1–6
4. Craye C, Rashwan A, Kamel MS, Karray F (2016) A multi-modal driver fatigue and distraction assessment system. Int J Intell Transp Syst Res 14(3):173–194
5. Cyganek B, Gruszczyński S (2014) Hybrid computer vision system for drivers eye recognition and fatigue monitoring. Neurocomputing 126:78–94
6. FLIR Instruments: Thermovision sdk user's manual, 2.6 sp2 edition (2010)
7. Forczmański P (2017) Human face detection in thermal images using an ensembleof cascading classifiers. Adv Intell Syst Comput 534:205–215. https://doi.org/10.1007/978-3-319-48429-7_19
8. Forczmański P (2018) Performance evaluation of selected thermal imaging-based human face detectors. Adv Intell Syst Comput 578:170–181. https://doi.org/10.1007/978-3-319-59162-9_18
9. Forczmański P, Kutelski K (2019) Driver drowsiness estimation by means of face depth map analysis. Adv Intell Syst Comput 889:396–407. https://doi.org/10.1007/978-3-030-03314-9_34
10. Forczmański P, Małecki K (2013) Selected aspects of traffic signs recognition: visual versus RFID approach. In: International conference on transport systems telematics. Springer, pp 268–274 (2013)
11. Jo J, Lee SJ, Park KR, Kim IJ, Kim J (2014) Detecting driver drowsiness using feature-level fusion and user-specific classification. Expert Syst Appl 41(4):1139–1152

12. Källhammer JE (2006) Night vision: requirements and possible roadmap for FIR and NIR systems. In: Photonics in the automobile II, vol 6198. International Society for Optics and Photonics, p 61980F (2006)
13. Kong W, Zhou L, Wang Y, Zhang J, Liu J, Gao S (2015) A system of driving fatigue detection based on machine vision and its application on smart device. J Sens 2015:11. Article ID 548602. https://doi.org/10.1155/2015/548602
14. Krishnasree V, Balaji N, Rao PS (2014) A real time improved driver fatigue monitoring system. WSEAS Trans Sig Process 10:146
15. Liu W, Anguelov D, Erhan D, Szegedy C, Reed S, Fu CY, Berg AC (2016) SSD: single shot multibox detector. In: European conference on computer vision. Springer, pp 21–37
16. Macioszek E (2017) Analysis of significance of differences between psychotechnical parameters for drivers at the entries to one-lane and turbo roundabouts in Poland. Adv Intell Syst Comput 505:149–161. https://doi.org/10.1007/978-3-319-43991-4_13
17. Makowiec-Dabrowska T, Siedlecka J, Gadzicka E, Szyjkowska A, Dania M, Viebig P, Kosobudzki M, Bortkiewicz A (2015) Work fatigue in urban bus drivers. Medycyna pracy 66(5):661–677
18. Małecki K, Nowosielski A, Forczmański P (2017) Multispectral data acquisition in the assessment of driver s fatigue. Commun Comput Inf Sci 715:320–332. https://doi.org/10.1007/978-3-319-66251-0_26
19. Małecki K, Watróbski J (2017) Mobile system of decision-making on road threats. Procedia Comput Sci 112:1737–1746
20. Mitas A, Czapla Z, Bugdol M, Ryguła A (2010) Registration and evaluation of biometric parameters of the driver to improve road safety. Scientific Papers of Transport, Silesian University of Technology, pp 71–79
21. Robert S, Adam W (2016) Mouth features extraction for emotion classification. In: Proceedings of the 2016 federated conference on computer science and information systems, FedCSIS, pp 1685–1692. https://doi.org/10.15439/2016F390
22. Staniek M (2017) Detection of cracks in asphalt pavement during road inspection processes. Sci J Silesian Univ Technol Ser Transp 96:175–184
23. Staniek M (2017) Stereo vision method application to road inspection. Baltic J Road Bridge Eng 12(1):38–47
24. Weller G, Schlag B (2007) Road user behavior model, deliverable D8 project RIPCORD-ISERET. In: 6th framework program of the European union
25. Zhang Y, Hua C (2015) Driver fatigue recognition based on facial expression analysis using local binary patterns. Optik-Int J Light Electron Opt 126(23):4501–4505

Image Contrast Enhancement Based on Laplacian-of-Gaussian Filter Combined with Morphological Reconstruction

Marcin Iwanowski(✉)

Institute of Control and Industrial Electronics, Warsaw University of Technology, ul.Koszykowa 75, 00-662 Warsaw, Poland
iwanowski@ee.pw.edu.pl

Abstract. In the paper, a method of contrast enhancement is presented. It combines linear and non-linear techniques. A non-linear approach based on morphological reconstruction is applied to select essential regions on the of Laplacian-of-Gaussian (LoG) of the input. The morphologically modified result of LoG is next added to the initial image. Thanks to the morphological processing, only the meaningful region boundaries are used to enhance the image contrast. The usage of morphological processing allows for selecting more precisely the image regions that are subject to contrast enhancement. The method performs well on textured images, allowing for adjusting the level of visible details in the output image while increasing the sharpness of meaningful image regions. A couple of examples illustrates the performance of the proposed method.

Keywords: Image contrast enhancement · Laplacian-of-Gaussian · Morphological image processing

1 Introduction

Contrast belongs to one of the most critical properties of digital images. Along with lightness, the (presence or absence of) noise, the contrast directly influences the perception of images. One may define it as a local difference of luminance that makes image details visible and distinguishable one from another and from the background. It is strictly related to the impression of sharpness of edges of image objects. To improve image contrast, one has to increase local pixel differences. Improving contrast allows for the better perception of the image content. Thus, the contrast improvement methods are of primary importance in digital imaging.

Classic contrast improvement methods make use of an estimate of the second image derivative. Modifying an image by subtracting (or adding, depending on the formula used) the estimate of the second derivative, results in an increase of the edge sharpness. These methods are usually based on the Laplacian or Laplacian-of-Gaussian convolution filters. The drawback of these filters is related

R. Burduk et al. (Eds.): CORES 2019, AISC 977, pp. 305–315, 2020.
https://doi.org/10.1007/978-3-030-19738-4_31

to the lack of selectiveness of the image details, the contrast of which is enhanced. In other words, no matter the image element it a real edge (contrast of which is to be adjusted) or noise, its sharpness is increased. Consequently, not only the contrast of meaningful details is increased, but also the noise.

In this paper, the method of contrast enhancement is proposed that allows for controlling the sharpness level of details the contrast of which is enhanced. It is based on the morphological processing of the result of the convolution-based second derivative estimate. In particular, the proposed processing scheme makes use of the morphological reconstruction that allows removing from the result of convolution filtering, the details of less importance. Moreover, this importance can be controlled by adjusting the set-up threshold.

The paper consists of 6 sections. In Sect. 2, previous works on the Laplacian-based contrast enhancement methods are shortly described. Section 3 is devoted to preliminaries referring to both linear and morphological tools used. In Sect. 4, the proposed method is described in details. Section 5 contains some results of experiments. Finally, Sect. 6 concludes the paper.

2 Related Works

An application of image derivatives to image edge processing that are estimated using the convolution filters belongs to standard well-established tools of image processing [2,4]. The fundamental work on edge-detection and its relation to an estimation of derivatives is the Marr and Hildreth paper [5]. The way of applying the Laplacian-based second-order derivative estimators to contrast enhancement was also discussed in [8]. Starting from basic Laplacian-based methods, some more sophisticated has been developed, e.g., based on the Laplacian Pyramid [1,14].

On the other hand, there exist several morphological approaches to image contrast enhancement. One of the most popular is the top-hat based operator [10]. In the paper [3], the morphological approach is applied to detect the image background and to improve image contrast. Issues of contrast improvement by means of the morphological image processing have also been discussed in [6,7,11,12].

3 Basic Notions

3.1 Laplacian-of-Gaussian-Based Contrast Enhancement

Let I be the input image – a 2D discrete function. The Laplacian operator is defined as a sum of second derivatives of an image function I along x and y axes:

$$\nabla^2 I = \frac{\partial^2 I}{\partial x^2} + \frac{\partial^2 I}{\partial y^2} \tag{1}$$

Laplacian of an image is usually computed using convolution. In general, the convolution filter is defined as a convolution of the image and a filter mask (kernel):

$$I' = I * F \Leftrightarrow I'(x,y) = \sum_{u,v \in F} I(x+u, y+v) \cdot F(u,v), \tag{2}$$

where I, I' are input and output images, respectively; p, q are image pixels and F is a mask of the filter (kernel). Laplacian is usually estimated using convolution filter with one of masks as defined by the Eq. 3.

$$L_1 = \begin{bmatrix} 0 & 1 & 0 \\ 1 & -4 & 1 \\ 0 & 1 & 0 \end{bmatrix} ; \ L_2 = \begin{bmatrix} 1 & 1 & 1 \\ 1 & -8 & 1 \\ 1 & 1 & 1 \end{bmatrix} \tag{3}$$

By subtracting the result of Laplacian from the initial image, one gets a result the classic contrast enhancement technique:

$$I' = I - \nabla^2 I = I - I * L. \tag{4}$$

The Laplacian computed using the convolution with the mask according to the Eq. 3 is considering the closest pixel neighborhood. To examine the variations of pixel values on a larger scale, while reducing the filter sensitivity to the noise, the original image if filtered, before applying the Laplacian, using the Gaussian filter. In the case of the Gaussian filter, the mask is defined as:

$$G(x,y) = \frac{1}{2\pi\sigma^2} e^{-\frac{x^2+y^2}{2\sigma^2}}, \tag{5}$$

where σ stands for the standard deviation, that in this case is the parameter defining the strength of a filter. Its combination with the Laplacian results in the Laplacian-of-Gaussian mask:

$$LoG(x,y) = \frac{1}{\pi\sigma^4} \left[1 - \frac{x^2+y^2}{2\sigma^2} \right] e^{-\frac{x^2+y^2}{2\sigma^2}}, \tag{6}$$

It is computed as linear filter with Gaussian mask followed by the one of Laplacian one or using the single LoG mask that is a combination of both (LG):

$$LoG(I) = (I * G) * L = I * LG \tag{7}$$

Following the Eq. 4, the LoG contrast enhancement is computed according to the following rule:

$$I' = I - LoG(I) = I - I * LG. \tag{8}$$

3.2 Morphological Processing

The morphological reconstruction [9,10,13][1] R of image J with marker image I, is defined as:

$$\mathcal{R}_J(I) = \delta_J^{(s)}(I), \tag{9}$$

[1] Formally saying, the reconstruction applied in the proposed method is the reconstruction by dilation. For the sake of clarity, since the reconstruction by erosion is not used here, the 'by dilation' words are omitted here.

where $\delta_J^{(n)}(I)$ is the morphological geodesic dilation of an image I with mask J of size n defined as:

$$\delta_{B,J}^{(n)}(I) = \underbrace{\delta_{B,J}(\delta_{B,J}(...\delta_{B,J}}_{n-times}(I)...))), \ \ \delta_{B,J}(I) = \delta_B(I) \wedge J, \tag{10}$$

where $\delta_B(I)$ stands for the morphological dilation of image I with structuring element B and $\wedge$ is a point-wise minimum operator. The size s of geodesic dilation in the Eq. 9 is defined as:

$$s = \min_n \left\{ \delta_{B,J}^{(n)}(I) = \delta_{B,J}^{(n+1)}(I) \right\}, \tag{11}$$

Morphological reconstruction allows for restoring image regions starting from given markers. In the current study, it is used to reconstruct image regions characterized by the gray level that is above a given threshold of t. In such a case, markers are defined as:

$$\mathcal{T}_t(I)(p) = \begin{cases} I(p) \ if \ I(p) \geq 0 \\ 0 \quad\ \ if \ I(p) < 0 \end{cases} \tag{12}$$

Simple thresholding according to the Eq. 12 results in removing all image regions that are below this threshold (cut-off). In case, however, an image contains connected areas that are only partially characterized by the graylevel above the threshold t, their lower parts are removed. To recover those parts, the morphological reconstruction is used:

$$\mathcal{RT}_t(I) = \mathcal{R}_{\ I}(\mathcal{T}_t(I)). \tag{13}$$

Figure 1 illustrates this concept. The input image (a) consists of multiple boxes lighter than the background with both uniform and non-uniform content. The binary thresholding at level 0.5 of the maximum pixel value produces a binary mask (b), while the thresholding using the Eq. 12 results in the multi-valued image (c), without complete visibility of boxes consisting of pixel values below and above the threshold – in their case only fragments above the threshold are present. Finally, an application of reconstruction allows for recovering the original content of boxes (d).

4 Proposed Approach

In the proposed approach, the non-linear method (morphological reconstruction from threshold) is applied to select essential regions on the result of high-pass convolution filtering using the Laplacian-of-Gaussian filter. The morphologically modified result of LoG is added to the initial image. Comparing to classic Laplacian-based contrast enhancement techniques, in this approach, an additional step of morphological processing of LoG is added. Thanks to that, LoG regions of high-amplitude (meaningful region boundaries) are separated

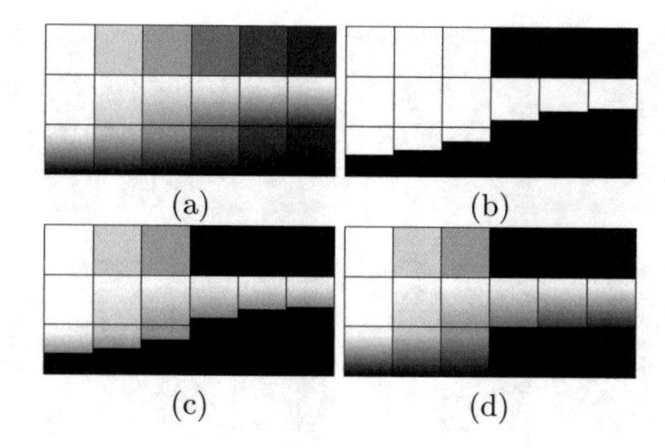

Fig. 1. Example of morphological reconstruction from threshold: (a) original image, (b) binary thresholding, (c) image masking from threshold – cut-off, (d) reconstruction from threshold.

from those of low-amplitude (noise), and the latter are removed before adding the Laplacian to the initial image. The combination of Gaussian filtering (the first part of the LoG) and the morphological processing allows for controlling more precisely image regions that are subject to contrast enhancement.

Since morphological processing applied in the proposed approach is defined exclusively on positive-valued images while the LoG result consists of both positive and negative values, the LoG image is split into two images, positive and negative:

$$\begin{aligned} LoG^{+}(I) &= \mathcal{T}_0(LoG(I)), \\ LoG^{-}(I) &= \mathcal{T}_0(-LoG(I)), \end{aligned} \tag{14}$$

where $\mathcal{T}$ the the threshold at level 0 defined by the Eq. 12.

The principal idea of the proposed method is to process separately LoG^+ and LoG^- images using the same processing scheme based on morphological reconstruction from threshold:

$$I' = I + \mathcal{RT}_t\left(LoG^{-}(I)\right) - \mathcal{RT}_t\left(LoG^{+}(I)\right), \tag{15}$$

where I and I' are initial and final image (with enhanced contrast). The flowchart of the method is shown in Fig. 3.

Morphological processing of LoG^+ and LoG^- removes their regions characterized by low pixel values while preserving high-valued ones. Finally, the morphologically processed LoG's are added to and subtracted from the initial image (as given by Eq. 15) to get the selective contrast improvement.

The morphological operator $\mathcal{RT}$ is, in fact, a *selection* operator that removes from the LoG image low-valued regions. The threshold t is a sensitivity parameter that allows controlling the level at which the LoG is cut-off and consequently the level of details the contrast of which is enhanced.

An example exhibiting detailed results of processing is shown in Fig. 2. The input image Fig. 2(a) is first processed using classic LoG contrast enhancement approach (b) and using the proposed (c). Comparing (b) and (c) one may observe that classic approach improves the contrast of all boundaries and a noise (lower

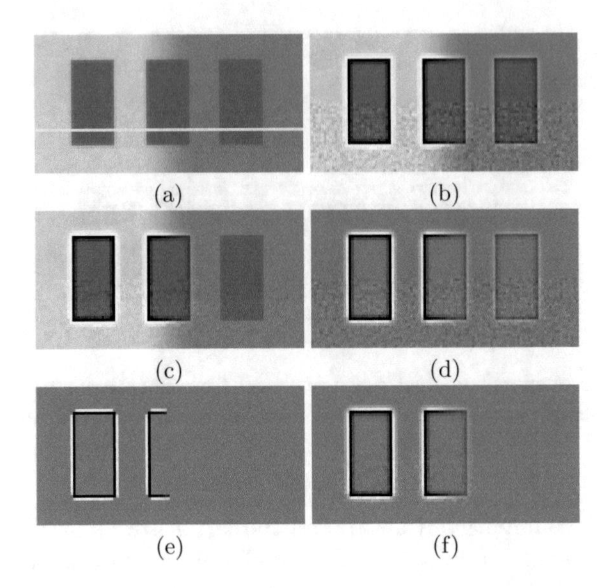

Fig. 2. Example image: original (a), simple LoG contrast enhancement result (b), filtered LoG result (c), simple LoG (d), $\mathcal{T}(LoG)$ (e), $\mathcal{RT}(LoG)$ (f).

part of the test image) while the proposed one introduces (controllable) selectivity. The proposed approach considers the amplitude of the LoG, that is shown in Fig. 2(d). Simple thresholding of the latter (Fig. 2(e)) does not recover the entire boundary of the middle object. The reconstruction (f) solves this problem, and the entire boundary is visible. Besides, the noise is filtered out and not taken into account when modifying the input image (Fig. 4).

5 Tests

The proposed methods have been tested on selected images from the Kodak database. The experiments proved that it allows for controlling the level of details, the contrast of which is enhanced. The control of details level is particularly useful for textures. Figure 5 exhibits the first test image along with the results of the proposed contrast enhancement scheme. In the experiments, the size of the mask of the LoG filter (Eq. 7) was equal to 5×5 with coefficients as given by the Eq. 16.

$$LG = \begin{bmatrix} 0.045 & 0.047 & 0.056 & 0.047 & 0.045 \\ 0.047 & 0.317 & 0.715 & 0.317 & 0.047 \\ 0.056 & 0.715 & -4.905 & 0.715 & 0.056 \\ 0.047 & 0.317 & 0.715 & 0.317 & 0.047 \\ 0.045 & 0.047 & 0.056 & 0.047 & 0.045 \end{bmatrix} \quad (16)$$

Test image contains non-uniform texture related mostly to bricks of the wall. The texture that model the brick is characterized by the regular shape of the

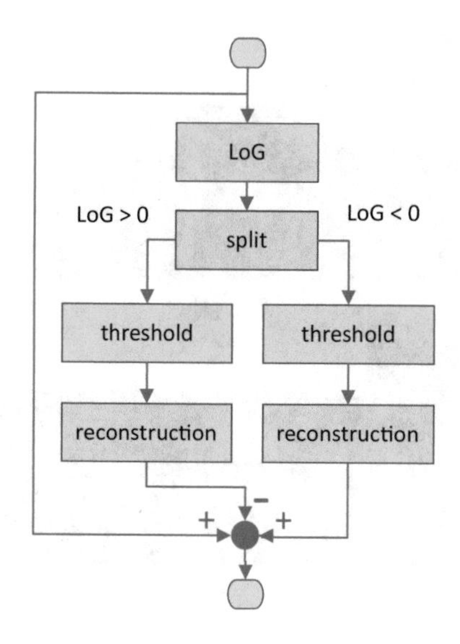

Fig. 3. Flowchart of the proposed method.

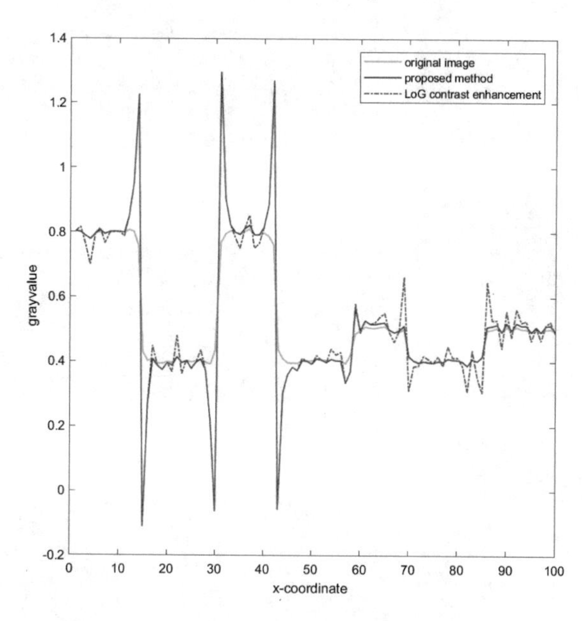

Fig. 4. Cross-section along horizontal line shown in Fig. 2(a).

brick itself filled by irregularities of the brick surface and the amplitude of the LoG of irregularities is varying. The human perception of image sharpness is correlated with the amplitude of the second derivative inside the brick. Consequently, the cut-off of the low-amplitude LoG components results in preserving the original texture of the brick on the output image. This relation may be

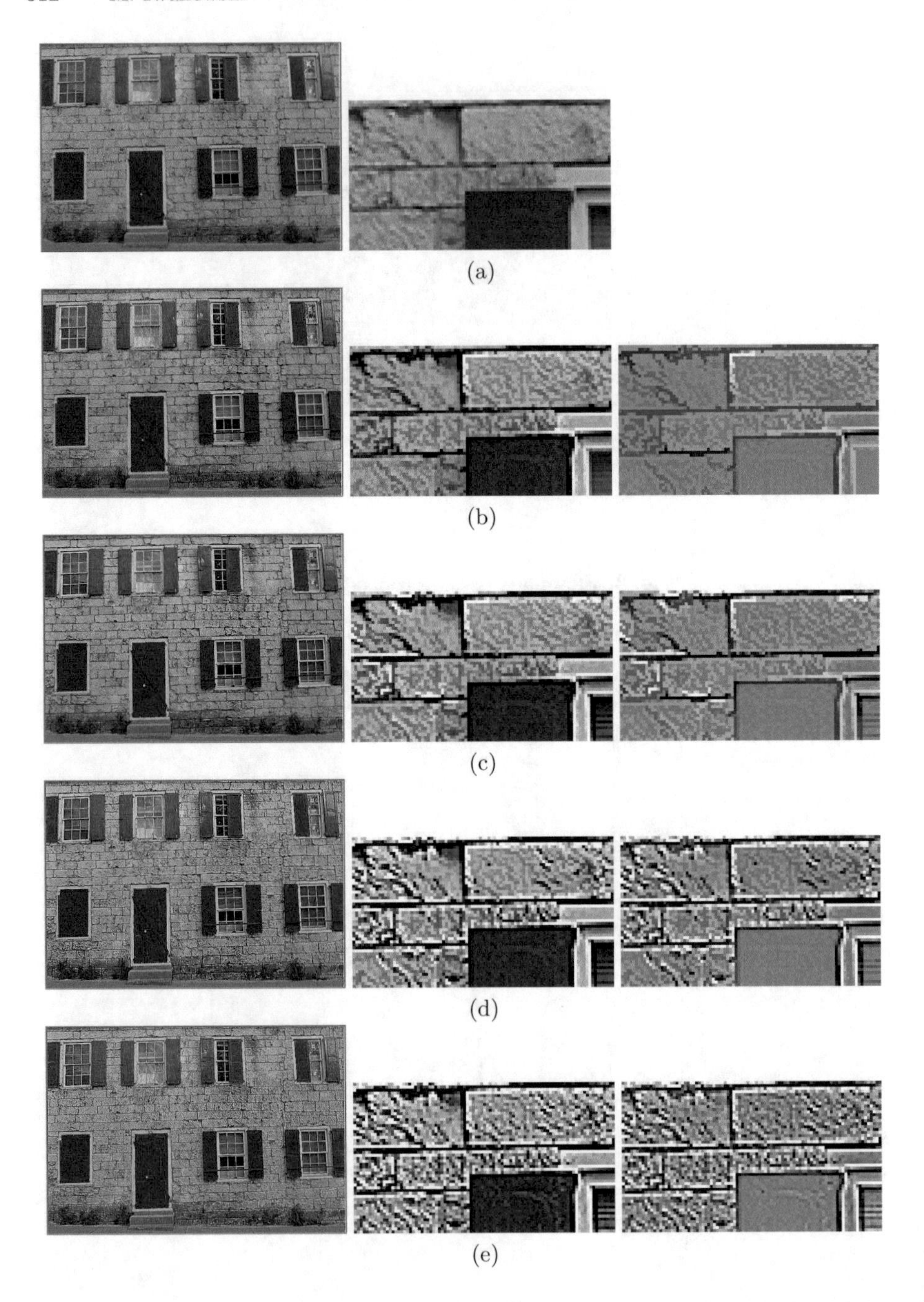

Fig. 5. The first test image, from left to right: full-size, zoomed part, filteres LoG for: original (a), $t = 1.5$ (b), $t = 1.0$ (c), $t = 0.5$ (d), $t = 0$ – simple LoG (e).

Fig. 6. Second test image (with zoomed part): original (a), $t = 2.0$ (b), $t = 1.5$ (c), $t = 1.0$ (d), $t = 0.5$ (e), $t = 0$ – simple LoG (f).

observed in Fig. 5 that present processing results obtained for increasing values of the threshold t. These values are scaled in such a way, that $t = 1$ stands for the threshold equal to the highest possible value of the input image. Values of the LoG image may be higher than one due to the obvious LoG filter properties.

The selectivity of the proposed approach concerning the control of the level of details preserved may also be observed on the second test image presented in Fig. 6. The textured area, in this case, consists of bricks, fence, sky, and trees. Here also when modifying the threshold t values, the level of preserved details is also varying.

6 Conclusions

In the paper, a method of controllable contrast enhancement has been presented. Non-linear method of morphological reconstruction from the parametrized threshold of the input image is applied to select essential regions on the result of the Laplacian-of-Gaussian filter. Finally, the morphologically modified result of LoG is added to the initial image. Thanks to the morphological processing, the meaningful region boundaries are separated from others, and not used to improve the image contrast. The combination of Gaussian filtering and morphological processing allows for controlling more precisely image regions that are subject to contrast enhancement. The method performs well on textured images, allowing for adjusting the level of visible details on the output image while increasing the sharpness of essential image regions. The proposed method has been tested on selected images from a Kodak image database. The experiments showed that the proposed approach, compared to the typical usage of the Laplacian of Gaussian, allows for more sophisticated control of the level of details the contrast of which is enhanced. Results obtained confirm the usefulness and applicability of the proposed method.

References

1. Dippel S, Stahl M, Wiemker R, Blaffert T (2002) Multiscale contrast enhancement for radiographies: Laplacian pyramid versus fast wavelet transform. IEEE Trans Med Imaging 21(4):343–353. https://doi.org/10.1109/TMI.2002.1000258
2. Gonzalez R, Woods R (2002) Digital image processing, 2nd edn. Prentice-Hall, Upper Saddle River
3. Jimenez-Sanchez AR, Mendiola-Santibanez JD, Terol-Villalobos IR, Herrera-Ruiz G, Vargas-Vazquez D, Garcia-Escalante JJ, Lara-Guevara A (2009) Morphological background detection and enhancement of images with poorlighting. IEEE Trans Image Process 18(3):613–623
4. Kak A, Rosenfeld A (1982) Digital picture processing. Academic Press, New York
5. Marr D, Hildreth E (1980) Theory of edge detection. Proc Roy Soc Lond B: Biol Sci 207(1167):187–217. https://doi.org/10.1098/rspb.1980.0020. http://rspb.royalsocietypublishing.org/content/207/1167/187
6. Meyer F, Serra J (1989) Contrasts and activity lattice. Sig Process 16:303–317

7. Mukhopadhyay S, Chanda B (2000) A multiscale morphological approach to local contrast enhancement. Sig Process 80(4):685–696
8. Neycenssac F (1993) Contrast enhancement using the Laplacian-of-a-Gaussian filter. CVGIP: Graph Models Image Process 55(6), 447–463. https://doi.org/10.1006/cgip.1993.1034. http://www.sciencedirect.com/science/article/pii/S1049965283710345
9. Salembier P, Serra J (1995) Flat zones filtering, connected operators, and filters by reconstruction. IEEE Trans Image Process 4(8):1153–1160
10. Soille P (2004) Morphological image analysis: principles and applications, corrected 2nd printing of the 2nd edn. Springer, New York
11. Terol-Villalobos IR (2004) Morphological connected contrast mappings based on top-hat criteria: a multiscale contrast approach. Opt Eng 43(7):1577–1595-19. https://doi.org/10.1117/1.1757456
12. Tsubai M, Nishimura T, Sasaki A (2004) Morphological image sharpening by double structuring elements for ultrasound images, pp. 1509–1512. https://www.scopus.com/inward/record.uri?eid=2-s2.0-11144332470&partnerID=40&md5=8686aee36a31768f655cc30d452e125e. Cited By 4
13. Vincent L (1993) Morphological grayscale reconstruction in image analysis: applications and efficient algorithms. IEEE Trans Image Process 2(2):176–201
14. Yun S, Kim JH, Kim S (2010) Image enhancement using a fusion framework of histogram equalization and Laplacian pyramid. IEEE Trans Consum Electron 56(4):2763–2771. https://doi.org/10.1109/TCE.2010.5681167

Lattice Auto-Associative Memories Induced Multivariate Morphology for Hyperspectral Image Spectral-Spatial Classification

Manuel Graña(✉)

Grupo de Inteligencia Computacional, University of the Basque Country (UPV/EHU), San Sebastian, Spain
manuelgrana@ehu.es

Abstract. The simultaneous use of spatial and spectral information for the classification-based analysis of hyperspectral images improves classification results and image segmentation quality. In this paper pixel spectra are individually classified by conventional support vector machines (SVM). The result of an innovative watershed transformation is used to postprocess the SVM result, imposing the homogeneity of the watershed region, i.e. classification disagreements inside the watershed region are solved by majority voting. This paper introduces several approaches to define reduced supervised orderings based on the recall distance of lattice auto-associative memories (LAAM). The automatic unsupervised selection of the foreground/background training sets from the hyperspectral image data is performed by the use of endmember induction algorithms (EIA). The proposed approach is compared with a recent state-of-the-art spectral-spatial approach.

Keywords: Mathematical morphology · Multivariate data · Hyperspectral images · Lattice auto-associative memories · Reduced orderings · Spectral-spatial classification

1 Introduction

The generation of thematic maps from hyperspectral images by classification of the pixel spectra has been a fruitful research line. Recently, there are increasingly successful proposals that combine spatial and spectral processing, allowed by the increasing on available computing power, for the generation of more robust algorithms. Some works propose extensions of neural networks classifiers [15] and support vector machines (SVM) [13] with spatial terms. Bayesian approaches allow the integration of spatial *a priori* modeled by Markov Random Fields with spectral posterior learnt by multinomial logistic regression [14], or SVM [27].

Mathematical morphology [4,11,20–22] has been very successful defining image operators and filters for binary and gray-scale images. Lattice theory

R. Burduk et al. (Eds.): CORES 2019, AISC 977, pp. 316–325, 2020.
https://doi.org/10.1007/978-3-030-19738-4_32

[2,9,10] gives the most general theoretical background for mathematical morphology [12,19]. We call lattice computing [5] an extension of mathematical morphology encompassing general data mining, neural computing, and machine learning applications.

We propose three novel h-supervised orderings based on lattice auto-associative memories (LAAM) [17,18], hereafter named as *LAAM-supervised orderings*. The proposed orderings keeps multivariate mathematical morphology defined using lattice algebra operators ($\vee$, $\wedge$ and $+$) in a coherent mathematical framework. Furthermore, foreground (F) and background (B) training sets need to be defined to run a supervised ordering method. Often sets B and F are defined manually by selecting data points from the ground-truth [25]. Here, we propose an automatic methodology for the definition of background and foreground sets based on the use of endmember induction algorithms (EIAs). A similar approach has been applied in [6] to neurimages looking for brain functional connections on resting state data.

2 Multivariate Mathematical Morphology

The extension of mathematical morphology to color and multivariate images is not straightforward since high dimensional pixels do not have an endowed total order that respects some natural properties expected from morphological operators, i.e. not creating new colors in the image. An approach to define a total order is the conditional ordering (C-ordering) which ranks the vector components so that the order of vector components is evaluated sequentially according to this rank, until any ambiguities are solved. *Lexicographical ordering* is the best known instance of C-ordering. For hyperspectral images, C-orderings are not useful, because it does not consider always all the vector components, and there is no inherent ranking inside the hyperspectral image bands. Finally, the reduced ordering approach [26] constructs a surjective mapping of the multivariate data space into a complete lattice $h : \mathbb{R}^n \to \mathbb{L}$ so that the order relation defined on the latter induces an order relation, the h-ordering $\leq_h$, in the former data space, formally defined as follows:

$$\mathbf{x} \leq_h \mathbf{y} \Leftrightarrow h(\mathbf{x}) \leq h(\mathbf{y}) ; \forall \mathbf{x}, \mathbf{y} \in \mathbb{R}^n. \quad (1)$$

h-orderings can be defined on the basis of a supervised classifier trained with some pixel values [26] to discriminate between a background and a foreground class. Discriminant function values or the estimated class *a posteriori* probabilities provide the surjective mapping h. The formal definition of these *h-supervised orderings* is as follows [26]:

Definition 2.1. *A h-supervised ordering for a non-empty set X is a h-ordering that satisfies the following conditions: $h(b) = \bot, \forall b \in B$, and $h(f) = \top, \forall f \in F$, where $B, F \subset X$ are subsets of X such that $B \cap F = \emptyset$.*

If h-functions are not injective an equivalence relation is induced on the input data space. The induced h-ordering $\leq_h$ is a well defined partial order relation if we collapse each equivalence class $\mathcal{L}[z] = \{\mathbf{c} \in \mathbb{R}^n | h(\mathbf{c}) = z\}$ to its canonical member z. If we need to define an order relation inside each class, other order must be defined, often it is the lexicographical order. This situation is clearly undesired because the pure mathematical building structure is lost, and the remaining is like a collage of methods, but in some practical situations is hard to avoid.

The supervised erosion of a multivariate image $\{I(p) \in \mathbb{R}^n\}_{p \in D_I}$, where D_I is the spatial domain of the image, with structuring object S, is defined as follows: $\varepsilon_{h,S}(I)(p) = I(q)$ s.t. $I(q) = \bigwedge_h \{I(s); s \in S_p\}$, where $\bigwedge_h$ is the infimum defined by the reduced ordering $\leq_h$ of Eq. (1), and S_p is the structuring element translated to the pixel position p. The supervised dilation is defined as $\delta_{h,S}(I)(p) = I(q)$ s.t. $I(q) = \bigvee_h \{I(s); s \in S_p\}$, where $\bigvee_h$ is the supremum defined by the order of Eq. (1).

3 Lattice Auto-Associative Memories (LAAM)

Given a set of input/output pairs of patterns $(X, Y) = \left\{\left(\mathbf{x}^\xi, \mathbf{y}^\xi\right)\right\}_{\xi=1}^k$, a linear heteroassociative neural network is built up as $W = \sum_\xi \mathbf{y}^\xi \cdot \left(\mathbf{x}^\xi\right)'$. Accordingly, lattice associative memories are proposed in [17,18], following dual construction $W_{XY} = \bigwedge_{\xi=1}^k \left[\mathbf{y}^\xi \times \left(-\mathbf{x}^\xi\right)'\right]$ and $M_{XY} = \bigvee_{\xi=1}^k \left[\mathbf{y}^\xi \times \left(-\mathbf{x}^\xi\right)'\right]$, where $\times$ is any of the $\boxtimes$ or $\boxdot$ operators, since $\mathbf{y}^\xi \boxtimes \left(-\mathbf{x}^\xi\right)' = \mathbf{y}^\xi \boxdot \left(-\mathbf{x}^\xi\right)'$. If $X = Y$ then W_{XX} and M_{XX} are called lattice auto-associative memories (LAAM). LAAMs present some surprising properties: (a) perfect recall for an unlimited number of stored patterns, i.e. $W_{XX} \boxtimes X = X = M_{XX} \boxdot X$, (b) convergence in one step for any input pattern, and (c) W_{XX} and M_{XX} share the same set of fixed points denoted by $\mathcal{F}(X)$. LAAMs have been applied to hyperspectral endmember induction [8] and brain functional magnetic resonance imaging data [7], and classification [23] based on the LAAM's recall error measured by the Chebyshev distance.

4 LAAM-Supervised Ordering

4.1 LAAM's h-Mapping

The LAAM's h-mapping is defined as the LAAM's recall error measured by the Chebyshev distance. Given a multivariate data vector $\mathbf{c} \in \mathbb{R}^n$ and a non-empty training set $X = \{\mathbf{x}_i\}_{i=1}^K$, $\mathbf{x}_i \in \mathbb{R}^n$ for all $i = 1, \ldots, K$, the LAAM h-mapping is given by:

$$h_X(\mathbf{c}) = \zeta\left(\mathbf{x}_M^\#, \mathbf{c}\right) \vee \zeta\left(\mathbf{x}_W^\#, \mathbf{c}\right), \tag{2}$$

where $\mathbf{x}^\# \in \mathbb{R}^n$ is the recall of vector $\mathbf{c}$ from either the erosive memory M_{XX}, i.e. $\mathbf{x}_M^\# = M_{XX} \boxdot \mathbf{c}$, or the dilative memory W_{XX}, i.e. $\mathbf{x}_W^\# = W_{XX} \boxtimes \mathbf{c}$. Function $\zeta(\mathbf{a}, \mathbf{b})$ denotes the Chebyshev distance between two vectors, given by the greatest absolute difference between the vectors' components: $\zeta(\mathbf{a}, \mathbf{b}) = \bigvee_{i=1}^n |a_i - b_i|$.

4.2 One-Side LAAM h-Supervised Ordering

Using the LAAM h-mapping of Eq. (2), a one-side LAAM h-supervised ordering, denoted $\leq_X$, is defined given a training set X as follows:

$$\forall \mathbf{x}, \mathbf{y} \in \mathbb{R}^n,\ \mathbf{x} \leq_X \mathbf{y} \Longleftrightarrow h_X(\mathbf{x}) \leq h_X(\mathbf{y}). \tag{3}$$

The one-side LAAM-supervised ordering generates a complete lattice $\mathbb{L}_X$, whose bottom element $\perp_X = 0$ corresponds to the set of fixed points of M_{XX} and W_{XX}, $h(\mathbf{x}) = \perp_X$ for $\mathbf{x} \in \mathcal{F}(X)$. On the other hand, the top element is $\top_X = +\infty$.

4.3 Background/Foreground LAAM h-Supervised Pseudo-orderings

In order to build background/foreground LAAM h-supervised ordering, we independently calculate the LAAM h-mapping of Eq. (2) on B and F, obtaining mappings h_B and h_F, respectively, composing them in different ways obtaining different orderings.

Relative LAAM h-pseudo-supervised Ordering. We define the relative LAAM h-supervised ordering combining both h_B and h_F into a h-mapping $h_r(\mathbf{x}) = h_F(\mathbf{x}) - h_B(\mathbf{x})$. This mapping is negative for $\mathbf{x} \in \mathcal{F}(B)$, and positive for $\mathbf{x} \in \mathcal{F}(F)$. Therefore, we assume that it defines a discriminant function such that $h_r(\mathbf{x}) > 0$ corresponds to pixels in the foreground class, and $h_r(\mathbf{x}) < 0$ to pixels in the background class. Points such that $h_r(\mathbf{x}) = 0$ correspond to the decision boundary. The induced ordering, denoted $\leq_r$, is defined as follows:

$$\forall \mathbf{x}, \mathbf{y} \in \mathbb{R}^n,\ \mathbf{x} \leq_r \mathbf{y} \Longleftrightarrow h_r(\mathbf{x}) \leq h_r(\mathbf{y}). \tag{4}$$

The relative LAAM h-supervised ordering generates a complete lattice, $\mathbb{L}_r$, whose bottom and top elements are $\perp = -\infty$ and $\top = +\infty$, respectively. However, this mapping does not follow Definition 2.1, because $h(\mathbf{b}) \neq \perp$ for $\mathbf{b} \in B$ and $h(\mathbf{f}) \neq \top$ for $\mathbf{f} \in F$. We nevertheless use this pseudo-supervised ordering and ensuing pseudo-morphological operators with good results.

Absolute LAAM h-pseudo-supervised Ordering. In order to define the absolute LAAM h-supervised ordering, denoted $\leq_a$, the data vectors are first classified into background and foreground classes, so that the ordering is defined depending on the data classes. A data point is classified as background, i.e. $\mathbf{x} \in B$, if $h_B(\mathbf{x}) \leq h_F(\mathbf{x})$. Conversely, it is classified as foreground, i.e. $\mathbf{x} \in F$, if $h_B(\mathbf{x}) > h_F(\mathbf{x})$. The absolute ordering is defined as follows:

$$\forall \mathbf{x}, \mathbf{y} \in \mathbb{R}^n,\ \mathbf{x} \leq_a \mathbf{y} \Longleftrightarrow \begin{cases} h_B(\mathbf{x}) \leq h_B(\mathbf{y}) & \text{if } \mathbf{x}, \mathbf{y} \in B \\ \mathbf{x} \in B \text{ and } \mathbf{y} \in F & \\ h_F(\mathbf{y}) \leq h_F(\mathbf{x}) & \text{if } \mathbf{x}, \mathbf{y} \in F \end{cases} \tag{5}$$

The absolute LAAM h-supervised ordering generates a complete lattice, $\mathbb{L}_a$, with bottom and top elements $\perp_a$ and $\top_a$, respectively. The mapping h_B maps the set of fixed points of the LAAM built with the background set into the bottom element: $\bot_a = h_B(\mathbf{b}); \mathbf{b} \in \mathcal{F}(B)$. Conversely, h_F maps the set of fixed points of the LAAM built with the foreground set into the top element:$\top_a = h_F(\mathbf{f}); \mathbf{f} \in \mathcal{F}(F)$. However, we do not have a single mapping perfectly complying with Definition 2.1, therefore we call pseudo- supervised ordering the resulting mapping.

4.4 Beucher Morphological Gradient

The Beucher morphological gradient [1] for scalar valued images is computed as the difference between the dilations and erosions of image I with structuring element S: $g_S(I) = \delta_S(I) - \varepsilon_S(I)$. For multivariate images, a straightforward approach is to compute a *component-wise* gradient averaging the independently computed scalar gradients per image band: $g(I) = \sum_{i=1}^{n}(\delta(I_i) - \varepsilon(I_i))$, where I_i denotes the i-th band of the image and n is the dimension of pixels in I.

The h-supervised Beucher gradient is defined on the h-supervised erosion $\varepsilon_{h,S}(I)$ and dilation $\delta_{h,S}(I)$ as follows:

$$g_{h,S}(I) = h(\delta_{h,S}(I)) - h(\varepsilon_{h,S}(I)), \tag{6}$$

where h can be a one side or relative mapping. When we consider the absolute LAAM h-supervised ordering as the defined in Eq. (5), we have designed a specific Beucher gradient $g_a(I)$ specified by Eq. (7).

$$\forall \mathbf{x} \in I,\ g_a(\mathbf{x}) = \begin{cases} h_B(\delta(\mathbf{x})) - h_B(\varepsilon(\mathbf{x})) & \text{if } \delta(\mathbf{x}), \varepsilon(\mathbf{x}) \in B \\ h_F(\varepsilon(\mathbf{x})) - h_F(\delta(\mathbf{x})) & \text{if } \delta(\mathbf{x}), \varepsilon(\mathbf{x}) \in F \\ h_F(\delta(\mathbf{x})) - h_B(\varepsilon(\mathbf{x})) + h_B(\delta(\mathbf{x})) - h_F(\varepsilon(\mathbf{x})) & \text{otherwise} \end{cases} \tag{7}$$

4.5 Unsupervised Selection of Training Sets

To avoid manual selection of foreground/background sets, we propose an automatic unsupervised methodology to obtain appropriate sets. Assume that an Endmember Induction algorithm (EIA) [7,8] provides a set of endmembers $E = \{\mathbf{e}_i\}_{i=1}^{p}$ from the image data. The matrix of distances $D = [d_{i,j}]_{i,j=1}^{p}$, where $d_{ij} = |\mathbf{e}_i, \mathbf{e}_j|$ is an appropriate distance between endmembers, i.e. the spectral angular mapping.

One-Side h-Supervised Ordering. The training data X consists of the endmember $\mathbf{e}_{k^*} \in E$ minimizing the average distance to the remaining endmembers: $k^* = \arg\min_k \left\{\frac{1}{p-1}\sum_{i\neq k} d_{ik}\right\}_{i=1}^{p}$.

Background/Foreground h-pseudo-supervised Orderings. The training sets F and B consist of the pair of endmembers $\mathbf{e}_{i^*}, \mathbf{e}_{j^*} \in E$ with maximum pairwise distance: $(i^*, j^*) = \arg\max_{i,j}\{(d_{ij})\}$. We arbitrarily set $F = \{\mathbf{e}_{i^*}\}$ and $B = \{\mathbf{e}_{j^*}\}$.

5 Experimental Results with Hyperspectral Images

To assess the value of the proposed LAAM h-supervised orderings, we perform spectral-spatial classification of benchmark hyperspectral images following the approach introduced in [24], which is also computed for comparison.

5.1 Methodology

Experimental Design. First, a spectral classification by SVM (multi-class, one-versus-all) is performed on the hyperspectral image obtaining the baseline thematic map. Second, a watershed transform is computed on the hyperspectral image giving an image segmentation into watershed regions. Third, watershed segmentation is combined with the baseline spectral SVM thematic map applying one of the two techniques proposed in [24], so-called WHEDS and NWHEDS. In both techniques, the majority class within each watershed region is computed, and pixels inside it are assigned to this majority class. For the boundary pixels defining the region watersheds, WHEDS assigns them to the neighboring watershed region with the closest median value; while NWHEDS keeps the class assigned by the spectral SVM classification. We have obtained results using the three proposed LAAM h-supervised orderings and a component-wise ordering to compute Beucher gradients and ensuing watershed segmentations.

Performance Indices. Classification performance is measured by the following performance indices [16]: (a) The overall accuracy (OA): $\mathrm{OA} = \frac{\sum_{i=1}^{C} n_i}{N}$, (b) the average accuracy (AA) $\mathrm{AA} = \sum_{i=1}^{C} \frac{n_i}{N_i}$, and (c) the Kappa coefficient, $\kappa = \frac{N\sum_k t_{kk} - \sum_k t_{k+}t_{+k}}{N^2 - \sum_k t_{k+}t_{+k}}$, as well as the class-specific sensitivity and specificity.

5.2 Pavia University Data

Dataset and Baseline Spectral Classification. The Pavia University hyperspectral image was taken by the ROSIS-03 sensor over the facilities of the University of Pavia in Italy. The hyperspectral data has been provided by Prof. Paolo Gamba from the Telecommunications and Remote Sensing Laboratory, Pavia University (Italy) [3].

Spatial-Spectral Classification Results. The quantitative evaluation of the classification performance is given in Tables 1, 2, and 3 showing the OA, AA and Kappa values obtained from the baseline thematic map, and the ones obtained with spectral-spatial classification using morphological operators with structuring elements of radius 1, 3 and 5, respectively. There is a systematic performance increase of all the spectral-spatial classification approaches relative to the baseline spectral SVM. Also, it can be appreciated a systematic improvement from

Table 1. Classification results of the Pavia University hyperspectral image: OA, AA, and Kappa (κ) values. Morphological structuring element disc shaped of radius $r = 1$.

Method		OA	AA	κ
Pixel-wise SVM		88.97	91.60	0.8565
SVM + NWHED	CW	91.42	93.73	0.8880
	LAAM_X	90.91	93.16	0.8815
	LAAM_a	91.09	93.32	0.8838
	LAAM_r	90.81	92.90	0.8801
SVM+WHED	CW	94.46	96.33	0.9274
	LAAM_X	93.40	95.27	0.9136
	LAAM_a	93.99	95.78	0.9213
	LAAM_r	93.77	95.46	0.9184

SVM-NWHEDS relative to SVM-WHEDS regardless of the ordering or structuring element size used. The performance of the SVM-NWHEDS and SVM-WHEDS classification has not significant differences between the component-wise ordering and the proposed LAAM h-supervised orderings with the unsupervised selection of training sets. Figures 1 and 2 show respectively the class-specific sensitivity and specificity computed on the confusion matrices of the baseline spectral SVM, the SVM-NWHEDS and the SVM-WHEDS spectral-spatial classification, for all orderings considered and a disk shaped structuring element with radius 3. The performance increase introduced by the spectral-spatial algorithms is evident in almost all the classes, except those with sensitivity or specificity values close to 100%, in the range 97–99%.

Table 2. Classification results of the Pavia University hyperspectral image: OA, AA, and Kappa (κ) values. Morphological structuring element disc shaped of radius $r = 3$.

Method		OA	AA	κ
Pixel-wise SVM		88.97	91.60	0.8565
SVM + NWHED	CW	92.87	94.83	0.9068
	LAAM_X	92.70	94.43	0.9045
	LAAM_a	92.81	94.46	0.9059
	LAAM_r	91.93	93.62	0.8944
SVM+WHED	CW	94.71	95.99	0.9306
	LAAM_X	94.90	96.27	0.9331
	LAAM_a	94.87	96.14	0.9326
	LAAM_r	94.69	95.83	0.9303

Table 3. Classification results of the Pavia University hyperspectral image: OA, AA, and Kappa (κ) values. Morphological structuring element disc shaped of radius $r = 5$.

Method		OA	AA	κ
Pixel-wise SVM		88.97	91.60	0.8565
SVM + NWHED	CW	93.41	94.39	0.9135
	$LAAM_X$	93.65	94.72	0.9167
	$LAAM_a$	93.09	94.16	0.9096
	$LAAM_r$	92.61	93.84	0.9034
SVM+WHED	CW	95.46	95.86	0.9403
	$LAAM_X$	95.27	96.11	0.9378
	$LAAM_a$	95.15	95.62	0.9364
	$LAAM_r$	94.91	95.71	0.9332

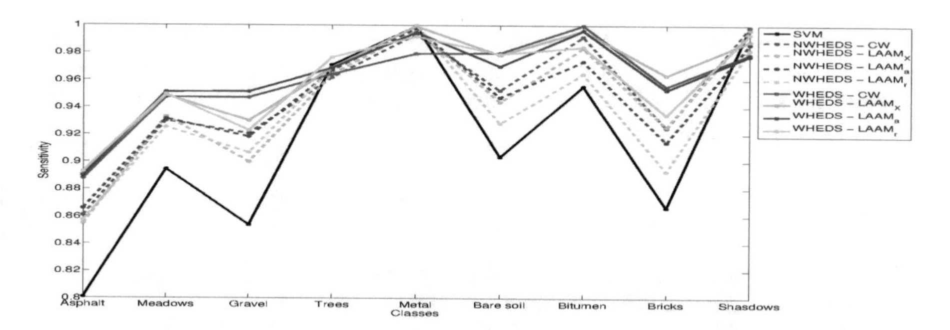

Fig. 1. Class-specific sensitivity results for the classification of the Pavia University hyperspectral image. Morphological results have been obtained using a disk shaped structuring element of radius $r = 3$.

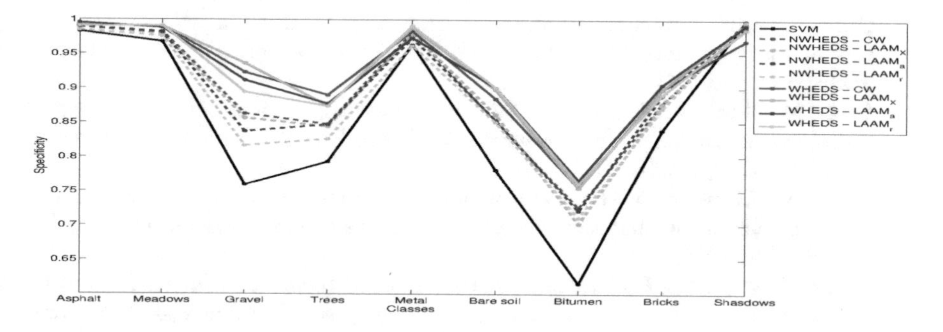

Fig. 2. Class-specific specificity results for the classification results of the Pavia University hyperspectral scene. Morphological results have been obtained using a disc shaped structuring element of radius $r = 3$.

6 Conclusions

This paper discusses the construction of some multivariate morphological operators using lattice computing techniques. Specifically, using the LAAM reconstruction error measured by the Chebyshev distance as a reduced ordering mapping it is possible to define some mathematical morphology operators based on the induced h-pseudo-supervised ordering, and its application to the segmentation of hyperspectral images.

Future work will focus in better exploiting the multivariate MM defined using LAAM-supervised orderings for spatial-spectral classification. Novel endmember induction algorithms can be used to improve the quality of the induced endmembers and more complex methodologies to define the training sets can be defined. Actually, the proposed automatic and unsupervised selection method defines the training sets by a single endmember, limiting the discriminative potential of the LAAMs. Another research avenue is to take advantage of the pixel-wise classification training samples to define the LAAM training sets.

Acknowledgements. M.A. Veganzones carried some of the computational experiments reported in this paper. The work has been partially funded by the Basque Government grant IT874-13 for the GIC research group and by FEDER funds through MINECO project TIN2017-85827-P.

References

1. Beucher S (1990) Segmentation d'images et morphologie mathematique. Ph.D. thesis, Ecole Nationale Superioure des Mines de Paris, Paris
2. Birkhoff G (1995) Lattice theory. AMS Bookstore
3. Dell'Acqua F, Gamba P, Ferrari A, Palmason J, Benediktsson J, Arnason K (2004) Exploiting spectral and spatial information in hyperspectral urban data with high resolution. IEEE Geosci Remote Sens Lett 1(4):322–326. https://doi.org/10.1109/LGRS.2004.837009
4. Goutsias J, Heijmans H (2000) Mathematical morphology. IOS Press, Amsterdam
5. Grana M (2008) A brief review of lattice computing. In: IEEE international conference on fuzzy systems, FUZZ-IEEE 2008 (IEEE world congress on computational intelligence), pp 1777–1781
6. Graña M, Chyzhyk D (2016) Image understanding applications of lattice autoassociative memories. IEEE Trans Neural Netw Learn Syst 27(9):1920–1932. https://doi.org/10.1109/TNNLS.2015.2461451
7. Graña M, Chyzhyk D, Garcia-Sebastian M, Hernandez C (2011) Lattice independent component analysis for functional magnetic resonance imaging. Inf Sci 181(10):1910–1928
8. Graña M, Villaverde I, Maldonado J, Hernández C (2009) Two lattice computing approaches for the unsupervised segmentation of hyperspectral images. Neurocomputing 72:2111–2120. https://doi.org/10.1016/j.neucom.2008.06.026
9. Gratzer G (2003) General lattice theory, 2 edn. Birkhauser Basel
10. Gratzer (2011) Lattice theory: foundation, 1st edn. Springer Basel

11. Haralick R, Sternberg S, Zhuang X (1987) Image analysis using mathematical morphology. IEEE Trans Pattern Anal Mach Intell PAMI-9(4):532–550. https://doi.org/10.1109/TPAMI.1987.4767941
12. Hawkes P, Heijmans H, Kazan B (1993) Morphological image operators. Academic Press, Cambridge
13. Li CH, Kuo BC, Lin CT, Huang CS (2012) A spatial-contextual support vector machine for remotely sensed image classification. IEEE Trans Geosci Remote Sens 50(3):784–799. https://doi.org/10.1109/TGRS.2011.2162246
14. Li J, Bioucas-Dias J, Plaza A (2012) Spectral-spatial hyperspectral image segmentation using subspace multinomial logistic regression and markov random fields. IEEE Trans Geosci Remote Sens 50(3):809–823. https://doi.org/10.1109/TGRS.2011.2162649
15. Prieto A, Bellas F, Duro R, López-Peña F (2010) An adaptive approach for the progressive integration of spatial and spectral features when training ground-based hyperspectral imaging classifiers. IEEE Trans Instrum Meas 59(8):2083–2093. https://doi.org/10.1109/TIM.2009.2030872
16. Richards J, Jia X (1999) Remote sensing digital image analysis: an introduction, 3rd edn. Springer, Heidelberg
17. Ritter G, Diaz-de-Leon J, Sussner P (1999) Morphological bidirectional associative memories. Neural Netw 12(6):851–867
18. Ritter G, Sussner P, Diaz-de-Leon J (1998) Morphological associative memories. IEEE Trans Neural Netw 9(2):281–293
19. Ronse C (1990) Why mathematical morphology needs complete lattices. Sig Process 21(2):129–154
20. Serra J (1984) Image analysis and mathematical morphology, vol 1, Image analysis & mathematical morphology series. Academic Press
21. Serra J (1988) Image analysis and mathematical morphology, vol 2, 1st edn. Theoretical advances. Academic Press, Cambridge
22. Soille P (2004) Morphological image analysis: principles and applications, 2nd edn. Springer, Heidelberg
23. Sussner P, Valle M (2006) Gray-scale morphological associative memories. IEEE Trans Neural Netw 17(3):559–570
24. Tarabalka Y, Chanussot J, Benediktsson J (2010) Segmentation and classification of hyperspectral images using watershed transformation. Pattern Recogn 43(7):2367–2379
25. Veganzones M, Graña M (2012, submitted) Lattice auto-associative memories induced supervised ordering defining a multivariate morphology on hyperspectral data. In: 2012 4th workshop on hyperspectral image and signal processing: evolution in remote sensing (WHISPERS). IEEE
26. Velasco-Forero S, Angulo J (2011) Supervised ordering in R^p: application to morphological processing of hyperspectral images. IEEE Trans Image Process 20(11):3301–3308
27. Zhang B, Li S, Jia X, Gao L, Peng M (2011) Adaptive markov random field approach for classification of hyperspectral imagery. IEEE Geosci Remote Sens Lett 8(5):973–977. https://doi.org/10.1109/LGRS.2011.2145353

Randomized Reference Classifier with Gaussian Distribution and Soft Confusion Matrix Applied to the Improving Weak Classifiers

Pawel Trajdos(✉) and Marek Kurzynski

Wroclaw University of Science and Technology, Wroclaw, Poland
{pawel.trajdos,marek.kurzynski}@pwr.edu.pl

Abstract. In this paper, an issue of building the RRC model using probability distributions other than beta distribution is addressed. More precisely, in this paper, we propose to build the RRR model using the truncated normal distribution. Heuristic procedures for expected value and the variance of the truncated-normal distribution are also proposed. The proposed approach is tested using SCM-based model for testing the consequences of applying the truncated normal distribution in the RRC model. The experimental evaluation is performed using four different base classifiers and seven quality measures. The results showed that the proposed approach is comparable to the RRC model built using beta distribution. What is more, for some base classifiers, the truncated-normal-based SCM algorithm turned out to be better at discovering objects coming from minority classes.

Keywords: Classification · Randomized reference classifier · Gaussian distribution · Multiclassifier systems

1 Introduction

Classification, in which one tries to assign a class label to an object, is one of the more common and well-known decision-making problems. Classification or pattern recognition tasks have been successfully applied in many areas including medicine, economy, agriculture, astronomy or defence. In the modern world, there is practically no field of human activity, where computer (automatic) classification methods would not be used. The large practical demand for computer-aided classification algorithms has resulted in the active development of object recognition methodologies in the last few decades.

Unfortunately, this variety of recognition methods does not always mean an acceptable quality of classification because each problem requires its individual approach and there is no easy solution which classifiers or algorithm should be applied.

R. Burduk et al. (Eds.): CORES 2019, AISC 977, pp. 326–336, 2020.
https://doi.org/10.1007/978-3-030-19738-4_33

If the built-in classifier does not meet the requirements of the quality of classification or is simply a weak classifier (is a bit better than the random guessing) one can use methods to improve the quality of classifiers. The most-known methods are the techniques related to the construction of an ensemble of classifiers on the basis of different training sets created from the original training set in the resampling process with a uniform distribution (bagging) [3] or uneven distribution and additionally adaptively changed (boosting) [7].

In shortly, bagging applies sampling with replacement to obtain independent training datasets for each individual classifier. Boosting modifies the input data distribution processed by each classifier in a sequence from the results of classifiers trained before, paying more attention to difficult samples. Both algorithms – even if they were applied to a weak classifier – lead to a powerful classifier in the form of a multiclassifier system.

In [18] authors introduced algorithm called Bayes metaclassifier (BMC) as a method for improving weak classifier in terms of its classification performance. In general, BMC constitutes the probabilistic generalization of any base classifier independent of its design paradigm and has the form of the Bayes scheme. Since BMC provides probabilistic interpretation for base classifier correct classification and misclassification, this method can be used in sequential classification or as a fusing mechanism in MC systems [15,16].

In [22] the original method of improving weak classifier was proposed which is based on the concept of soft confusion matrix (SCM), built using validation set. Soft confusion matrix gives a picture of local properties (for a given test object x) of base classifier including empirical probabilities of class-dependent correct and incorrect classifications. The high value of class dependent correct (incorrect) classification probability for a given object denotes that classifier is capable of the correct classification of the object x coming - let say - from the ith class (it tends to misclassify object x from ith class to - lets say - jth class). This knowledge can be directly used to correct classifying functions of the base classifier and to improve its quality.

The developed method additionally requires the formal procedure for calculation of the probability of correct and misclassification of base classifier at the point x. For this purpose, the concept of randomized reference classifier (RRC) was used, which originally was proposed in [25] as a method for calculation of competence of base classifier in the combining procedure of multiclassifier systems. This approach assumes, that the most natural measure of the classifier's competence at a given point of feature space is its probability of correct classification at this point. Unfortunately, this probability is equal to 1 or 0, unless we adopt a probabilistic model of the recognition task or assume that the classifier works in a random manner. However, both cases are difficult to accept. First, the competence should be neutral to the base classifier models, and many concepts of classifiers use the probabilistic approach. Second, in MC systems, deterministic base classifiers are generally used. For these reasons, the authors developed an indirect method. In the proposed approach the base classifier is modelled by a hypothetical classifier called randomized reference classifier (RRC). The RRC is defined by a set of random variables, which observed values are class support produced for the object to be classified. Since expected values of random variables are equal to the supports produced by the modelled base classifier, the RRC can be considered - on average - as the equivalent of this classifier. Consequently, the probability of correct classification of RRC at any point of feature space can be used as the competence of modelled base classifier at this point. The concept of RRC proved to be very effective, as it enabled the construction of MC systems, which in experimental research outperformed different state-of-the-art methods. It also turned out that with the help of RRC it is possible to determine other properties of modelled classifiers: the class-dependent probability of correct/incorrect classification and diversification of two classifiers [17,22,24] and these RRC capabilities were used in the improved of weak classifier via SCM concept.

The key problem in the construction of RRC is the choice of the probability distribution of random variables. This choice is not unique and the values of probability of correct classification of RRC and consequently the probabilities of correct/incorrect classification of modelled classifier depend on the definition of the distributions. In the original proposition of the RRC, beta probability distributions have been used. Such a choice follows from the specific definition of the class supports produced by the RRC as a random division of the unit interval. From the theory of order, statistics results that then the supports must be beta distributed [5]. As it seems the use of beta distribution in the draw process is the main disadvantage of the RRC concept. The only justification, related to the geometrical interpretation of the support vector is not a substantive justification. In particular, the proposed distribution does not strictly associate the RRC with the modelled base classifier in the context of its properties which are observable for the validation objects.

In this study, the concept of RRC based on Gaussian distribution is developed. Because in Gaussian distribution we can tune not only the expected value as in the beta distribution but also the variance, therefore the proposed RRC classifier is more flexible and can better adapt to the properties of the modelled base classifier.

The paper is divided into four sections and organized as follows. Section 2 introduces the formal notation used in the paper and provides a description of the proposed approach. The experimental setup is given in Sect. 3. In Sect. 4 experimental results are given and discussed. Section 5 concludes the paper.

2 Proposed Method

2.1 Preliminaries

In the single-label classification approach, a $d-$ dimensional vector $\boldsymbol{x} \in \mathbb{X} = \mathbb{R}^d$ is assigned a class $m \in \mathbb{M}$, where $\mathbb{M} = \{0, 1, 2, \cdots, M\}$ is a set of available classes. The classifier $\psi : \mathbb{X} \mapsto \mathbb{M}$ is an approximation of an unknown mapping $f : \mathbb{X} \mapsto \mathbb{M}$ which assigns the classes to the instances. The classification methods analyzed in this paper follow the statistical classification framework. Hence, a feature vector $\boldsymbol{x}$ and its label m are assumed to be realisations of random variables $\mathbf{X}$ and $\mathbf{M}$, respectively. The random variables follow the joint probability distribution $P(\mathbf{X}, \mathbf{M})$. Given the zero-one loss, the optimal decision is made using the maximum *a posteriori* rule:

$$\psi^*(\boldsymbol{x}) = \underset{\boldsymbol{k} \in \mathbb{M}}{\operatorname{argmax}}\, P(\mathbb{M} = k|\mathbb{X} = x), \tag{1}$$

where $P(\mathbb{M} = k|\mathbb{X} = x)$ is the conditional probability that the object x belongs to class k.

In this paper, the so-called soft output of the classifier $\nu : \mathbb{X} \mapsto [0,1]^2$ is also defined. The soft output vector ν contains values proportional to the conditional probabilities. Consequently, the following conditions need to be satisfied:

$$\nu_i \approx P(\mathbb{M} = i|\mathbb{X} = x), \tag{2}$$

$$\nu_i(\boldsymbol{x}) \in [0,1], \tag{3}$$

$$\sum_{i=1}^{M} \nu_i(\boldsymbol{x}) = 1. \tag{4}$$

2.2 Soft Confusion Matrix

The SCM approach is based on an assessment of the probability of classifying an object $\boldsymbol{x}$ into the class $s \in \mathbb{M}$ using the classifier ψ. It also provides an extension of the Bayesian model in which the object's description $\boldsymbol{x}$ and its true label $m \in \mathbb{M}$ are realizations of random variables $\mathbf{X}$ and $\mathbf{M}$, respectively. In the SCM approach, classifier ψ predicts randomly based on the probabilities $P(\boldsymbol{\Psi}(\boldsymbol{x}) = s) = P(s|\boldsymbol{x})$ [1]. Hence, the outcome of the classification s is a realization of the random variable $\boldsymbol{\Psi}(\boldsymbol{x})$. Unfortunately, for deterministic classifiers, these probabilities would be zero or one. The problem may be dealt with using a randomized classifier equivalent to the given one (RRC).

According to the extended Bayesian model, the posterior probability $P(m|\boldsymbol{x})$ of label m can be defined as:

$$P(m|\boldsymbol{x}) = \sum_{s \in \mathcal{M}} P(s|\boldsymbol{x}) P(m|s, \boldsymbol{x}). \tag{5}$$

where $P(m|s, \boldsymbol{x})$ denotes the probability that an object $\boldsymbol{x}$ belongs to the class m given that $\boldsymbol{\Psi}(\boldsymbol{x}) = s$. This probability is estimated using local soft confusion matrix. The locality of the matrix is defined using Gaussian potential function with β parameter. The detailed procedure of obtainging this matrix is given in [22].

Unfortunately, the assumption that base classifier assigns labels in a stochastic way is rather impractical, since most real-life classifiers are deterministic. This issue was addressed by implementation of deterministic binary classifiers in which their statistical properties were modelled using the RRC procedure, as described in Sect. 2.3.

2.3 Randomized Reference Classifier

In above-mentioed approach, the behaviour of a base classifier ψ was modeled using a stochastic classifier defined by a probability distribution over the set of labels $\mathbb{M}$. In this study, the randomized reference classifier (RRC) proposed by Woloszynski and Kurzynski [25] was used. The RRC is a hypothetical classifier that allows a randomised model of a given deterministic classifier to be built.

We assumed that for a given instance $\boldsymbol{x}$, the randomised classifier $\psi^{(R)}$ generates a vector of class supports ν being observed values of random variables $\Delta_i(\boldsymbol{x})$. The chosen probability distribution of random variables needs to satisfy the following conditions:

$$\Delta_i(\boldsymbol{x}) \in [0, 1], \tag{6}$$

$$\sum_{i=1}^{M} \Delta_i(\boldsymbol{x}) = 1, \tag{7}$$

$$\mathbf{E}\left[\Delta_i(\boldsymbol{x})\right] = \nu_i(\boldsymbol{x}),\ i \in \{0, 1\}, \tag{8}$$

where $\mathbf{E}$ is the expected value operator. Conditions (6) and (7) follow from the normalisation properties of class supports, whereas condition (8) provides the equivalence of the randomized model $\psi^{(R)}$ and base classifier ψ. Based on the latter condition, the RRC can be used to provide a randomised model of any classifier that returns a vector of class-specific supports $\nu(\boldsymbol{x})$.

The probability of classifying an object $\boldsymbol{x}$ into the class i using the RRC can be calculated from the following formula:

$$P(\boldsymbol{\Psi} = m|\mathbf{X} = \boldsymbol{x}) = Pr\left[\Delta_m(\boldsymbol{x}) > \Delta_{\mathbb{M}\setminus m}(\boldsymbol{x})\right], \tag{9}$$

where $Pr\left[\Delta_m(\boldsymbol{x}) > \Delta_{\mathbb{M}\setminus m}(\boldsymbol{x})\right]$ is the probability that the value obtained by the realisation of random variable Δ_m is greater than the realisation of the remaining random variables.

In this paper, we propose using the normal distribution truncated to the interval $[0, 1]$ [13] instead of the beta distribution suggested by Woloszynski et al. [25]. The expected value for each random variable is simply determined using formula (8). The standard deviation is determined using a rescaled variance of the beta distribution:

$$sd_i = \left(\frac{\nu_i(1-\nu_i)}{M+1}\right)^{\gamma}, \tag{10}$$

where γ is a parameter that should be tuned in order to achieve the best classification quality of the SCM method.

3 Experimental Setup

The goal of this paper is to determine how changing the underlying distribution of RRC classifier affect the classification quality of algorithms built using RRC model. To do so the experimental evaluation, which setup is described below, is performed.

The following base classifiers were employed:

- ψ_{NB} – Naive Bayes classifier with kernel density estimation [12].
- ψ_{KNN} – nearest neighbours classifier [4].
- ψ_{J48} – Weka implementation of the C4.5 algorithm [20]. Laplace smoothing is used to produce estimation of conditional probability [19].
- ψ_{NC} – nearest centroid (Nearest Prototype) [14].

The classifiers implemented in WEKA framework [10] were used. If not stated otherwise, the classifier parameters were set to their defaults. For the KNN classifier, the number of neighbours was selected from the following values $K \in \{1, 3, 5, \ldots, 11\}$.

During the experimental evaluation the following classifiers were compared:

1. ψ_{R} – unmodified base classifier,
2. ψ_{B} – SCM classifier with beta distribution,
3. ψ_{N} – SCM classifier with truncated normal distribution.

The size of the neighborhood, expressed as β coefficient and the variance-related coefficient γ, were chosen using a fiveefold cross-validation procedure and the grid search technique. The following values of β and γ were considered: $\beta \in \{1, 2, 3, \cdots, 21\}$, $\gamma \in \{0.1, 0.2, 0.3, \cdots, 1.0\}$. The values were chosen in such a way that minimizes macro-averaged F_1 loss function.

The experimental code was implemented using WEKA framework [10]. The source code of the algorithms is available online[1].

To evaluate the proposed methods the following classification-loss criteria are used [21]: Zero-one loss (1-Accuracy); Macro-averaged FDR (1-precision), FNR (1-recall), F_1; Micro-averaged FDR, FNR, F_1.

Following the recommendations of [6] and [9], the statistical significance of the obtained results was assessed using the two-step procedure. The first step is to perform the Friedman test [8] for each quality criterion separately. Since the multiple criteria were employed, the familywise errors (FWER) should be controlled [26]. To do so, the Bergman-Hommel [2] procedure of controlling FWER of the conducted Friedman tests was employed. When the Friedman test shows that there is a significant difference within the group of classifiers, the pairwise tests using the Wilcoxon signed-rank test [6,23] were employed. To control FWER of the Wilcoxon-testing procedure, the Bergman-Hommel approach was employed [2]. For all tests the significance level was set to $\alpha = 0.05$.

Table 1 displays the collection of the 64 benchmark sets that were used during the experimental evaluation of the proposed algorithms. The table is divided into three columns. Each column is organized as follows. The first column contains the names of the datasets. The remaining ones contain the set-specific characteristics of the benchmark sets: The number of instances in the dataset ($|S|$); dimensionality of the input space (d); the number of classes (C); average imbalance ratio (IR). Benchmark datasets are available online[2].

[1] https://github.com/ptrajdos/rrcBasedClassifiers/tree/develop.
[2] https://github.com/ptrajdos/MLResults/blob/master/data/slDataFull.zip.

To reduce the computational burden and remove irrelevant features, the correlation-based feature selection described in [11] was applied.

Table 1. The characteristics of the benchmark sets

Name	$\|S\|$	d	C	IR	Name	$\|S\|$	d	C	IR	Name	$\|S\|$	d	C	IR
appendicitis	106	7	2	2.52	housevotes	435	16	2	1.29	shuttle	57999	9	7	1326.03
australian	690	14	2	1.12	ionosphere	351	34	2	1.39	sonar	208	60	2	1.07
balance	625	4	3	2.63	iris	150	4	3	1.00	spambase	4597	57	2	1.27
banana2D	2000	2	2	1.00	led7digit	500	7	10	1.16	spectfheart	267	44	2	2.43
bands	539	19	2	1.19	lin1	1000	2	2	1.01	spirals1	2000	2	2	1.00
Breast Tissue	105	9	6	1.29	lin2	1000	2	2	1.83	spirals2	2000	2	2	1.00
check2D	800	2	2	1.00	lin3	1000	2	2	2.26	spirals3	2000	2	2	1.00
cleveland	303	13	5	5.17	magic	19020	10	2	1.42	texture	5500	40	11	1.00
coil2000	9822	85	2	8.38	mfdig fac	2000	216	10	1.00	thyroid	7200	21	3	19.76
dermatology	366	34	6	2.41	movement libras	360	90	15	1.00	titanic	2201	3	2	1.55
diabetes	768	8	2	1.43	newthyroid	215	5	3	3.43	twonorm	7400	20	2	1.00
Faults	1940	27	7	4.83	optdigits	5620	62	10	1.02	ULC	675	146	9	2.17
gauss2DV	800	2	2	1.00	page-blocks	5472	10	5	58.12	vehicle	846	18	4	1.03
gauss2D	4000	2	2	1.00	penbased	10992	16	10	1.04	Vertebral Column	310	6	3	1.67
gaussSand2	600	2	2	1.50	phoneme	5404	5	2	1.70	wdbc	569	30	2	1.34
gaussSand	600	2	2	1.50	pima	767	8	2	1.44	wine	178	13	3	1.23
glass	214	9	6	3.91	ring2D	4000	2	2	1.00	winequality-red	1599	11	6	20.71
haberman	306	3	2	1.89	ring	7400	20	2	1.01	winequality-white	4898	11	7	82.94
halfRings1	400	2	2	1.00	saheart	462	9	2	1.44	wisconsin	699	9	2	1.45
halfRings2	600	2	2	1.50	satimage	6435	36	6	1.66	yeast	1484	8	10	17.08
hepatitis	155	19	2	2.42	Seeds	210	7	3	1.00					
HillVall	1212	100	2	1.01	segment	2310	19	7	1.00					

4 Results and Discussion

To compare multiple algorithms on multiple benchmark sets the average ranks approach [6] is used. In the approach, the winning algorithm achieves rank equal '1', the second achieves rank equal '2', and so on. In the case of ties, the ranks of algorithms that achieve the same results, are averaged. To provide a visualisation of the average ranks, the radar plots are employed. In the plots, the data is visualised in such way that the lowest ranks are closer to the centre of the graph. The radar plots related to the experimental results are shown in Figs. 1a–d.

Due to the page limit, the full results are published online[3].

The numerical results are given in Table 2. The table is structured as follows. The table is divided into-base-classifier-specific sections and each section has it's own header containing base classifier name. The first row of each section contains names of the investigated algorithms. Then the table is divided into seven sections – one section is related to a single evaluation criterion. The first

[3] https://github.com/ptrajdos/MLResults/tree/master/RandomizedClassifiers/RRC_NormalDistribution_CORES2019.

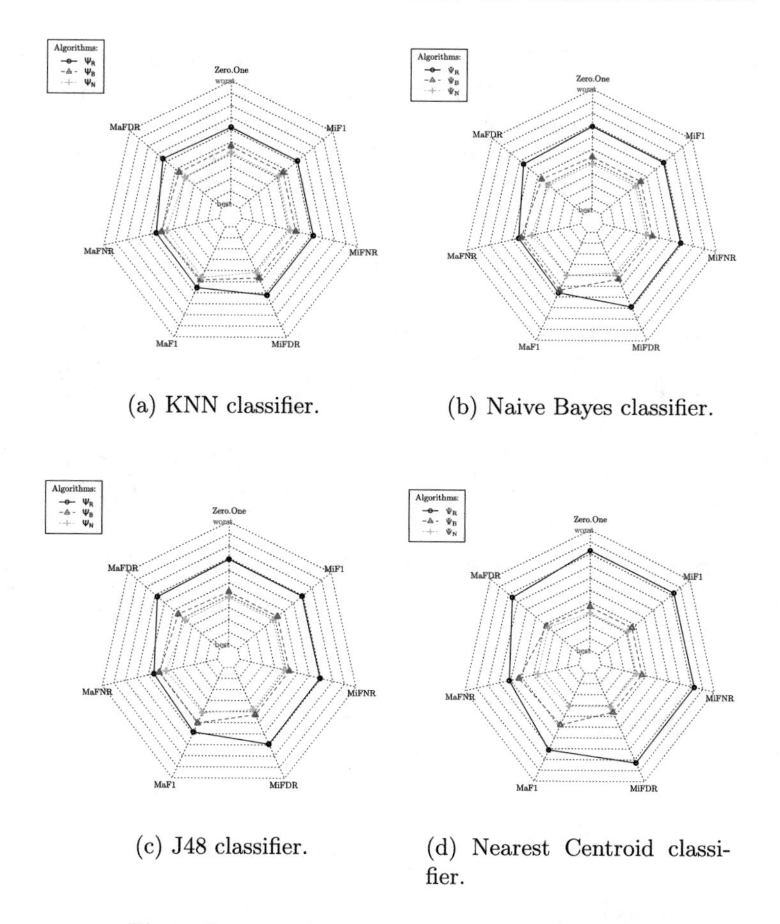

(a) KNN classifier.

(b) Naive Bayes classifier.

(c) J48 classifier.

(d) Nearest Centroid classifier.

Fig. 1. Radar plots for the investigated classifiers.

row of each section is the name of the quality criterion investigated in the section. The second row shows the p-value of the Friedman test. The third one shows the average ranks achieved by algorithms. The following rows show p-values resulting from pairwise Wilcoxon test. The p-value equal to 0.000 informs that the p-values are lower than 10^{-3} and p-value equal to 1.000 informs that the value is higher than 0.999.

Let us begin with the analysis of results for zero-one loss and micro-averaged criteria which are known to be biased towards majority classes [21]. For all investigated base classifiers the results are pretty consistent. That is, SCM-based classifiers are significantly better than the unmodified classifier. And there are no significant differences between ψ_{B} and ψ_{N} classifiers.

For macro-averaged criteria, on the other hand, the results are a bit different. Generally, the average ranks suggest that truncated-normal-based SCM classifiers may be a bit better than the beta-based SCM. However, not all differences are significant. For ψ_{KNN} classifier there are no significant differences between the investigated methods. For ψ_{NB} and ψ_{J48} SCM-based classifiers are significantly better according to FDR criterion. For ψ_{NC} base classifier (the weakest one) the ψ_{N} classifier outperforms the remaining classifiers. These results suggest that truncated-normal-based SCM classifier is a bit better than beta-based SCM method in discovering objects coming from minority classes.

Table 2. Statistical evaluation. Wilcoxon test results.

ψ_{KNN}

	ψ_R	ψ_B	ψ_N	ψ_R	ψ_B	ψ_N	ψ_R	ψ_B	ψ_N	ψ_R	ψ_B	ψ_N	ψ_R	ψ_B	ψ_N	ψ_R	ψ_B	ψ_N	ψ_R	ψ_B	ψ_N
Nam	Zero-One			MaFDR			MaFNR			MaF1			MiFDR			MiFNR			MiF1		
Frd	7.667e-02			2.360e-02			8.487e-01			8.487e-01			7.667e-02			7.667e-02			7.667e-02		
Rnk	2.24	1.93	1.83	2.28	1.92	1.80	2.09	2.00	1.91	2.11	1.97	1.92	2.24	1.93	1.83	2.24	1.93	1.83	2.24	1.93	1.83
ψ_R		.043	.043		.236	.236		1.000	1.000		1.000	1.000		.043	.043		.043	.043		.043	.043
ψ_B			.432			.270			1.000			1.000			.432			.432			.432

ψ_{NB}

	ψ_R	ψ_B	ψ_N	ψ_R	ψ_B	ψ_N	ψ_R	ψ_B	ψ_N	ψ_R	ψ_B	ψ_N	ψ_R	ψ_B	ψ_N	ψ_R	ψ_B	ψ_N	ψ_R	ψ_B	ψ_N
Nam	Zero-One			MaFDR			MaFNR			MaF1			MiFDR			MiFNR			MiF1		
Frd	1.960e-03			9.098e-03			3.327e-01			2.205e-01			1.960e-03			1.960e-03			1.960e-03		
Rnk	2.38	1.87	1.75	2.31	1.92	1.77	2.09	2.05	1.86	2.12	2.08	1.80	2.38	1.87	1.75	2.38	1.87	1.75	2.38	1.87	1.75
ψ_R		.000	.000		.003	.003		.183	.069		.089	.083		.000	.000		.000	.000		.000	.000
ψ_B			.151			.131			.069			.083			.152			.152			.152

ψ_{J48}

	ψ_R	ψ_B	ψ_N	ψ_R	ψ_B	ψ_N	ψ_R	ψ_B	ψ_N	ψ_R	ψ_B	ψ_N	ψ_R	ψ_B	ψ_N	ψ_R	ψ_B	ψ_N	ψ_R	ψ_B	ψ_N
Nam	Zero-One			MaFDR			MaFNR			MaF1			MiFDR			MiFNR			MiF1		
Frd	4.727e-04			7.379e-04			3.987e-01			1.646e-01			4.727e-04			4.727e-04			4.727e-04		
Rnk	2.38	1.85	1.76	2.35	1.90	1.75	2.10	2.01	1.89	2.17	2.01	1.82	2.38	1.85	1.76	2.38	1.85	1.76	2.38	1.85	1.76
ψ_R		.000	.000		.010	.003		.220	.220		.118	.118		.000	.000		.000	.000		.000	.000
ψ_B			.078			.206			.220			.118			.078			.078			.078

ψ_{NC}

	ψ_R	ψ_B	ψ_N	ψ_R	ψ_B	ψ_N	ψ_R	ψ_B	ψ_N	ψ_R	ψ_B	ψ_N	ψ_R	ψ_B	ψ_N	ψ_R	ψ_B	ψ_N	ψ_R	ψ_B	ψ_N
Nam	Zero-One			MaFDR			MaFNR			MaF1			MiFDR			MiFNR			MiF1		
Frd	1.750e-09			4.146e-06			1.263e-02			3.022e-05			1.750e-09			1.750e-09			1.750e-09		
Rnk	2.65	1.72	1.62	2.52	1.76	1.72	2.22	2.05	1.72	2.42	1.96	1.61	2.65	1.72	1.62	2.65	1.72	1.62	2.65	1.72	1.62
ψ_R		.000	.000		.000	.000		.440	.055		.019	.000		.000	.000		.000	.000		.000	.000
ψ_B			.102			.453			.003			.002			.102			.102			.102

5 Conclusions

In this paper, the issue of building the RRC model using truncated-normal distribution has been investigated. During the experimental evaluation, promising results have been obtained. Despite a naive-heuristic method has been applied to estimate the variance of the underlying normal distribution, the proposed method is comparable to the original beta-distribution-based approach. What is more, for some classifiers the results show that the proposed approach is better at discovering objects coming from minority classes. We believe that applying a better method of variance estimation will improve the results. Consequently, our further research will explore this issue.

Acknowledgments. This work was supported by the statutory funds of the Department of Systems and Computer Networks, Wroclaw University of Science and Technology.

References

1. Berger JO (1985) Statistical decision theory and Bayesian analysis. Springer, New York. https://doi.org/10.1007/978-1-4757-4286-2
2. Bergmann B, Hommel G (1988) Improvements of general multiple test procedures for redundant systems of hypotheses. In: Multiple hypothesenprüfung/multiple hypotheses testing. Springer, Heidelberg, pp 100–115.
3. Breiman L (1996) Bagging predictors. Mach Learn 24(2):123–140. https://doi.org/10.1007/bf00058655
4. Cover T, Hart P (1967) Nearest neighbor pattern classification. IEEE Trans Inform Theory 13(1):21–27. https://doi.org/10.1109/tit.1967.1053964
5. David HA, Nagaraja HN (2003) Order Statistics. Wiley, Hoboken. https://doi.org/10.1002/0471722162
6. Demšar J (2006) Statistical comparisons of classifiers over multiple data sets. J Mach Learn Res 7:1–30
7. Freund Y, Schapire RE (1997) A decision-theoretic generalization of on-line learning and an application to boosting. J Comput Syst Sci 55(1):119–139. https://doi.org/10.1006/jcss.1997.1504
8. Friedman M (1940) A comparison of alternative tests of significance for the problem of m rankings. Ann Math Stat 11(1):86–92. https://doi.org/10.1214/aoms/1177731944
9. Garcia S, Herrera F (2008) An extension on "statistical comparisons of classifiers over multiple data sets" for all pairwise comparisons. J Mach Learn Res 9:2677–2694
10. Hall M, Frank E, Holmes G, Pfahringer B, Reutemann P, Witten IH (2009) The WEKA data mining software. SIGKDD Explor Newsl 11(1):10. https://doi.org/10.1145/1656274.1656278
11. Hall, M.A.: Correlation-based feature selection for machine learning. Ph.D. thesis, The University of Waikato (1999)
12. Hand DJ, Yu K (2001) Idiot's bayes: not so stupid after all? Int Stat Rev/Revue Internationale de Statistique 69(3):385. https://doi.org/10.2307/1403452
13. Johnson N (1994) Continuous Univariate Distributions. Wiley, New York
14. Kuncheva L, Bezdek J (1998) Nearest prototype classification: clustering, genetic algorithms, or random search? IEEE Trans Syst Man Cybern Part C (Appl Rev) 28(1):160–164. https://doi.org/10.1109/5326.661099
15. Kurzynski M, Majak M (2016) Meta-bayes classifier with Markov model applied to the control of bioprosthetic hand. In: Intelligent decision technologies 2016: proceedings of the 8th KES international conference on intelligent decision technologies (KES-IDT 2016) – part II. Springer, Cham, pp 107–117
16. Kurzynski M, Majak M, Zolnierek A (2016) Multiclassifier systems applied to the computer-aided sequential medical diagnosis. J Biocybern Biomed Eng 36:619–625
17. Lysiak R, Kurzynski M, Woloszynski T (2014) Optimal selection of ensemble classifiers using measures of competence and diversity of base classifiers. Neurocomputing 126:29–35. https://doi.org/10.1016/j.neucom.2013.01.052
18. Majak M, Kurzynski M (2018) On a new method for improving weak classifiers using bayes metaclassifier. In: Proceedings of the 10th international conference on computer recognition systems, CORES 2017. Springer, Cham, pp 258–267.
19. Provost F, Domingos P (2003) Tree induction for probability-based ranking. Mach Learn 52(3):199–215. https://doi.org/10.1023/a:1024099825458

20. Quinlan JR (1993) C4.5: programs for machine learning. Morgan Kaufmann Publishers Inc., San Francisco
21. Sokolova M, Lapalme G (2009) A systematic analysis of performance measures for classification tasks. Inf Process Manag 45(4). https://doi.org/10.1016/j.ipm.2009.03.002
22. Trajdos P, Kurzynski M (2016) A dynamic model of classifier competence based on the local fuzzy confusion matrix and the random reference classifier. Int J Appl Math Comput Sci 26(1). https://doi.org/10.1515/amcs-2016-0012
23. Wilcoxon F (1945) Individual comparisons by ranking methods. Biometrics Bull 1(6):80. https://doi.org/10.2307/3001968
24. Woloszynski T, Kurzynski M (2009) On a new measure of classifier competence applied to the design of multiclassifier systems. In: International conference on image analysis and processing. Springer, pp 995–1004
25. Woloszynski T, Kurzynski M (2011) A probabilistic model of classifier competence for dynamic ensemble selection. Pattern Recogn 44(10–11):2656–2668. https://doi.org/10.1016/j.patcog.2011.03.020
26. Yekutieli D, Benjamini Y (2001) The control of the false discovery rate in multiple testing under dependency. Ann Stat 29(4):1165–1188. https://doi.org/10.1214/aos/1013699998

Hybrid Algorithm for the Detection and Recognition of Railway Signs

Ewelina Choodowicz, Pawe Lisiecki, and Piotr Lech(✉)

Szczecin Faculty of Electrical Engineering,
Department of Signal Processing and Multimedia Engineering,
West Pomeranian University of Technology, 26 Kwietnia 10, 71-126 Szczecin, Poland
cholodowicz.ewelina@gmail.com, pawel.lisiecki.priv@gmail.com,
piotr.lech@zut.edu.pl

Abstract. This paper presents an application of a hybrid algorithm for detection and recognition of railway signalling. The proposed algorithm has been implemented in a safety application which supports train drivers work. The algorithm is dedicated to performing the tasks of early warning about events related to signalling at railway infrastructure. Simulation tests have been performed in order to find an optimal solution of the problem with the use of classic digital image processing algorithms, the Haar cascade classifier and neural networks. As a result of the conducted research, a hybrid algorithm is developed that uses the You Only Look Once (YOLO) neural network to detect the position of semaphores in the image and the classic methods of digital image processing to recognize the colour light signals. As a part of the research, a database of test images was prepared along with test applications. Research methodology is based on Lane Startup method.

Keywords: Neural network · YOLO · Haar · Object recognition

1 Introduction

The achievements of modern technologies associated with Industry 4.0 are increasingly supporting (or replacing) the work of people which leads to overall safety improvement. One of such novel solutions may be an early warning system based on events. These events can be the result of a variety of activities. A good example of such applications can be the system which indicates and signalizes the potential threat or location on which the operator has to focus. An alternative path of technological development of the Industrial Internet of Things (IIoT) is to achieve full autonomy and eliminate human interference from the decision-making process because it is the most frequently encountered source of threats. Such solutions can be found, for example, in autonomous cars. The IIoT areas supporting this type of activities include machine vision, databases and analytics, as well as artificial intelligence [8].

R. Burduk et al. (Eds.): CORES 2019, AISC 977, pp. 337–347, 2020.
https://doi.org/10.1007/978-3-030-19738-4_34

This problem is particularly important in monotonous and tedious works. One of them is the work of the train driver. In this case, even the mere indication of the place (assisted by, for example, a sound signal), on which to focus, can be helpful and increase the level of safety. For example, a generated warning that a semaphore is in the field of view allows one to find it, or makes it possible to find more time to make decisions and take necessary actions. In this paper, the main aim is to develop the algorithm which provides detection and recognition of the occurrence of events while driving train. The event is associated with semaphore occurrence and its colour light signals. The application provides almost real-time assistance along with recognition and visualization of colour light determination in the daytime.

The problem of signalling detection in the railway system is similar to road traffic signalling, but it is more complicated due to the greater number of potential light signalling devices and a more diverse environment in which the semaphore can be located. Urban signalling usually occurs in heavily urbanized places whereas in the railway infrastructure such a generalization cannot be accepted. The problem of detection and recognition of specific objects cannot be considered as a simple task [5]. The main difficulties associated with these tasks stem from the imperfections of the image acquisition process. In the digital image, there are noises, optical and chromatic distortions, etc. Another problem is a large amount of data to be processed, which impacts the speed of operation and the accuracy of image processing algorithms.

2 Research Methodology, Initial Analysis, Implementations and Tests

Scientific research in this work has been carried out in accordance with the Lean Startup [2] methodology. The Lean Startup approach focuses on the rapid creation of prototypes to test the assumptions. Feedback from the user is available to gather the information how to make reviews and necessary corrections. The application of the Lean Startup methodology to the innovation process or research and development works can result in the products delivered to customers earlier and at a lower cost per iteration, thereby reducing costs and generating a gradual return that leads to increased profits. This methodology helps to "lose" (achieve failure, bring the project to a downfall) faster with a less negative impact on the implemented project, to minimize the risk in the long term and to create a process for gradual improvement. You can easily retreat from the wrong path, change or modify it as well as make a pivot, as shown in Fig. 1. In this publication, the research is presented in accordance with the iterations, but taking into account the assumptions of the Lean Startup method, we do not focus on algorithms which lead to a failure (except for the necessary minimum proving that this approach will probably lead to a "fall").

On the basis of the literature, the potential algorithms were found to solve the problem. Initially, three paths (corresponding to learn phase in Lean Startup

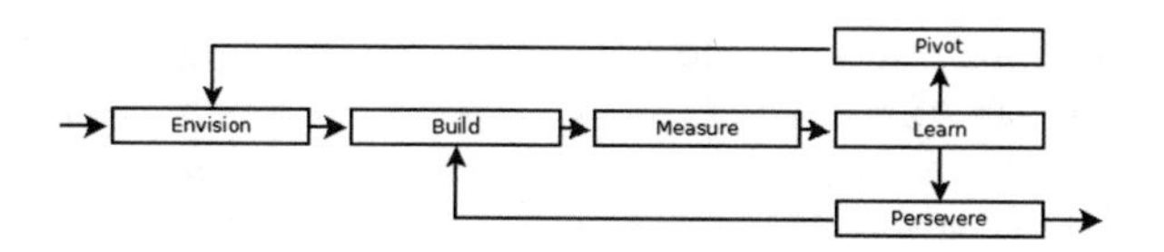

Fig. 1. The illustration of the Lean Startup idea.

methodology) were selected that could be used to accomplish the stated problem: classic methods of image processing, cascade Haar classifier and YOLO convolutional neural network. The main goal of the research is to find a solution with the highest efficiency of detection. For research purposes, a test application was written in Python, utilising Open CV libraries and support for the YOLO neural network. The designed application allows for testing the algorithms in the development phase and to load graphics or video files. Its main purpose is to detect and locate colour light railway signals and then determine which light is visible on the semaphore head. The developed application is able to detect and locate objects in real-time or nearly real-time.

In this research, the images used for analysis are pictures acquired by a digital camera with a resolution of 1920 × 1080 pixels. The camera was placed just in front of a train windshield in such a way as to cover the field of view of the entire railway infrastructure. In Fig. 2, an exemplary image captured by the utilised camera is presented. The images from the camera were not obtained in the permanent lighting conditions and the image illumination changes very dynamically, being correlated with the speed of the train.

Fig. 2. Example view from the camera (left) and application preprocessing window (right).

2.1 Classic Image Processing Methods

Designing a semaphore detector algorithm using digital image analysis methods would be time-consuming. To overcome this issue a graphical user interface was developed for the purpose of initial image processing. The user interface allows loading any image or video file for processing. Loaded image (or video frame) can be divided into smaller fragments and then processed locally using any sequence of operations. Additionally, user can check the colour distribution of each fragment. The results of local processing can be seen in almost real-time. Basic operations such as: Gaussian blur, thresholding, gradient, average local thresholding, dilation, erosion, opening, closing and Canny edge filtering were implemented in the graphical user interface. Users can see a preview of object contours after applying a sequence of operations. These contours are also visible in the original image, which is shown in a separate window. During image analysis, it was noticed that semaphores usually appear in the upper half of the image, therefore all operations that are performed in the image can be limited to its upper part, shortening the time needed to perform a series of operations.

As a result of the performed tests, the detection algorithm was designed that gives the best effectiveness in the developed environment. The algorithm consists of the following steps:

1. An original image is divided into rectangular areas in which further algorithm steps are performed locally. In the next steps, the original image fragment is simply called the image.
2. All fragments of the image on which green predominates are removed, at this stage, the image is blurred using the Gaussian filter and then masked. The mask is created from the image in such a way that the pixel from the original image, which has a green value greater than half of the average of the green channel values in the original image, the red colour value greater than half of the average red channel values in the original image and the blue colour value greater than half of the average blue channel values becomes a pixel with the value of 0. The remaining pixels of the mask are set to 255.
3. Image background is removed as a result of local segmentation operation based on average pixel values preceded by Gaussian blur of the image.
4. All objects contours are specified using OpenCV function: *findContours()*.
5. The semaphore is detected based on the determined contours. For a contour to be classified as a semaphore, the ratio of its width to height must be less than 0.55 but greater than 0.15. These values were calculated experimentally based on the analysis of the geometry of real semaphores.

The presented algorithm has several significant disadvantages. The first problem is local analysis within the fragment of the original image, which results in correct detection only for the semaphores containing fully within a single block. In case of occurrence of a semaphore covering more fragments, it may not be detected due to the conditions regarding the ratio of the width to height of the found contour. The second problem is the occurrence of dark trees in the images. A large diversity of the environment causes that on some images there are large, dark green areas which have a colour scheme and brightness very similar to the brightness and colour of the semaphore. A semaphore located on the background of such area will not be detected. The assumed criterion for semaphore detection is greatly simplified, and therefore the algorithm is characterized by a high percentage of incorrect indications. The advantage of the detector is the speed of its operation. For an image divided into 20 fragments, detection takes about 33 ms for a computer with Intel Core i7-8750H processor (algorithm working without the GPU acceleration).

An attempt to generalize the task of eliminating false positive detections could be difficult or impossible with the use of classical digital image processing methods. The reason for this is the immense diversity of images, particularly the diversity of lighting, location and number of semaphores in images. The implemented detector cannot cope with the detection of semaphores in images with poor lighting and low contrast between the semaphore and its background, and in images where the semaphore is on the background of trees or dark green areas.

The proposed method of semaphore detections can be used only for static images in which semaphores appear against the bright background (for example clear sky). As a result of the tests, it was decided to reject this solution and not to place this method in the target application.

2.2 Haar Algorithm

The second research path was related to the cascade Haar classifier [1,7]. The Haar classifier uses a cascade of Haar-like features which contain information about the change of contrast value for an individual group of pixels. In the design process of the semaphore detector, some tools for designing and training a cascade Haar classifier available in the OpenCV library have been used.

The process of creating a cascade Haar classifier consists of the following steps:

1. Creating a set of photos without any semaphores; so-called background images/negative images. For this purpose, 3500 images taken from inside of a locomotive have been selected from the image database. Each of the images was converted to greyscale and resized to the size of 300×300 pixels.
2. Creating a descriptor file for images that do not contain semaphores. The descriptor file is a *txt* file in which each line consists of a path to a single background image.
3. Creating a database of images containing the semaphores. These images were created automatically using OpenCV tool: *opencv_createsamples*. The tool takes a single image of the object to be found, transforms it and place it on the background images. In addition, a descriptor file (*info.lst*) is created, which describes the location of the semaphore on the generated images.
4. Training the cascade Haar classifier. The training was performed using the OpenCV *opencv_traincascade* tool with parameters set to the default values.

The learning process took about 4 h. The result of the training was the XML file, which was later used in the implementation of the Haar classifier test application. The implementation of Haar classifier was created as the Python script. The key to efficient object detection implementation using previously defined Haar cascade is the use of the *detectMultiScale()* function from the OpenCV library. The function takes several parameters, e.g.: input image, image scaling factor and the minimum number of neighbouring detections so that the given area is classified as a detected object. Unfortunately, the proposed algorithm demonstrated low detection efficiency (Fig. 3).

Fig. 3. Incorrect results of the cascade Haar classifier (cyan rectangles) compared with YOLO detection algorithm (purple rectangles).

The Haar classifier leads to many incorrect detections and in more than 70% of cases, it does not detect semaphores which are detected by the YOLO neural network. It has been noticed that the algorithm tends to search for dark objects, such as the window of a building or dark elements next to the semaphore pole. In this case, the only advantage of the cascade classifier is its execution speed, it works almost twice as fast as the implemented neural network. The detection time was on average about 18 ms. As a result of the test, it was decided to reject this solution and not to place it in the final application.

2.3 YOLO

YOLO is the convolutional neural network [3] designed for object detection and localization in real time. The network structure allows achieving very high operating speed, on average up to 45 frames per second (the operating speed given for the Titan X GPU) while maintaining high detection accuracy. The network uses characteristic features of the whole image to create frames in which objects can potentially be found. In addition to each frame, the class of the given object is assigned.

The YOLO network is an example of SSD (Single Shot Detector) [4] - it means that the network works globally, on entire image, not just its fragments. The network training process, conducted with the use of *Darknet* software, has been divided into several stages:

1. Creating a descriptor file for training images. The file consists of paths to each training image (*train_images.txt*).
2. Creating a descriptor file for test images. The file consists of paths to all the test images used to check the network's operation (*test_images.txt*).

3. Creating a descriptor file for object class names a file in which all the object class names, which should be found during the network execution, are located (*objects.names*). In this paper, only one class name (semaphore) was placed in the descriptor file.
4. Creating files which describe the location of semaphores in the training set of images. Each *txt* file contains the position and class of the presented object in the training image. For this step an external tool *LabelIMG* was used. In this way, a database of training images was manually created along with files describing the location of the objects.
5. Modifying the network configuration file describing the network architecture and training process (**.cfg* file). Modifications regarding the size of filters in the YOLO layers are necessary. These changes depend on the number of object classes to be detected in the YOLO layers. The filter size can be calculated using the following formula: $Number\_of\_filters = anchors \times 5 + number\_of\_object\_classes$. The anchors are the relationships between the width and the height of the frame in which the detected object can be found.
6. Creating the file that consists of paths to descriptor files, the path to folder where training results will be saved and the path to the network configuration file (*train_yolo.data*).
7. Neural network training using the *Darknet* tool, which supports operations on the CPU and GPU. During training, a pre-trained model was used and the target network was trained on it.

The trained neural network has been implemented as a part of an application with a graphical user interface. The interface has implemented the possibility of adjusting the detection parameters in real-time. The detection parameters include the confidence coefficient threshold and the non-maximum suppression threshold [6]. The first one is responsible for the minimum confidence coefficient at which the object is detected. The higher the parameter, the less false positive detections are at the expense of not detecting some objects. The second parameter is responsible for reducing the number of detections inside a single area. During the detection, the neural network often finds many objects inside an area where there is only a single one, however in an ideal case, the neural network should detect only one object. To overcome this problem, the non-maximum suppression algorithm is used. However, this method does not provide the best possible detection and, in the presence of many objects close to each other, makes it impossible to detect them. Increase of the value of the second parameter reduces the number of objects detected closely to a single object, but also reduces the chances of detecting semaphores that are near to each other in the image. While testing the network, the detection parameters were adjusted and then the preformance of network operation on 100 different images was calculated. The results of this experiment are presented in Table 1.

Detection accuracy is calculated according to the following principles and criteria:

1. In order for the semaphore to be classified as possible to be detected, there should be at least 30% of its area visible in the image. Any semaphores which are less visible are classified as very difficult to detect, thus they are not included in the accuracy indicator calculations.
2. The semaphore head has to be facing the locomotive. Hence, all detected semaphores, which are in a backward position, are wrongly classified.
3. The accuracy indicator is calculated using the following formula:

$$j(i) = \frac{sem_{ok}(i) - sem_{bad}(i)}{sem(i)}, \tag{1}$$

$$J = \frac{1}{100}\sum_{i=1}^{100} j(i), \tag{2}$$

where: $sem_{ok}(i)$ - number of correctly detected semaphores in the i-th image, $sem_{bad}(i)$ - number of incorrectly detected semaphores in the i-th image, $sem(i)$ - the actual number of semaphores of the i-th image (this number does not include semaphores in a backward position and with less than 30% of their areas being visible).

Table 1. Performance results obtained for the set of 100 images.

Number of cases	True positives	False positives	True negatives	False negatives
100	79	0	0	6

It is worth noting that the images, which cause non-full coverage, were often images taken from interpolated frames of the video stream. Interpolation of the video frame gives the image distortion effect that affects proper detection. The best (semaphore detection accuracy approximates 86%) and worst scenarios of semaphore detection in the image were noted and the results are shown in Table 2.

Table 2. Detection scenarios recorded in the test.

Best case scenarios		Worst case scenarios	
Number of semaphores	Semaphores found	Number of semaphores	Semaphores found
8	8	4	1
4	4	3	1

The average duration of semaphore detection (without displaying the result on the screen) in a single image was 35 ms. As a result, this solution was selected for the implementation in the final application.

3 Hybrid Concept

On the basis of the article posted by Jonathan Hui[1], it can be concluded that the YOLO neural network has some issues with the classification of small objects and objects located close to each other. For the purpose of creating a colour detection system, image fragments that were detected by YOLO neural network were analysed. It has been noticed that the detected areas have several similar features. The most important feature is the high coverage of a given fragment of the image with a dark colour that appears on the entire surface of the semaphore. It was also noticed that the colours of lights on a given semaphore are well separated from the rest of the image fragment due to high contrast. During the analysis of images from the database, it was assumed that the algorithm should detect several light configurations: no light, red light, green light, green light along with orange light, double orange light.

The algorithm of colour detection (using classic methods of image processing described earlier) consists of the following steps:

1. Blurring the image using Gaussian filter with 5×5 pixels mask and standard deviation $\sigma = 3$ (parameters were determined experimentally).
2. Removing the image background using a local segmentation based on the average pixel values.
3. Increasing the contrast between the light and dark pixels of the image fragment to highlight the bright colours of the lights on the semaphore head.
4. Detecting a colour for a given colour (described with soft or hard constraints) an image mask was created to remove pixels of a different colour. In the next step a logical product operation on an image with created masks was performed, thus obtaining the fragment of the image itself that contains a masked colour. The result image is subjected to the dilation operation using the 7×7 pixels structural element.
5. Detection of a light contour using *find_contours()* function from OpenCV library.
6. Classification of lighting configuration on the basis of a simple classifier implemented with conditional statements.
7. Displaying lighting classification on the screen.

As a result of the algorithms operation in the 100 images from a database, 80% of correct colour classifications were obtained. Example result of the algorithm can be seen in Fig. 4. The algorithm operated on fragments of images found using the YOLO neural network. Taking into account the simplicity of the implemented algorithm and the large diversity of input images, the obtained results can be considered as good.

[1] "What do we learn from single shot object detectors SSD YOLO fpn focal loss" available at: https://medium.com.

Fig. 4. Result of the colour detection using the proposed method.

4 Concluding Remarks

The analysis and testing of the proposed methods for recognizing semaphores have shown that classical image processing methods are the least useful. The use of the Haar classifier has turned out to be a better approach to the problem, but this method was definitely less accurate than the implemented YOLO convolutional neural network. The implementation of the YOLO network made it possible to reach a compromise between the high accuracy of detection, the correct localization of objects and relatively high speed of computation. On the contrary, classical methods could be successfully used to detect the colour of semaphore lights. Separation of the area associated with the semaphore with the use of the YOLO network and colour detection by classical image processing is the basis of the hybrid algorithm proposed in this paper. In the research process utilising the Lean Startup methodology, three iterations were carried out (classic methods, Haar and YOLO), two of them ended with failures. The last one, combined with partially positive conclusions from the first iteration, led to the working product - the final hybrid method.

References

1. Bradski G, Kaehler A (2008) Learning OpenCV: computer vision with the OpenCV library. O'Reilly Media, Inc., Newton
2. Frederiksen DL, Brem A (2017) How do entrepreneurs think they create value? Ascientific reflection of Eric Ries' lean startup approach. Int Entrep Manage J 13(1):169–189. https://doi.org/10.1007/s11365-016-0411-x
3. Karn U (2016) An intuitive explanation of convolutional neural networks. The Data Science Blog
4. Liu W, Anguelov D, Erhan D, Szegedy C, Reed S, Fu CY, Berg AC (2016) SSD: single shot multibox detector. In: Leibe B, Matas J, Sebe N, Welling M (eds) Computer vision - ECCV 2016. Springer, Cham, pp 21–37
5. Ritika S, Mittal S, Rao D (2017) Railway track specific traffic signal selection using deep learning. arXiv preprint arXiv:1712.06107
6. Rothe R, Guillaumin M, Van Gool L (2015) Non-maximum suppression for object detection by passing messages between windows. In: Cremers D, Reid I, Saito H, Yang MH (eds) Computer vision - ACCV 2014. Springer, Cham, pp 290–306

7. Viola P, Jones M (2001) Rapid object detection using a boosted cascade of simple features. In: Proceedings of the 2001 IEEE computer society conference on computer vision and pattern recognition, CVPR 2001, vol 1. IEEE
8. Ye T, Wang B, Song P, Li J (2018) Automatic railway traffic object detection system using feature fusion refine neural network under shunting mode. Sensors 18(6):1916

Combination of Linear Classifiers Using Score Function – Analysis of Possible Combination Strategies

Pawel Trajdos and Robert Burduk(✉)

Department of Systems and Computer Networks,
Wroclaw University of Science and Technology,
Wybrzeze Wyspianskiego 27, 50-370 Wroclaw, Poland
{pawel.trajdos, robert.burduk}@pwr.edu.pl

Abstract. In this work, we addressed the issue of combining linear classifiers using their score functions. The value of the scoring function depends on the distance from the decision boundary. Two score functions have been tested and four different combination strategies were investigated. During the experimental study, the proposed approach was applied to the heterogeneous ensemble and it was compared to two reference methods – majority voting and model averaging respectively. The comparison was made in terms of seven different quality criteria. The result shows that combination strategies based on simple average, and trimmed average are the best combination strategies of the geometrical combination.

Keywords: Binary classifiers · Linear classifiers · Geometrical space · Potential function

1 Introduction

The combination of multiple base classifiers has been an important issue in machine learning for about twenty years [8,35]. The ensembles of classifiers (EoC) or multiple classifiers systems (MCSs) [5,11,21,25,34] are popular in supervised classification algorithms where single classifiers are often unstable (small changes in input data may result in creation of very different decision boundaries) or are often more accurate than any of the base classifiers.

The task of constructing MCSs can be generally divided into three steps: generation, selection and integration [2]. In the first step a set of base classifiers is trained using manipulation of the training patterns, manipulation of the training parameters or manipulation of the feature space.

The second phase of building EoCs is related to the choice of a set or one classifier from the whole available pool of base classifiers. It is popular to use the diversity measure to select one classifier or a subset of all base classifiers. In the literature, there are many approaches to the selection phase of building EoCs [3,17,27,28].

R. Burduk et al. (Eds.): CORES 2019, AISC 977, pp. 348–359, 2020.
https://doi.org/10.1007/978-3-030-19738-4_35

The integration process is the last stage of constructing EoCs and it is widely discussed in the pattern recognition literature [24,32]. Generally, supervised learning methods produce a classifier whose output is represented as a score function. This function is mapping to a function that is interpreted as a posteriori probability, rank level function or directly as a class label. Depending on the type of mapping, many methods for integrating base classifiers can be distinguished [19,26,31].

In this paper we propose the concept of the classifier integration process which uses score functions without their further transformation. In this paper we examined two forms of the score function that is called the potential function and four different combination strategies were investigated.

The remainder of this paper is organized as follows. Section 2 presents the proposed method of EoC integration using two types of the potential function. The experimental evaluation is presented in Sect. 3. The discussion and conclusions from the experiments are presented in Sect. 4.

2 Proposed Method

In this section, the proposed approach is explained. Additionally, this section introduces the notation used in this paper.

2.1 Linear Binary Classifiers

In this paper, it is assumed that the input space $\mathbb{X}$ is a $d-$dimensional Euclidean space $\mathbb{X} = \mathbb{R}^d$. Each object from the input space $x \in \mathbb{X}$ belongs to one of two available classes, so the output space is: $\mathbb{M} = \{-1; 1\}$. It is assumed that there exists an unknown mapping $f : \mathbb{X} \mapsto \mathbb{M}$ that assigns each input space coordinates into a proper class. A classifier $\psi : \mathbb{X} \mapsto \mathbb{M}$ is a function that is designed to provide an approximation of the unknown mapping f. A linear classifier makes its decision according to the following rule:

$$\psi(x) = \operatorname{sign}(\omega(x)), \tag{1}$$

where $\omega(x) = \langle n; x\rangle + b$ is the so called *discriminant function* of the classifier ψ [19], n is a unit normal vector of the decision hyperplane ($\|n\| = 1$), b is the distance from the hyperplane to the origin and $\langle\cdot;\cdot\rangle$ is a dot product defined as follows:

$$\langle a; b\rangle = \sum_{i=1}^{d} a_i b_i,\ \forall a, b \in \mathbb{X}. \tag{2}$$

In this paper, we use a norm of the vector x defined using the dot product:

$$\|x\| = \sqrt{\langle x; x\rangle}. \tag{3}$$

When the normal vector of the plane is a unit vector, the absolute value of the discriminant function equals to the distance from the decision hyperplane to point x. The sign of the discriminant function depends on the site of the plane where the instance x lies.

Now, let us define an ensemble classifier:

$$\Psi = \left\{\psi^{(1)}, \psi^{(2)}, \cdots, \psi^{(N)}\right\} \tag{4}$$

that is a set of N classifiers that work together in order to produce a more robust result [19]. In this paper, it is assumed that only linear, binary classifiers are employed. There are multiple strategies to combine the classifiers constituting the ensemble. The simplest strategy to combine the outcomes of multiple classifiers is to apply the majority voting scheme [19]:

$$\omega_{\mathrm{MV}}(x) = \sum_{i=1}^{N} \mathrm{sign}(\omega^{(i)}(x)), \tag{5}$$

where $\omega^{(i)}(x)$ is the value of the discriminant function provided by the classifier $\psi^{(i)}$ for point x. However, this simple yet effective strategy completely ignores the distance of the instance x from the decision planes.

Another strategy is model averaging [29]. The output of the averaged model may be calculated by simply averaging the values of the discriminant functions:

$$\omega_{\mathrm{MA}}(x) = \frac{1}{N}\sum_{i=1}^{N} \omega^{(i)}(x) \tag{6}$$

After combining the base classifiers, the final prediction of the ensemble is obtained according to the rule (1).

2.2 The Proposed Method

In this paper, an approach similar to the softmax [19] normalization is proposed. Contrary to the softmax normalization, our goal is not to provide a probabilistic interpretation of the linear classifier but to provide a fusion technique that works in the geometrical space. The idea is to span a potential field around the decision plane. The potential field may be constructed by applying a transformation on the value of the discriminant function. The transformation must meet the following properties:

$$\mathrm{sign}(g(\omega^{(i)}(x))) = \mathrm{sign}(\omega^{(i)}(x))\forall x \in \mathbb{X}, \tag{7}$$

$$g(\omega^{(i)}(x)) \in [-1;1]\,\forall z \in \mathbb{R}, \tag{8}$$

$$g(0) = 0. \tag{9}$$

Property (7) assures that the crisp decision based on the transformed value is the same as the decision based on the unmodified discriminant function. Property (8) bounds g in interval $[-1;1]$. However, contrary to the softmax normalization the transformation does not have to be a sigmoid function. Property (9)

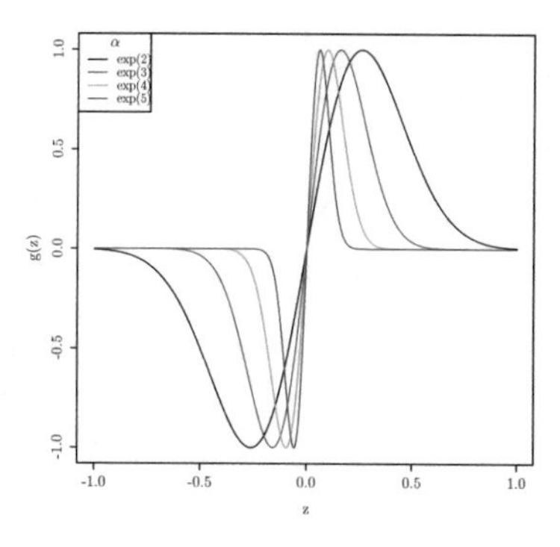

Fig. 1. Potential function g.

assures that the potential is 0 at the surface of the decision plane. In this paper, the following transformation function is used:

$$g(z) = z\exp(-\gamma z^2 + 0.5)\sqrt{2\gamma}, \tag{10}$$

where γ is a coefficient that determines the position and steepness of the peak. The translation constant 0.5 and the scaling factor $\sqrt{2\gamma}$ guarantee that the maximum and minimum values are 1 and -1 respectively. The function is visualised in the Fig. 1.

All models in the ensemble share the same shape coefficient γ. The shape coefficient is tuned in order to achieve the best quality of the entire ensemble.

After transforming the values of discriminant functions for the entire ensemble, there is a need to combine the outcomes to produce the final decision. In this paper, we analyze four different combination rules. The first one is a simple average of the transformed values of discriminant functions:

$$\omega_{\mathrm{TA}}(x) = \frac{1}{N}\sum_{i=1}^{N} g(\omega^{(i)}(x)). \tag{11}$$

The other one is to apply the trimmed mean approach:

$$\omega_{\mathrm{TME}}(x) = \frac{1}{N-2}\sum_{i=1}^{N}\left[g(\omega^{(i)}(x)) - \max_{i\in\{1,2,\cdots,N\}}\omega^{(i)}(x) - \min_{i\in\{1,2,\cdots,N\}}\omega^{(i)}(x)\right]. \tag{12}$$

Before the remaining combination rules are defined, let us introduce subsets of negative and positive values of the transformed ensemble outcomes:

$$\mathcal{G}_{-}(x) = \left\{g(\omega^{(i)}(x)) \mid g(\omega^{(i)}(x)) < 0\right\}, \tag{13}$$

$$\mathcal{G}_{+}(x) = \left\{g(\omega^{(i)}(x)) \mid g(\omega^{(i)}(x)) \geq 0\right\}. \tag{14}$$

Then, the remaining rules are as follows:

$$\omega_{\mathrm{MAX}}(x) = \max(\mathcal{G}_{+}(x)) + \min(\mathcal{G}_{-}(x)), \tag{15}$$

$$\omega_{\mathrm{MIN}}(x) = \min(\mathcal{G}_{+}(x)) + \max(\mathcal{G}_{-}(x)), \tag{16}$$

$$\omega_{\mathrm{GME}}(x) = \left(\prod_{z \in \mathcal{G}_{+}(x))} |z|\right)^{|\mathcal{G}_{+}(x))|^{-1}} - \left(\prod_{z \in \mathcal{G}_{-}(x))} |z|\right)^{|\mathcal{G}_{-}(x))|^{-1}}, \tag{17}$$

where $|\mathcal{G}_{-}(x))|$ and $|z|$ are cardinality of set $\mathcal{G}_{-}(x))$ and the absolute value of z respectively.

The proposed algorithm is able to deal only with the binary classification problems. However, any multi-class problem can be decomposed into multiple binary problems. In the experimental stage the One-vs-One strategy was used [16]. This strategy builds a separate binary classifier for each pair of classes. In our method, a single pair-specific is replaced by the above-described ensemble classifier.

3 Experimental Setup

In the conducted experimental study, the proposed approach was used to combine classifiers in the heterogeneous ensemble of classifier. The following base classifiers were employed:

- ψ_{FLDA} – Fisher LDA [22]
- ψ_{MLP} – single layer MLP classifier [12]
- ψ_{NC} – nearest centroid (Nearest Prototype) [18,20]
- ψ_{SVM} – SVM classifier with linear kernel (no kernel) [4],
- ψ_{LR} – logistic regression classifier [7].

The classifiers implemented in WEKA framework [13] were used. The classifier parameters were set to their defaults. The multi-class problems were dealt with using One-vs-One decomposition [16]. The experimental code was implemented using WEKA framework [13]. The source code of the algorithms is available online[1]. The heterogeneous ensemble employs one copy of each of the above-mentioned base classifiers. Each classifier is learned using the entire dataset.

During the experimental evaluation the following combination methods were compared:

1. Ψ_{MV} – the ensemble combined using the majority voting approach,
2. Ψ_{MA} – the ensemble combined using the model averaging approach,
3. Ψ_{TA} – the ensemble combined using the rule described in (11).
4. Ψ_{MAX} – the ensemble combined using the rule described in (15).
5. Ψ_{MIN} – the ensemble combined using the rule described in (16).

[1] https://github.com/ptrajdos/piecewiseLinearClassifiers/tree/master.

6. Ψ_{TME} – the ensemble combined using the rule described in (17).
7. Ψ_{GME} – the ensemble combined using the rule described in (17).

The coefficient γ for transformation and g was tuned using the grid search approach. The following set of parameter values were investigated:

$$\{\gamma = \exp(i)|i \in \{2, \cdots, 10\}\}.$$

The parameter is chosen in such a way that it provides the maximum value of the macro-averaged F_1 criterion.

To evaluate the proposed methods the following classification-quality criteria are used [30]: Zero-one loss (Accuracy); Macro-averaged FDR, FNR, F_1;Micro-averaged FDR, FNR, F_1.

Following the recommendations of [6] and [10], the statistical significance of the obtained results was assessed using the two-step procedure. The first step is to perform the Friedman test [9] for each quality criterion separately. Since the multiple criteria were employed, the familywise errors (FWER) should be controlled [36]. To do so, the Bergman-Hommel [1] procedure of controlling FWER of the conducted Friedman tests was employed. When the Friedman test shows that there is a significant difference within the group of classifiers, the pairwise tests, which use the Wilcoxon signed-rank test [6,33] were employed. To control FWER of the Wilcoxon-testing procedure, the Bergman-Hommel approach was employed [15]. For all the tests the significance level was set to $\alpha = 0.05$.

Table 1 displays the collection of the 64 benchmark sets that were used during the experimental evaluation of the proposed algorithms. The table is divided into two columns. Each column is organized as follows. The first column contains the names of the datasets. The remaining ones contain the set-specific characteristics of the benchmark sets: the number of instances in the dataset ($|S|$); dimensionality of the input space (d); the number of classes (C); average imbalance ratio (IR).

The datasets come from the Keel[2] repository or are generated by us. The datasets are available online[3].

During the dataset-preprocessing stage, a few transformations on datasets were applied. That is, features are selected using the correlation-based approach [14]. Then, the PCA method was applied [23] and the percentage of variance was set to 0.95. The attributes were also scaled to fit the interval [0; 1]. Additionally, in order to ensure the dot product to be in the interval $[-1; 1]$, vectors in each dataset were scaled using the factor $\frac{1}{d^2}$. This normalization makes it easier to find proper γ.

4 Results and Discussion

To compare multiple algorithms on multiple benchmark sets the average ranks approach [6] is used. In the approach, the winning algorithm achieves rank equal

[2] https://sci2s.ugr.es/keel/category.php?cat=clas.
[3] https://github.com/ptrajdos/MLResults/blob/master/data/slDataFull.zip.

Table 1. The characteristics of the benchmark sets

Name	$\lvert S\rvert$	d	C	IR	Name	$\lvert S\rvert$	d	C	IR	Name	$\lvert S\rvert$	d	C	IR
appendicitis	106	7	2	2.52	housevotes	435	16	2	1.29	shuttle	57999	9	7	1326.03
australian	690	14	2	1.12	ionosphere	351	34	2	1.39	sonar	208	60	2	1.07
balance	625	4	3	2.63	iris	150	4	3	1.00	spambase	4597	57	2	1.27
banana2D	2000	2	2	1.00	led7digit	500	7	10	1.16	spectfheart	267	44	2	2.43
bands	539	19	2	1.19	lin1	1000	2	2	1.01	spirals1	2000	2	2	1.00
Breast Tissue	105	9	6	1.29	lin2	1000	2	2	1.83	spirals2	2000	2	2	1.00
check2D	800	2	2	1.00	lin3	1000	2	2	2.26	spirals3	2000	2	2	1.00
cleveland	303	13	5	5.17	magic	19020	10	2	1.42	texture	5500	40	11	1.00
coil2000	9822	85	2	8.38	mfdig fac	2000	216	10	1.00	thyroid	7200	21	3	19.76
dermatology	366	34	6	2.41	movement libras	360	90	15	1.00	titanic	2201	3	2	1.55
diabetes	768	8	2	1.43	newthyroid	215	5	3	3.43	twonorm	7400	20	2	1.00
Faults	1940	27	7	4.83	optdigits	5620	62	10	1.02	ULC	675	146	9	2.17
gauss2DV	800	2	2	1.00	page-blocks	5472	10	5	58.12	vehicle	846	18	4	1.03
gauss2D	4000	2	2	1.00	penbased	10992	16	10	1.04	Vertebral Column	310	6	3	1.67
gaussSand2	600	2	2	1.50	phoneme	5404	5	2	1.70	wdbc	569	30	2	1.34
gaussSand	600	2	2	1.50	pima	767	8	2	1.44	wine	178	13	3	1.23
glass	214	9	6	3.91	ring2D	4000	2	2	1.00	winequality-red	1599	11	6	20.71
haberman	306	3	2	1.89	ring	7400	20	2	1.01	winequality-white	4898	11	7	82.94
halfRings1	400	2	2	1.00	saheart	462	9	2	1.44	wisconsin	699	9	2	1.45
halfRings2	600	2	2	1.50	satimage	6435	36	6	1.66	yeast	1484	8	10	17.08
hepatitis	155	19	2	2.42	Seeds	210	7	3	1.00					
HillVall	1212	100	2	1.01	segment	2310	19	7	1.00					

'1', the second achieves rank equal '2', and so on. In the case of ties, the ranks of algorithms that achieve the same results, are averaged. To provide a visualisation of the average ranks, the radar plots are employed. In the plots, the data is visualised in such a way that the lowest ranks are closer to the centre of the graph. The radar plots related to the experimental results are shown in Fig. 2.

Due to the page limit, the full results are published online[4].

The numerical results are given in Table 2. The table is structured as follows. The first row contains names of the investigated algorithms. Then, the table is divided into seven sections – one section is related to a single evaluation criterion. The first row of each section is the name of the quality criterion investigated in the section. The second row shows the p-value of the Friedman test. The third one shows the average ranks achieved by algorithms. The following rows show p-values resulting from pairwise Wilcoxon test. The p-value which is equal to 0.000 informs that the p-values are lower than 10^{-3} and p-value is equal to 1.000 informs that the value is higher than 0.999.

The analysis of the radar plot suggests that two groups of classification criteria can be distinguished. The first group contains micro-averaged criteria and the zero-one criterion, the second one contains macro-averaged criteria. Evaluation of the classifiers carried out with the use of criteria belonging to a specific group reveals different relationships between classifiers. These differences are a consequence of the properties of the quality criteria used. This means that the zero-one

[4] https://github.com/ptrajdos/MLResults/blob/master/Boundaries/bounds_hetero_15.01.2019E4_m_R.zip.

criterion and micro-averaged criteria give us information related to the classification quality for the majority classes. On the other hand, the macro-averaged criteria put more emphasis on classification quality for minority classes [30].

Table 2. Statistical evaluation. Wilcoxon test for the heterogeneous ensemble – p-values for paired comparisons of the investigated methods.

	Ψ_{MV}	Ψ_{MA}	Ψ_{TA}	Ψ_{MAX}	Ψ_{MIN}	Ψ_{TME}	Ψ_{GME}
Nam	Zero-One						
Frd	5.729e-14						
Rnk	2.98	3.78	3.36	3.73	5.72	3.56	4.87
Ψ_{MV}		.007	.091	.002	.000	.161	.000
Ψ_{MA}			.968	.968	.000	.968	.007
Ψ_{TA}				.080	.000	.968	.000
Ψ_{MAX}					.000	.846	.000
Ψ_{MIN}						.000	.000
Ψ_{TME}							.000

	Ψ_{MV}	Ψ_{MA}	Ψ_{TA}	Ψ_{MAX}	Ψ_{MIN}	Ψ_{TME}	Ψ_{GME}
Nam	MaFDR						
Frd	2.873e-04						
Rnk	3.76	4.45	3.41	3.93	4.75	3.31	4.39
Ψ_{MV}		.016	.295	.969	.155	.279	.673
Ψ_{MA}			.001	.025	.878	.002	.295
Ψ_{TA}				.056	.013	.878	.025
Ψ_{MAX}					.028	.056	.155
Ψ_{MIN}						.004	.295
Ψ_{TME}							.003

	Ψ_{MV}	Ψ_{MA}	Ψ_{TA}	Ψ_{MAX}	Ψ_{MIN}	Ψ_{TME}	Ψ_{GME}
Nam	MaFNR						
Frd	1.791e-08						
Rnk	4.27	5.32	3.32	3.58	4.00	3.09	4.42
Ψ_{MV}		.000	.003	.505	1.00	.000	1.00
Ψ_{MA}			.000	.000	.018	.000	.002
Ψ_{TA}				.049	.139	1.00	.008
Ψ_{MAX}					.601	.049	.016
Ψ_{MIN}						.139	1.00
Ψ_{TME}							.001

	Ψ_{MV}	Ψ_{MA}	Ψ_{TA}	Ψ_{MAX}	Ψ_{MIN}	Ψ_{TME}	Ψ_{GME}
Nam	MaF1						
Frd	2.641e-09						
Rnk	3.96	5.10	3.23	3.59	4.81	2.96	4.35
Ψ_{MV}		.000	.017	.548	.117	.000	.340
Ψ_{MA}			.000	.000	.315	.000	.017
Ψ_{TA}				.017	.002	.454	.001
Ψ_{MAX}					.007	.014	.011
Ψ_{MIN}						.000	.185
Ψ_{TME}							.000

	Ψ_{MV}	Ψ_{MA}	Ψ_{TA}	Ψ_{MAX}	Ψ_{MIN}	Ψ_{TME}	Ψ_{GME}
Nam	MiFDR						
Frd	5.729e-14						
Rnk	2.98	3.78	3.36	3.73	5.72	3.56	4.87
Ψ_{MV}		.007	.091	.002	.000	.161	.000
Ψ_{MA}			.968	.968	.000	.968	.007
Ψ_{TA}				.080	.000	.968	.000
Ψ_{MAX}					.000	.846	.000
Ψ_{MIN}						.000	.000
Ψ_{TME}							.000

	Ψ_{MV}	Ψ_{MA}	Ψ_{TA}	Ψ_{MAX}	Ψ_{MIN}	Ψ_{TME}	Ψ_{GME}
Nam	MiFNR						
Frd	5.729e-14						
Rnk	2.98	3.78	3.36	3.73	5.72	3.56	4.87
Ψ_{MV}		.007	.091	.002	.000	.161	.000
Ψ_{MA}			.968	.968	.000	.968	.007
Ψ_{TA}				.080	.000	.968	.000
Ψ_{MAX}					.000	.846	.000
Ψ_{MIN}						.000	.000
Ψ_{TME}							.000

	Ψ_{MV}	Ψ_{MA}	Ψ_{TA}	Ψ_{MAX}	Ψ_{MIN}	Ψ_{TME}	Ψ_{GME}
Nam	MiF1						
Frd	5.729e-14						
Rnk	2.98	3.78	3.36	3.73	5.72	3.56	4.87
Ψ_{MV}		.007	.091	.002	.000	.161	.000
Ψ_{MA}			.968	.968	.000	.968	.007
Ψ_{TA}				.080	.000	.968	.000
Ψ_{MAX}					.000	.846	.000
Ψ_{MIN}						.000	.000
Ψ_{TME}							.000

For the zero-one criterion and micro-averaged criteria, three main groups of classifiers can be seen. The first group contains Ψ_{MIN} and Ψ_{GME} classifiers that perform significantly worse than the other analysed classifiers. What is more, classifier Ψ_{MIN} is significantly worse than Ψ_{GME} for all quality criteria belonging to the investigated group. The second group contains only one classifier – Ψ_{MV}. According to average ranks, this classifier is the best performing one for the investigated set of quality criteria. According to the statistical analysis, this classifier outperforms the remaining classifiers except for Ψ_{TA} and Ψ_{TME}. The third group consisted of classifiers Ψ_{MA}, Ψ_{TA}, Ψ_{MAX}, and Ψ_{TME}. There are no significant differences between the classifiers within this group.

For macro-averaged measures, the situation changes significantly. First of all, it may be noticed that average ranks of reference methods (Ψ_{MV} and Ψ_{MA}) increase, whereas the average ranks of the proposed methods decrease. That is, the model-averaging classifier Ψ_{MA} becomes the worst one except for Ψ_{MIN} according to macro-averaged F_1 and FNR criteria. The majority voting classifier Ψ_{MV} also deteriorates significantly. Now it is comparable to Ψ_{MAX}, Ψ_{MIN} and Ψ_{GME} classifiers. What is more, Ψ_{MV} classifier is outperformed by Ψ_{TA} and Ψ_{TME} classifiers in terms of macro-averaged FNR and F_1 criteria. The reason for the above-mentioned deterioration of the reference methods is the fact that they are not tuned to perform better on minority classes, whereas the investigated methods were tuned to do so.

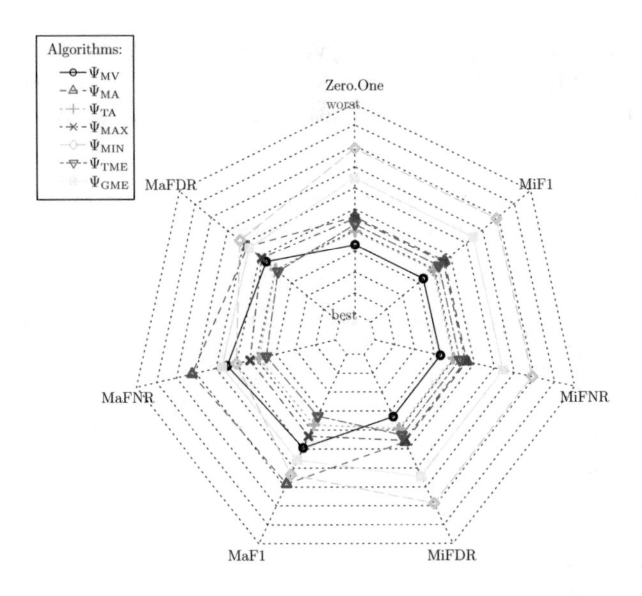

Fig. 2. Average ranks of for the heterogeneous ensemble.

Now let us investigate the differences inside the group of the proposed combination criteria. First of all, classifiers Ψ_{TA} and Ψ_{TME} offer the best classification quality under macro-averaged F_1 measure. It means that these classifiers offer the best trade-off between macro-averaged precision and recall. Under macro-averaged FDR (1 – precision) measure, these algorithms outperform only Ψ_{MIN} and Ψ_{GME} classifiers. For macro-averaged FNR (1 – recall) the investigated classifiers outperform all but Ψ_{MIN} classifiers. On the other hand, under the macro-averaged measures, there are no significant differences between Ψ_{TA} and Ψ_{TME}.

5 Conclusions

In this paper, a geometric combination scheme was proposed. Four different methods of producing the final output of EoC were investigated. The goal of this paper is to determine the best combination strategy for the given potential-function-induced geometrical space. The experimental comparison shows that Ψ_{TA} and Ψ_{TME} algorithms are the best choice. This is because under macro-averaged measures they are outperforming the other proposed strategies and reference methods. What is more, under the micro-averaged criteria they are comparable to the majority voting procedure. According to the outcome of the statistical evaluation, these algorithms perform equally well. However, under macro-averaged measures, Ψ_{TME} achieves a slightly lower average rank. This suggests that Ψ_{TME} may be slightly better since the truncated mean combination rule removes extreme values of the potential function so it may be less influenced by outliers.

The obtained results are very interesting, so we are willing to continue our research in the field of combining classifiers in the geometrical space. An interesting direction to explore may be the application of the potential function whose shape is not given arbitrary but is created considering data distribution.

Acknowledgments. This work was supported in part by the National Science Centre, Poland under the grant no. 2017/25/B/ST6/01750.

References

1. Bergmann B, Hommel G (1988) Improvements of general multiple test procedures for redundant systems of hypotheses. In: Multiple hypothesenprüfung/multiple hypotheses testing. Springer, Heidelberg, pp 100–115. https://doi.org/10.1007/978-3-642-52307-6_8
2. Britto AS, Sabourin R, Oliveira LE (2014) Dynamic selection of classifiers—a comprehensive review. Pattern Recogn 47(11):3665–3680
3. Burduk R, Walkowiak K (2015) Static classifier selection with interval weights of base classifiers. In: Asian conference on intelligent information and database systems. Springer, pp 494–502
4. Cortes C, Vapnik V (1995) Support-vector networks. Mach Learn 20(3):273–297. https://doi.org/10.1007/bf00994018
5. Cyganek B (2012) One-class support vector ensembles for image segmentation and classification. J Math Imaging Vis 42(2–3):103–117
6. Demšar J (2006) Statistical comparisons of classifiers over multiple data sets. J Mach Learn Res 7:1–30
7. Devroye L, Györfi L, Lugosi G (1966) A probabilistic theory of pattern recognition. Springer, New York. https://doi.org/10.1007/978-1-4612-0711-5
8. Drucker H, Cortes C, Jackel LD, LeCun Y, Vapnik V (1994) Boosting and other ensemble methods. Neural Comput 6(6):1289–1301
9. Friedman M (1940) A comparison of alternative tests of significance for the problem of m rankings. Ann Math Stat 11(1):86–92. https://doi.org/10.1214/aoms/1177731944
10. Garcia S, Herrera F (2008) An extension on "statistical comparisons of classifiers over multiple data sets" for all pairwise comparisons. J Mach Learn Res 9:2677–2694
11. Giacinto G, Roli F (2001) An approach to the automatic design of multiple classifier systems. Pattern Recogn Lett 22:25–33
12. Gurney K (1997) An introduction to neural networks. Taylor & Francis, London. https://doi.org/10.4324/9780203451519
13. Hall M, Frank E, Holmes G, Pfahringer B, Reutemann P, Witten IH (2009) The WEKA data mining software. SIGKDD Explor Newsl 11(1):10. https://doi.org/10.1145/1656274.1656278
14. Hall MA (1999) Correlation-based feature selection for machine learning. Ph.D. thesis, The University of Waikato
15. Holm S (1979) A simple sequentially rejective multiple test procedure. Scand J Stat 6(2):65–70. https://doi.org/10.2307/4615733
16. Hüllermeier E, Fürnkranz J (2010) On predictive accuracy and risk minimization in pairwise label ranking. J Comput Syst Sci 76(1):49–62. https://doi.org/10.1016/j.jcss.2009.05.005

17. Ko AH, Sabourin R, Britto AS Jr (2008) From dynamic classifier selection to dynamic ensemble selection. Pattern Recogn 41(5):1718–1731
18. Kuncheva L, Bezdek J (1998) Nearest prototype classification: clustering, genetic algorithms, or random search? IEEE Trans Syst Man Cybern: Part C (Appl Rev) 28(1):160–164. https://doi.org/10.1109/5326.661099
19. Kuncheva LI (2004) Combining pattern classifiers: methods and algorithms, 1st edn. Wiley-Interscience
20. Manning CD, Raghavan P, Schutze H (2008) Introduction to information retrieval. Cambridge University Press, New York. https://doi.org/10.1017/cbo9780511809071
21. Markiewicz A, Forczmański P (2015) Detection and classification of interesting parts in scanned documents by means of adaboost classification and low-level features verification. In: International conference on computer analysis of images and patterns. Springer, pp 529–540
22. McLachlan GJ (1992) Discriminant analysis and statistical pattern recognition. Wiley series in probability and mathematical statistics: applied probability and statistics. A Wiley-Interscience Publication. https://doi.org/10.1002/0471725293
23. Pearson K (1901) LIII. On lines and planes of closest fit to systems of points in space. The London, Edinburgh, and Dublin Philos Mag J Sci 2(11):559–572. https://doi.org/10.1080/14786440109462720
24. Ponti MP Jr (2011) Combining classifiers: from the creation of ensembles to the decision fusion. In: 2011 24th SIBGRAPI conference on graphics, patterns and images tutorials (SIBGRAPI-T). IEEE, pp 1–10
25. Przybyła-Kasperek M et al (2019) Three conflict methods in multiple classifiers that use dispersed knowledge. Int J Inf Tech Decis Making (IJITDM) 18(02):555–599
26. Przybyła-Kasperek M, Wakulicz-Deja A (2017) Comparison of fusion methods from the abstract level and the rank level in a dispersed decision-making system. Int J Gener Syst 46(4):386–413
27. Reif M, Shafait F, Goldstein M, Breuel T, Dengel A (2014) Automatic classifier selection for non-experts. Pattern Anal Appl 17(1):83–96
28. Rejer I, Burduk R (2017) Classifier selection for motor imagery brain computer interface. In: IFIP international conference on computer information systems and industrial management. Springer, pp 122–130
29. Skurichina M, Duin RP (1998) Bagging for linear classifiers. Pattern Recogn 31(7):909–930. https://doi.org/10.1016/s0031-3203(97)00110-6
30. Sokolova M, Lapalme G (2009) A systematic analysis of performance measures for classification tasks. Inf Process Manag 45(4). https://doi.org/10.1016/j.ipm.2009.03.002
31. Trawiński B, Lasota T, Kempa O, Telec Z, Kutrzyński M (2017) Comparison of ensemble learning models with expert algorithms designed for a property valuation system. In: International conference on computational collective intelligence. Springer, pp 317–327
32. Tulyakov S, Jaeger S, Govindaraju V, Doermann D (2008) Review of classifier combination methods. In: Machine learning in document analysis and recognition. Springer, pp 361–386
33. Wilcoxon F (1945) Individual comparisons by ranking methods. Biometrics Bull 1(6):80. https://doi.org/10.2307/3001968
34. Woźniak M, Graña M, Corchado E (2014) A survey of multiple classifier systems as hybrid systems. Inf Fusion 16:3–17

35. Xu L, Krzyzak A, Suen CY (1992) Methods of combining multiple classifiers and their applications to handwriting recognition. IEEE Trans Syst Man Cybern 22(3):418–435
36. Yekutieli D, Benjamini Y (2001) The control of the false discovery rate in multiple testing under dependency. Ann Stat 29(4):1165–1188. https://doi.org/10.1214/aos/1013699998

Multi Sampling Random Subspace Ensemble for Imbalanced Data Stream Classification

Jakub Klikowski(✉) and Michał Woźniak(✉)

Wrocław University of Science and Technology, Wrocław, Poland
{jakub.klikowski,michal.wozniak}@pwr.edu.pl

Abstract. The classification of data streams is a frequently considered problem. The data coming in over time has a tendency to change its characteristics over time and usually we also encounter some difficulties in data distributions as inequality of the number of learning examples from considered classes. The combination of these two phenomena is an additional challenge. In this article, we propose a novel MSRS (*Multi Sampling Random Subspace Ensemble*) a chunk-based ensemble method for imbalanced non-stationary data stream classification. The proposed algorithm employs random subspace approach and balancing data using various sampling methods to ensure an appropriate diversity of the classifier ensemble. MSRS has been evaluated on the basis of the computer experiments carried out on the diverse pool of the non-stationary imbalanced data streams.

Keywords: Ensemble learning · Imbalanced data · Concept drift · Data stream

1 Introduction

Considering real-life data, there is often a problem of imbalanced data. This phenomenon appears when solving tasks in areas such as fraud detection [1], medical diagnosis [2] or spam filtering [3]. Skew distribution is a situation when one of the class is represented by less instances than other classes. Imbalanced learning problem is reviewed by He and Garcia in survey [4] and by Branco et al. in another survey [5]. Imbalance data may negatively impact on the performance of the canonical classifiers that tend to falsely classify a minority class as a majority class. Visa and Ralescu also show in [6] the main issues and distinguish three factors: accuracy, class distribution and error costs. Additionally, the generalization abilities of classifier learning methods may be inhibited by small, insufficient fraction of minority examples and local data characteristics could make some observations harder to classify than the others. Imbalanced data methods need different metrics to measure an efficiency in the proper way. Brodersen et al. in [7] proposed a balanced accuracy metric which is appropriate for classification problems in imbalanced datasets. Bifet et al. [8] proposed

R. Burduk et al. (Eds.): CORES 2019, AISC 977, pp. 360–369, 2020.
https://doi.org/10.1007/978-3-030-19738-4_36

an another evaluation metric - $Kappa - m$ statistic which is the modification of Kohen-Cappa statistic [9]. Krawczyk [10] presents open challenges and future directions in imbalanced learning. One of the mentioned topic is imbalanced data stream classification.

There are several propositions how to deal with imbalanced data. Two basic groups may be distinguished: data preprocessing and methods adapted to the classification of imbalanced data. *Data preprocessing* tries to balance data distributions by the reduction of objects belonging to the majority class known (*undersapmling*) or by *oversampling*, i.e., the creation of additional synthetic minority class examples. It is also possible to combine aforementioned methods. The next known approach are ensemble methods that use some data balancing mechanics based on classifiers' committees.

Another approach is to design specified method based on rules which avoid result bias towards majority class. In most of the cases this is an ensemble method. The main idea is to split complex problems into easier to solve problems. This operation increases the efficiency of classification. When a new object comes, it is classified by each classifier in ensemble and decisions are made according a combination rule. An important thing is to achieve diversity. Ensemble methods are very popular techniques to classify stream data which is very well reviewed by Krawczyk et al. in this survey [11]. Ensembles owe this popularity to their efficient performance and scalability. They can be relatively easy deployable to work in real world applications.

The main contribution of this work are listed below:

- Proposition of the novel MSRS (*Multi Sampling Random Subspace Ensemble*) method that combines various preprocessing techniques into a uniform method based on classifiers' committees.
- Merging different oversampling techniques and random subspace approach into one ensemble forming step.
- Detailed performance evaluation of the MSRS on the basis of the wide range of computer experiments using benchmark data streams and comparing the proposed methods with the well-known benchmark solutions.

2 Related Works

The most important issue in the imbalanced data stream is the skewed class distribution. This phenomenon causes that regular classification methods tend to favor the majority class which was described by Sun et al. in [12]. There are two different ways to deal with imbalanced problem. One of it is to try rebalancing data distribution by using different algorithms in preprocessing phase. Data balancing is split into four different techniques. Under-sampling which removes excessive examples from majority set. *Condensed Nearest Neighbour* [13] method uses the *1-Nearest Neighbor*'s rule to attractively decide if the sample should be removed. *Near Miss* algorithm uses 3 different types of heuristic rules based on *Nearest Neighbors* algorithm. Yen and Lee [14] proposed a method of under-sampling based on centroids created during majority data clustering.

Over-sampling, where new objects are created based on existing examples to increase the minority set. He et al. [15] proposed *ADASYN*. This method generates samples next to the original samples which are wrongly classified using a k-Nearest Neighbors algorithm. Chawla et al. [16] developed *SMOTE* method, which creates new synthetic data samples based on the *k-Nearest Neighbors* algorithm. The next solution is to mix over and under sampling into the combine sampling algorithm. Batista et al. [17] invented *SMOTE ENN* method. It uses *SMOTE* to create new samples and the *Edited Nearest Neighbours* method to clean whole data set. Batista et al. [18] proposed also *SMOTETomek* method. It also uses SMOTE to create new samples but Tomek Links method to clean the data. The last category is to use ensemble balanced methods to process data. Liu et al. [19] proposed two methods. *Balance Cascade* creates an ensemble of balanced sets by attractively under-sampling the imbalanced data set using an estimator, while *Easy Ensemble* also creates an ensemble of data set, but using randomly under-sampling the original set of data.

Methods designed for data stream classification are split into two main categories according to reading data manner. Online methods, where data is processed one object after another and chunk based approach, where data comes in certain portions of objects. This work focuses on chunk-based ensembles. Gao et al. [20] proposed an *OUSEnsemble* (Over Under Sampling Ensemble) method. This method based on the combination of over-sampling by collecting minority examples from older chunks. Then it is combined with under-sampling of excessive majority data. Ditzler and Polikar [21] proposed two methods based on *Learn++* [22] algorithm. *Learn++.CDS* combines *Learn++.NSE* [23] for concept drift and *SMOTE* sampling method to deal with the imbalance data. *Learn++.NIE* [21] is modification of *Learn++.CDS* which replace *SMOTE* with their own bagging variation method for balancing data. Wang et al. [24] proposed *KMeanClustering* ensemble method which undersamples imbalance data with the *k-mean clustering* algorithm. The main idea is to use the clustering algorithm on the majority data and to resample it using the created centroids. Chen and He proposed *REA* [25] method which is an extension of *SERA* [26] and *MuSeRa* [27] algorithms. *REA* ensemble based on collecting minority examples and selection using *k-Nearest Neighbors* algorithm.

3 Multi Sampling Random Subspace Ensemble

The proposed MSRS (*Multi Sampling Random Subspace Ensemble*) assumed that data stream is available in the form of data chunks. Each portion of data is used to learn a new model. Each model is built on the basis of a given learning method. Chunk of data is properly processed before passing it to classifier. The learning process can be divided into two stages Algorithm 1.

The first stage is the selection of features to create a data subspace. The ensemble is based on the several weaker classifiers combined into one committee. This strategy allows obtaining better results than using only one classifier. Diversification of classifiers is a very important factor when creating an ensemble. In this solution, the method of random subspace of features is used. For each classifier, random attributes from the data set are selected. This technique allows to reduce correlations in the classifiers committee.

The second stage is data preprocessing aimed at reducing the imbalance ratio. Most well-known ensemble methods applied to imbalanced data classification use one sampling solution [21,24,25]. Main idea and innovation in this proposition is to use a few sampling methods instead of one. Then they can be combine into one "multi-sampling" ensemble. That combination allows to find in a dynamic way the best sampling algorithm in actual data. Different methods of data resampling are used for this. At each iteration, a random balancing method is selected. The resampling method pool can be passed as an parameter. By default, the following methods are selected. *Condensed Nearest Neighbour* and *TomekLinks* from under-sampling methods. *SMOTE* and *ADASYN* from over-sampling methods. *SMOTEENN* and *SMOTETomek* from combine methods. Then the new model is added to the ensemble. When the maximum ensemble capacity is reached, the new models replace those with the lowest weight. The weights are calculated using the selected metric, in this case it is a BAC (*balance accuracy score*) [28].

4 Experiments

The main goal of experiments is to empirically compare whether the method proposed in this work is as efficient as other literature methods. Planned experiments may be divided into several groups in order to verify:

- The impact of the features number in data on the classification quality.
- The impact of the imbalance ratio in data on the classification quality.
- Comparison of the classification quality for tested methods in imbalanced synthetic stationary and non-stationary streams.
- Comparison of the classification quality of the tested methods in imbalanced real data.

All tested classifiers committees have been implemented in the Python programming language according to the descriptions provided in the articles. Selected methods that were used during testing to compare the quality of the proposed solution - *KMeanClustering* (KMC) [24], *Learn++CDS* (L++CDS) [21], *Learn++NIE* (L++NIE) [21], *REA* [25], *Over Under Sampling Ensemble* (SE) [20]. The method proposed in this article appears in the abbreviated name MSRS. Each ensemble consists of ten base classifiers. As the base classifier, the implementation of k-Neighbors Classifier from the Scikit-learn [29] library was used. In addition, the Multilayer Perceptron Classifier as a baseline was used during the tests. The implementation used from the scikit-learn library on the default parameter settings. The entire program code is publicly available and placed in the repository[1].

[1] https://github.com/JakubKlik/msrs.

Algorithm 1. Multi Sampling Random Subspace ensemble - pseudocode

```
Data:
  DS: data stream
  PM: pool of sampling methods
  BC: base classifier
  N: maximum number of classifiers
  WM: weight method
1  foreach data chunks DC in DS do
2      foreach classifiers C in ensemble E do
3          calculate weight of C using WM;
4          update value in classifiers weights CW;
5      end
6      select random subspace SD from DC;
7      randomly select sampling method SM from PM;
8      create balance data chunk BD balancing SD by SM;
9      build new hypothesis C based on BD using BC;
10     if |E| ≥ N then
11         find hypothesis CM with minimum weight in CW;
12         replace CM with new hypothesis C;
13     end
14     else
15         add new hypothesis C to ensemble E;
16     end
17     set weight to 1 for new hypothesis C in CW;
18 end
```

4.1 Data Streams

The synthetic streams were created as a result of the generation of multidimensional data in the normal distribution from own generator. For all streams, it was assumed that these are two-class data and having only floating-point values. In addition, data on various degrees of class unbalance were used for the study. When generating synthetic data, there is a problem of determining the difficulty for a given set. A common solution is to use a certain classifier as a baseline. In this work it has also been applied, but in addition a measure of the difficulty of a given stream has been proposed. Assuming that the data is generated in the form of normal distributions, the Kolmogorov-Smirnov test is used to determine the difficulty. This test is performed on each dimension separately for the positive and negative class. Then, the geometric mean is used, which is used as a number to determine the difficulty of the stream being generated. In the case when the difficulty is calculated for data streams with sudden drift, tests are performed separately for objects before and after the drift.

4.2 Evaluation

All tests are executed in the *test then train* manner [30]. This way allows to evaluate methods based on data chunks. Each chunk is used twice. First, it is used to evaluate the tested method. Then the same chunk is used for training. In this way, there is no data skipping because it is not necessary to divide the data into a training set and test set. The exception to this rule is the first step where the boot chunk is used. It is characterized with more data than normal and is only used to learn the ensemble. Using the selected metrics the result in the given chunk is calculated. In the case of these experiments, known metrics such as the precision score and recall score are taken into account. Instead of accuracy score, the balanced accuracy score is used. The next measured metric is the Km statistic. It is the modification of Kohen-Cappa statistic proposed by Bifet et al. [9].

5 Results

5.1 Features Test

The first test was to check whether the proposed algorithm has a dependence of quality on the number of attributes in the data stream. Additional methods have been tested to determine a certain reference point to other literature solutions. Tests were performed on data streams with no drift concept. All streams had 20,000 objects, 2 classes, 20% of imbalance ratio. In addition, these streams were with a different number of attributes. It was a range from 10 to 100 increased every 5 attributes and the results are showed in Fig. 1.

The conducted tests showed no dependence of the proposed algorithm on the number of attributes. The results on the chart show that with the increase in the number of attributes, the balance accuracy score Fig. 1 is not significantly deteriorated. In comparison to others, the proposed method does not achieve the worst result. The MLPC method achieved the worst result. It's worth considering a different way of selecting features to get better results with a large number of attributes in the stream.

5.2 Imbalance Ratio Test

The next test was to check the dependence of the data imbalance on the quality of the classification in the proposed method. The tests were also carried out for the remaining methods to obtain a proper comparison. Streams used for this did not have a drift concept. They had 20,000 objects, 2 classes, 10 attributes, and a difficulty level of 30. They differed in the degree of unbalancing, which took values from 0.01 to 0.05 every 0,01 and from 0.05 to 0.40 every 0,05. The results of this experiment are presented in Fig. 2.

The conducted tests showed no major impact on the quality of the algorithm classification by changing the degree of unbalancing. The results in the charts indicate that increasing the unbalancing does not affect the BAC. The proposed method achieves comparable results to other methods.

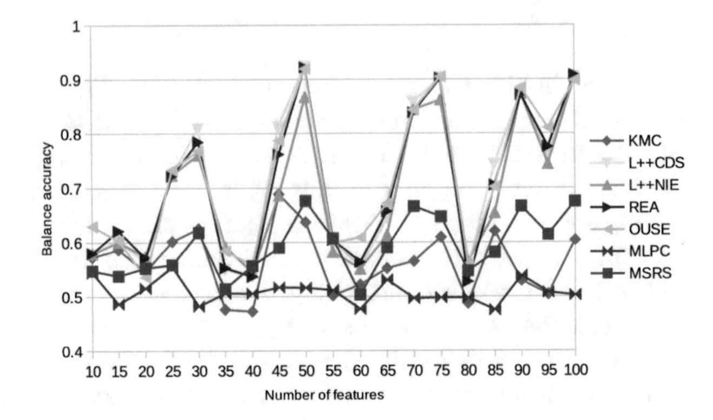

Fig. 1. Balanced accuracy for methods on sudden drift streams with different number of features.

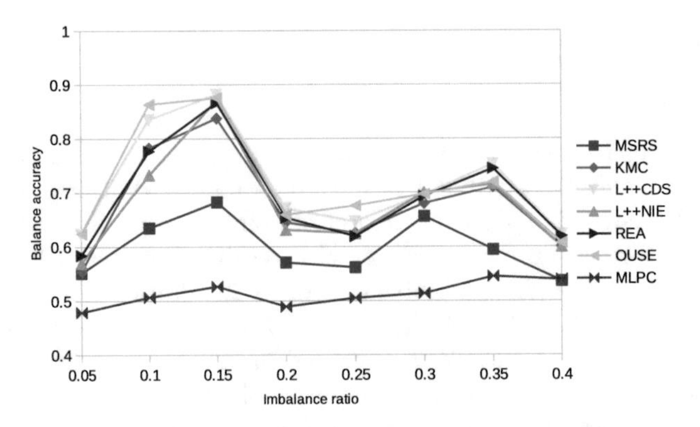

Fig. 2. Balance accuracy for methods on sudden drift streams with different imbalance ratio.

5.3 Ranking Tests

Synthetic Data. Another experiment in this article is the general test on drift streams and without drift. The purpose of this test is to check how the proposed method handles when classifying non-stationary imbalanced data streams compared to other methods. The obtained results were accumulated using the ranking method based on non-parametric tests. The whole procedure consisted in comparing all methods with pairs using the Wilcoxon Signed-Ranks test. For this experiment, data streams were used that had 20,000 objects and 2 classes. The streams had 5 or 10 attributes, the unbalancing at 10 or 20%, stationary or with sudden drift, and the difficulty level between 5 and 50 increased by 5 points. Rank tests have shown that the method achieves results that place the method at almost last place (Table 1). Such a bad result says that the method did not succeed in comparison tests with other literature methods. It is clearly visible that *OUSE* is the best method in this ranking tests.

Table 1. Summed Wilcoxon Signed-Ranks test

Synthetic imbalanced streams.

Rank	Method	Points
1	OUSE	227.41
2	L++CDS	147.56
3	REA	104.29
4	KMC	104.13
5	L++NIE	95.5
6	MSRS	-98.85
7	MLPC	-580.05

Real imbalanced data.

Rank	Method	Points
1	REA	169.38
2	L++CDS	164.99
3	OUSE	55.84
4	MSRS	16.77
5	KMC	13.2
6	L++NIE	-54.27
7	MLPC	-365.9

Real Data. The last experiment that has been carried out is to check how the method deals with real sets. In this case, the data sets from the KEEL [31] repository were used. Following data sets where used: yeast1, vehicle0, segment0, page-blocks0, shuttle-2vs5. Data sets were treated as streams from which data chunks of a certain size were downloaded. As in the previous experiment, the results were collected using Wilcoxon Signed-Ranks test. The obtained results indicate a better operation of the proposed method when using real data (Table 1). This is not the best result, but there is a better ability to deal with real data. The *REA* method was the best in this case, which in previous experiments was also at the forefront of the best methods.

6 Conclusions

The experiments conducted showed that this method has no potential for further development. The main idea of the proposed method was based on ensuring diversity through the use of various oversampling techniques. Results of experiments shows that this idea does not provide significant diversification of classifiers. In addition, the sampling, which is used on every data chunk, may cause the classification to become unstable. This results in the fact that the proposed method obtains fairly poor results compared to other literature methods. However, this method deals quite better with the classification of real data. MSRS is not dependent on the number of features or the degree of unbalancing, which allows for stable operation in the case of more difficult data.

It is worth considering some changes in assumptions. One of such ideas may be replacing random sub-spaces with a selection of features. This would allow maintaining more diversity of classifiers and could significantly increase the quality of the whole ensemble classification. One should also consider how to choose methods of balancing. This could be replaced with some heuristic, which would try to dynamically choose more valuable preprocessing methods for a given data stream. Another possible solution is to reflect on whether better results could be balancing the data, but not until the full compensation. Limited data balancing can lead to more stable ensemble.

Acknowledgement. This work was supported by the Polish National Science Centre under the grant No. 2017/27/B/ST6/01325.

References

1. Phua C, Alahakoon D, Lee V (2004) Minority report in fraud detection: classification of skewed data. Acm Sigkdd Explor Newslett 6(1):50–59
2. Krawczyk B, Galar M, Jeleń Ł, Herrera F (2016) Evolutionary undersampling boosting for imbalanced classification of breast cancer malignancy. Appl Soft Comput 38:714–726
3. Alqatawna J, Faris H, Jaradat K, Al-Zewairi M, Adwan O (2015) Improving knowledge based spam detection methods: the effect of malicious related features in imbalance data distribution. Int J Commun Netw Syst Sci 8(05):118
4. He H, Garcia EA (2009) Learning from imbalanced data. IEEE Trans Knowl Data Eng 21(9):1263–1284
5. Branco P, Torgo L, Ribeiro R (2015) A survey of predictive modelling under imbalanced distributions. arXiv preprint arXiv:1505.01658
6. Visa S, Ralescu A (2005) Issues in mining imbalanced data sets-a review paper. In: Proceedings of the sixteen midwest artificial intelligence and cognitive science conference, vol 2005, pp 67–73
7. Brodersen KH, Ong CS, Stephan KE, Buhmann JM (2010) The balanced accuracy and its posterior distribution. In: 2010 20th international conference on Pattern recognition (ICPR), pp 3121–3124. IEEE
8. Bifet A, de Francisci Morales G, Read J, Holmes G, Pfahringer B (2015) Efficient online evaluation of big data stream classifiers. In: Proceedings of the 21th ACM SIGKDD international conference on knowledge discovery and data mining, pp 59–68. ACM
9. Cohen J (1960) A coefficient of agreement for nominal scales. Educ Psychol Measur 20(1):37–46
10. Krawczyk B (2016) Learning from imbalanced data: open challenges and future directions. Progress Artif Intell 5(4):221–232
11. Krawczyk B, Minku LL, Gama J, Stefanowski J, Woźniak M (2017) Ensemble learning for data stream analysis: a survey. Inf Fusion 37:132–156
12. Sun Y, Wong AK, Kamel MS (2009) Classification of imbalanced data: a review. Int J Pattern Recogn Artif Intell 23(04):687–719
13. Hart P (1968) The condensed nearest neighbor rule (corresp.). IEEE Trans Inf Theory 14(3):515–516
14. Yen SJ, Lee YS (2009) Cluster-based under-sampling approaches for imbalanced data distributions. Expert Syst Appl 36(3):5718–5727
15. He H, Bai Y, Garcia EA, Li S (2008) ADASYN: adaptive synthetic sampling approach for imbalanced learning. In: IEEE international joint conference on neural networks, IJCNN 2008. (IEEE World Congress on Computational Intelligence), pp 1322–1328. IEEE
16. Chawla NV, Bowyer KW, Hall LO, Kegelmeyer WP (2002) SMOTE: synthetic minority over-sampling technique. J Artif Intell Res 16:321–357
17. Batista GE, Prati RC, Monard MC (2004) A study of the behavior of several methods for balancing machine learning training data. ACM SIGKDD Explor Newslett 6(1):20–29
18. Batista GE, Bazzan AL, Monard MC (2003) Balancing training data for automated annotation of keywords: a case study. In: WOB, pp 10–18

19. Liu XY, Wu J, Zhou ZH (2009) Exploratory undersampling for class-imbalance learning. IEEE Trans Syst Man Cybern Part B (Cybern) 39(2):539–550
20. Gao J, Ding B, Fan W, Han J, Philip SY (2008) Classifying data streams with skewed class distributions and concept drifts. IEEE Internet Comput 12(6):37–49
21. Ditzler G, Polikar R (2013) Incremental learning of concept drift from streaming imbalanced data. IEEE Trans Knowl Data Eng 25(10):2283–2301
22. Polikar R, Upda L, Upda SS, Honavar V (2001) Learn++: an incremental learning algorithm for supervised neural networks. IEEE Trans Syst Man Cybern Part C (Appl Rev) 31(4):497–508
23. Elwell R, Polikar R (2009) Incremental learning of variable rate concept drift. In: International workshop on multiple classifier systems, pp 142–151. Springer
24. Wang Y, Zhang Y, Wang Y (2009) Mining data streams with skewed distribution by static classifier ensemble. In: Opportunities and challenges for next-generation applied intelligence, pp 65–71. Springer
25. Chen S, He H (2011) Towards incremental learning of nonstationary imbalanced data stream: a multiple selectively recursive approach. Evolving Syst 2(1):35–50
26. Chen S, He H (2009) Sera: selectively recursive approach towards nonstationary imbalanced stream data mining. In: International joint conference on neural networks, IJCNN 2009, pp 522–529. IEEE
27. Chen S, He H, Li K, Desai S (2010) Musera: multiple selectively recursive approach towards imbalanced stream data mining. In: 2010 international joint conference on neural networks (IJCNN), pp 1–8. IEEE
28. Branco P, Torgo L, Ribeiro RP (2017) Relevance-based evaluation metrics for multi-class imbalanced domains. In: Proceedings of advances in knowledge discovery and data mining - 21st Pacific-Asia conference, Part I, PAKDD 2017, Jeju, South Korea, 23–26 May 2017, pp 698–710
29. Pedregosa F, Varoquaux G, Gramfort A, Michel V, Thirion B, Grisel O, Blondel M, Prettenhofer P, Weiss R, Dubourg V, Vanderplas J, Passos A, Cournapeau D, Brucher M, Perrot M, Duchesnay E (2011) Scikit-learn: machine learning in Python. J Mach Learn Res 12:2825–2830
30. Bifet A, Holmes G, Kirkby R, Pfahringer B (2010) MOA: massive online analysis. J Mach Learn Res. 11:1601–1604
31. Alcalá-Fdez J, Sánchez L, Garcia S, del Jesus MJ, Ventura S, Garrell JM, Otero J, Romero C, Bacardit J, Rivas VM et al (2009) KEEL: a software tool to assess evolutionary algorithms for data mining problems. Soft Comput 13(3):307–318

Author Index

R. Burduk et al. (Eds.): CORES 2019, AISC 977, pp. 371–372, 2020.
https://doi.org/10.1007/978-3-030-19738-4

Printed in the United States
By Bookmasters